**Mann-Whitney U Test**

$$U = n_1 n_2 + \frac{n_1(n_1 + 1)}{2.0} - R$$

312

**McNemar Test**

$$\chi^2(1) = \frac{(|a - d| - 1.0)^2}{a + d}$$

299

**Mean**

$$\bar{X} = \frac{\sum X}{n}$$

45

**One-Way ANOVA:**
**Correction Term for the Mean**

$$C = \frac{(\sum\sum X_{ij})^2}{N}$$

232

**Omega Squared**

$$\omega^2 = \frac{SS_A - (k - 1)MS_{S/A}}{SS_{TOT} + MS_{S/A}}$$

234

**SS Contrast**

$$\frac{(\sum_j p_j T_{.j})^2}{n\sum p_j^2}$$

237

**SS Groups (Equal n)**

$$SS_A = \frac{\sum T_{.j}^2}{n} - C$$

233

**SS Groups (Unequal n)**

$$SS_A = \sum \frac{T_{.j}^2}{n_j} - C$$

233

**SS Persons Within Groups**

$$SS_{S/A} = SS_{TOT} - SS_A$$

233

**SS Total**

$$SS_{TOT} = \sum\sum X_{ij}^2 - C$$

233

**Paired t Test**

$$t(n - 1) = \frac{\bar{X}_D}{\sqrt{\dfrac{\sum D^2 - (\sum D)^2/n}{n(n - 1)}}}$$

211

# Statistics for the Social and Behavioral Sciences

# Statistics for the Social and Behavioral Sciences

David A. Kenny

*University of Connecticut*

**Little, Brown and Company**

*BOSTON    TORONTO*

Library of Congress Cataloging-in-Publication Data

Kenny, David A., 1946–
 Statistics for the social and behavioral sciences.

 Bibliography: p.
 Includes index.
 1. Social sciences—Statistical methods.   2. Psychology
—Statistical methods.   3. Statistics.   I. Title.
 HA29.K4295   1986       300'.1'5195       86-20909
 ISBN 0-316-48915-8

Library of Congress Catalog Card Number 86-20909

ISBN 0-316-48915-8

9  8  7  6  5  4  3  2  1

MV

Published simultaneously in Canada
by Little, Brown and Company (Canada) Limited

Printed in the United States of America

## Acknowledgments

Table 13.1, Table 13.2, Table 16.1, and Table 16.2: From Jacob Cohen, *Statistical Power for the Behavioral Sciences,* Rev. ed. (New York: Academic Press, 1977). Adapted by permission.

Problem 13.14 and Problem 18.16. From Randy L. Diehl, Keith R. Kluender, and Ellen M. Parker, "Are Selective Adaptation and Contrast Effects Really Distinct?" *Journal of Experimental Psychology: Human Perception and Performance* 11 (1985), Table 1, p. 215. Copyright 1985 by the American Psychological Association. Adapted by permission of the author.

Appendix C and Appendix H: From Donald B. Owen, *Handbook of Statistical Tables* (Reading, MA: Addison-Wesley, 1962), Table 1.1, pp. 3–10, Table 11.4, pp. 349–353. © 1962 by Addison-Wesley Publishing Company, Inc. Adapted with permission.

Appendix D: From Enrico T. Federighi, "Extended Tables of the Percentage Points of Students' *t*-Distribution," *Journal of the American Statistical Association* 54 (1959): 683–688. Adapted by permission.

Appendix E: From Egon S. Pearson and Herman O. Hartley, *Biometrika Tables for Statisticians,* Vol. 1, 3rd ed. (Cambridge: The University Press, 1972), Table 18, pp. 169–175. Adapted with the kind permission of Egon S. Pearson and the Biometrika Trustees.

Appendix G: From Table IV of Ronald A. Fisher and Frank Yates, *Statistical Tables for Biological, Agricultural and Medical Research,* 6th ed. (London: Longman Group Ltd., 1974; previously published by Oliver & Boyd Ltd., Edinburgh). Adapted by permission of the authors and publishers.

Appendix I: From Gerald J. Glasser and Robert F. Winter, "Critical Values of the Coefficient of Rank Correlation for Testing the Hypothesis of Independence," *Biometrika* 48 (1961), Table 2, p. 446. Adapted by permission.

Pp. 39, 125, 126, 143: Life-expectancy data is from U.S. Bureau of the Census, *Statistical Abstract of the United States 1984,* 104th ed. (Washington, DC: United States Government Printing Office, 1984), p. 861.

P. 142: Hurricane data is from *Information Please Almanac* ® 1983 (Boston: Houghton Mifflin Co., 1983), p. 431.

*To my children,*
*Katherine, Deirdre, and David*

# *Preface*

A few years ago, I found myself in the emergency room of the local hospital. While painting my house, I had fallen through a window and cut both my arms. As a nurse busily worked to stop the bleeding, she asked me what I did for a living. I told her I taught psychology. My response apparently interested her, and she eagerly asked me what area of psychology I taught. When I said statistics, she was quite disappointed. I told her that I found statistics to be fun and exciting; she thought I must be crazy. Probably many of the student readers of this text (and perhaps a few of their instructors as well) feel the same way about statistics as the nurse did. It is my hope that this text will make students and instructors fear the topic less and even make a few of you enjoy the topic as much as I do.

This textbook is written for students majoring in the social and behavioral sciences. It provides a first course in data analysis. Students should learn the basic methods that social and behavioral scientists use in analyzing data to test hypotheses. The book is intended to be comprehensible to students who are not planning to go on to postgraduate study, but I have also included material to prepare students for graduate school. Even the active researcher may find the book a useful resource, because I have covered many practical issues that are not typically included in textbooks.

The book begins with a general introduction to the major terms in data analysis. The next seven chapters present procedures that have been developed to describe data. Measures of central tendency, variability, and association are presented. Chapters 9, 10, and 11 introduce the key concepts that underlie the drawing of statistical conclusions. Presented are sampling, the normal and binomial distributions, and sampling distributions. The final seven chapters discuss statistical models, and the standard tests of statistical significance are presented here as well.

Much of the computation of statistics is no longer an unpleasant and laborious task. With the easy access to computers and calculators today, an undergraduate can do the statistical computation in five minutes that used to take a Ph.D statistician weeks to do. This book recognizes this fact and emphasizes interpretation and understanding as opposed to computation.

Some students understandably feel intimidated by statistics and its formulas and nomenclature. One way to lessen their fear is to make the style of presentation informal. I have avoided numbering sections and formulas and have tried to use words instead of symbols whenever possible. In fact, many of the formulas are stated in terms of words instead of abstract symbols.

Another way to increase comprehensibility is to use many examples. I have included examples from the areas of nonverbal communication, teacher expectancies, vandalism, age and short-term memory, obedience to authority, voting in Congress, and many others. Where possible, I present actual data instead of made-up numbers. Examples are drawn from psychology, education, genetics, public policy, business, sociology and anthropology, medicine, and meteorology.

I have also attempted to make the book very practical and to discuss topics that professional researchers face with real data. Many of the classical topics in statistics (e.g., selecting balls from urns) are of interest to the statistician, but their abstract discussion is of little value to the undergraduate who is struggling with the topic for the first time.

A related goal of mine was to include important topics that are not covered in enough detail in many statistics texts. I have incorporated much more material on the effect of unusual data points, issues of data transformation, repeated measures designs, and model testing than is found in most contemporary texts.

Modern data analysis, whether the discipline is psychology or economics, or whether the design is experimental or observational, uses primarily correlational techniques. This book has four chapters on the subject of correlation and regression and contains much more detail than most of the statistics texts that are currently available. The purpose of the text is to feature data analysis techniques that are being used now and will be increasingly used in the future, and to avoid the discussion of techniques that are no longer used but were important many years ago.

The task of writing a book is always a collective effort that extends beyond the listed authors. This was certainly true of my endeavor. I would first like to acknowledge the assistance of my reviewers: David Chizar, the University of Colorado, Boulder; Jon A. Christopherson, formerly of the U.S. Coast Guard Academy; Charles M. Judd, the University of Colorado, Boulder; Katherine W. Klein, North Carolina State University; Thomas E. Nygren, the Ohio State University; Mike Raulin, State University of New York at Buffalo; Howard M. Sandler, Peabody College of Vanderbilt University; Robert Seibel, the Pennsylvania State University; Joseph B. Thompson, Washington and Lee University; and Anthony A. Walsh, Salve Regina–the Newport College. They patiently educated me about a number of important issues.

Special thanks are due to those at Little, Brown who guided this project. Tom Pavela convinced me that I could write a book in this area that would be new and exciting. Molly Faulkner took what was only an outline and developed it into a plan for a book. And Mylan Jaixen took charge of the difficult task of bringing the book into production. Barbara Breese of Little, Brown and Melinda Wirkus turned the manuscript into a book.

A number of my colleagues and students read various chapters and provided important feedback. In the early stages, Cindi Zagiebroylo and Lisa Cassady provided me with helpful comments. Later, Thomas Malloy and especially Claire Harrison carefully read each number and word.

I also wish to thank colleagues who provided me with data. In particular, Bella DePaulo gave me a fascinating data set that is described in Chapter 2. Also, Starkey Duncan and Don Fiske are thanked for the data described in Chapters 5 and 7. Finally, the data in Chapter 6 relating age and short-term memory were gathered by the late Dennis Ilchisin.

A project of this length requires extensive clerical support from many people. I want to acknowledge their crucial assistance. In the early stages, Mary Ellen Kenny and Robyn Ireland typed numerous drafts of the chapters. It was Robyn Ireland who put the manuscript onto a word processor. Over the last year, Claire Harrison handled all the clerical details. Without her assistance, the book would have been delayed considerably.

I am grateful to the literary executor of the late Sir Ronald Fisher, F.R.S.; to Dr. Frank Yates, F.R.S.; and to Longman Group–London, Ltd., for permission to reprint Table IV from *Statistical Tables for Biological, Agricultural and Medical Research* (6th ed., 1974).

Finally, my home institution, the University of Connecticut, provided me with the resources and support to undertake this effort. I completed the project at the Psychology Department of Arizona State University.

# *Brief Contents*

# Contents

# Statistics for the Social and Behavioral Sciences

PART *1*

# *Getting Started*

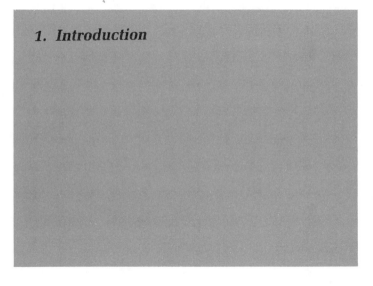

1. Introduction

# 1 | *Introduction*

Numbers are very much a part of modern life. "The average high temperature in New York City for January is 37 degrees Fahrenheit." "The president's popularity is 45%, subject to an error of 3%." "After the institution of the 55 mph speed limit, highway fatalities decreased by 16.4%." "Prices are reduced in this sale by over 50%." "There were 12.5 million people enrolled in college during 1983." Numbers are just as much a part of modern life as pollution, rock and roll, miracles of medicine, and nuclear weapons. If the quality of life is to be improved in this modern world, its citizens must understand how to make sense out of numbers.

Numbers are not important in and of themselves. They are important because they help us make decisions. Decisions can be made without numbers, but if the right numbers are used, in the right way, the quality of decisions can be improved. In the purchase of a home computer, for example, one could guess which computer model is the most reliable, but one would be better informed if the statistics for repair rate were available. Knowing the numbers can help the consumer make the decision, but other factors besides repair rate are important. Ease of use, styling, and cost also contribute to sensible decision making. Numbers help people make all kinds of decisions in everyday life—what courses to take, what stereo to purchase, whether to have a surgical operation.

Numbers are essential in helping us as a society make decisions. Changes in the economy alert business and government to the need for changes in investment and tax laws. Increases in the rate of cancer point to potential causes and indicate new environmental legislation. Also, numbers obtained from Scholastic Aptitude Tests can be used to indicate how good a job schools are doing. As with individual decisions, societal decisions are not made by the numbers but rather are guided by them. The political process sets the priorities and the values. Numbers are a means to an end and not the end unto themselves.

Numbers are not only used by human beings, but they come from human beings. People attempt to monitor reality and attach numbers to objects. For instance, today's temperature is 65 degrees, dinner cost $10.41, or the last movie I saw was a "10." All these numbers refer to an object (a day, a meal or a movie), but the numbers are attached to the objects by people. Someone decided to charge $10.41 for a meal, scientists have agreed that, in a fever thermometer, so many millimeters of mercury correspond to a temperature of 98.6 degrees. Thus, although numbers seem to be cold and impersonal, they are by necessity personal. Numbers attached to objects only seem to be objective, but they are actually based on a set of social conventions. A number referring to an object is given a meaning by persons. Humans use machines and computers to assist them in assigning numbers to objects, but it is a human being, not the machine, that assigns the numbers.

Many of today's numbers come from computers. Phone bills, class registration forms, library cards, and cash registers continually bombard us with computer-generated numbers. It is becoming common practice to blame the impersonal computer for all of society's ills. People fume when a computer makes a mistake, but the error is usually caused by the person who entered the data into the computer and not by the machine itself. The numbers from a computer only *seem* impersonal. They are actually products of human thought and action, although it may be difficult to see the hand of the person who programmed the computer.

If numbers are to be used intelligently, their meaning must first be comprehended. To most, numbers are not as enjoyable as a day at the beach, but if we are to survive and thrive we must learn how to make sense out of them. That is the purpose of this book: making sense out of numbers.

# *Essential Definitions*

Social and behavioral scientists are busy measuring intelligence, recall, conformity, fear, and social status. *Measurement* is the assignment of numbers to objects using an operational definition in order to measure a variable. The following terms must be carefully distinguished: *number,* an *object,* a *variable,* and a variable's *operational definition.* A *number,* a *score,* or a *datum* is a numeric value given to an object. If someone has a temperature of 102 degrees, the number is 102. More than one number is called *data.* The *object* is the entity to which the number is attached. So for the previous example of a person with a 102 degree temperature, the object is the person. The *variable* is the construct that the researcher is attempting to measure. For this example the variable is temperature. Measurement always requires performing a series of steps. These steps are called the variable's *operational definition.* For temperature, one operational definition is the level of mercury in a thermometer that has been in the mouth of a person for two minutes.

Another example might help clarify the distinctions between number, object, variable, and operational definition. Jane receives a score of 98 on her midterm examination. The number is 98, the object is Jane, the variable is midterm examination grade, and the operational definition is the number correct on the midterm examination divided by the total number possible and then multiplied by 100. Still yet another example is that Paul's car travels at 63 miles per hour. The number or datum is 63, the object is Paul's car, the variable is speed, and the operational definition is the number of miles traveled during a time period divided by the length of the time period.

## Sample and Population

Knowing that your score is 109 on an examination tells you little or nothing. If 109 is the lowest grade on the test, however, you know you are in trouble. Thus it is evident that numbers make little sense in isolation. Only when they are in groups do they have any meaning. This is the reason that numbers come in batches. In this book a set of numbers is called a sample. The term *sample* is used because the numbers are assumed to be a subset of scores from a larger group of scores. The term for the larger group of scores is *population*. So if a researcher studies the performance of rats running a maze, he or she may examine closely the behavior of a sample of ten rats. But the behaviors of the ten rats are assumed to be representative of behaviors of the larger population of laboratory rats.

From the sample data, the researcher computes quantities that are used to summarize the data. A quantity computed from sample data is called a *statistic*. In statistical work, the data are analyzed for two very different purposes. A statistic can be used either to describe the sample or to serve as a basis for drawing inferences about the population. *Descriptive statistics* concern ways of summarizing the scores of the sample. *Inferential statistics* concern using the sample data to draw conclusions about the population. So descriptive statistics refer to the description of the sample, and inferential statistics have to do with inferences about the population.

Consider a survey of 1000 voters that is to be used to predict a national election. The 1000 voters form the sample, and all those voting in the national election form the population. The characterization of the preferences of the 1000 voters in the sample involves descriptive statistics. Using these descriptions to infer about the results of the national election involves inferential statistics.

Descriptive statistics always precede inferential statistics. First the sample data are carefully analyzed and then the researcher is in position to draw inferences about the population. In Chapters 2 through 8 the methods used in descriptive statistics are presented. Chapters 9 through 18 discuss the procedures used in inferential statistics.

## Level of Measurement

Numbers can provide three different types of information. First, they can be used simply to *differentiate* objects. The numbers on the players' backs when they play football or baseball are there so that the spectators can know who scored the touchdown or hit the home run. Second, numbers can be used to *rank* objects. Those objects with lower numbers have *more* (or sometimes, less) of some quantity. In that case the numbers tell how objects rate relative to each other. It is a common practice to rank order the participants after the finish of a race: first, second, third, and so on. Third, numbers can be used to *quantify* the relative difference between persons. This is done when height and weight are measured. When you lose weight you want to know the *number* of pounds that you lost and not just that you weigh less. So, there are three major uses for numbers:

1. to differentiate objects,
2. to rank objects, and
3. to quantify objects.

The *nominal level of measurement* serves only to differentiate objects or persons. Variables for which persons are only differentiated are called *nominal* variables. Examples of nominal variables are ethnicity, gender, and psychiatric diagnostic category. Consider the categorization of persons' religious affiliations. It is possible to assign a one to those who are Protestants, a two to those who are Jewish, and a three to the Catholics. These numbers are used to differentiate the religions, but they do not rank or quantify them. Sometimes each individual receives a unique number, such as a social security number, and other times many individuals share a common number (zip code). Both zip codes and social security numbers are nominal variables.

The *ordinal level of measurement* not only differentiates the objects but also ranks them. For instance, students may be asked to rank-order a set of movies from most to least enjoyable. Records are often ranked from the least to the most popular. The "top 40" musical hits illustrate the ordinal level of measurement. At the ordinal level of measurement, the numbers show which objects have *more* of the variable than other objects, but the numbers do not say how much more.

The *interval level of measurement* presumes not only that objects can be differentiated and ranked but also that the differences can be quantified. Thus, if John weighs 198 pounds, Jim 206, and Sam 214, the amount that Sam weighs more than Jim, 8 pounds, is equal to the amount that Jim weighs more than John. The interval level of measurement differentiates, ranks, and quantifies. Table 1.1 summarizes the ways in which the three levels of measurement differ.

For variables measured at the interval level of measurement, the numbers

*TABLE 1.1*  **Properties of Different Levels of Measurement**

| Level | Differentiate | Rank | Quantify |
|-------|---------------|------|----------|
| Nominal | Yes | No | No |
| Ordinal | Yes | Yes | No |
| Interval | Yes | Yes | Yes |

are said to be in a particular unit of measurement. The *unit of measurement* defines the meaning of a difference of one point in the variable. So if a researcher measures weight, it is important to know whether the unit of measurement is pound, gram, or kilogram.

For the runner in a race, the number on his or her back is at the nominal level of measurement, the place the runner comes in is at the ordinal level, and the time it takes to complete the race is at the interval level. Another example might help in understanding the differences between the three levels of measurement. You and your classmates have different numbers. Your social security numbers differ, and those are nominal differences. You will have class ranks on the midterm exam, and those are ordinal differences. Finally, you are different ages, and those are interval differences.

Determining the level of measurement of many variables can be accomplished by common sense. It is fairly obvious that a variable such as age in years is at the interval level of measurement and that gender (male versus female) is a nominal variable. Generally, the best way to determine the level of measurement is by the procedures used to measure the variable. For instance, if subjects are asked to rank order the stimuli, then the stimuli are at the ordinal level of measurement. Most problematic is establishing that a variable is at the interval level of measurement. One can safely assume that a variable that is measured in physical units—such as time, size, and weight—is at the interval level of measurement. However, for variables whose units are quite subjective, for example, a rating of how much a person likes a movie on a scale from 1 to 10, it is not clear whether the level of measurement is the ordinal or the interval level. Some researchers prefer to be quite conservative, and claim that the level of measurement is only at the ordinal level. Most researchers, however, are willing to assume that the variable is at the interval level.

The decision concerning the level of measurement has very important consequences for the statistical analysis. The valid interpretation of many of the commonly used statistical techniques requires that a variable be measured at the interval level of measurement. There are also techniques that can be used if the variables are measured at the nominal and the ordinal levels of measurement, but these techniques tend not to be nearly as informative as those that were developed for the interval level. If one wishes to be conservative, one can always assume that the variables are at the nominal or ordinal level of measurement.

In some very special cases the same variable can be at one level for one purpose and at another level for a second purpose. Consider the variable of first letter of a person's last name. Ordinarily this would be considered a nominal variable. That is, the first letter of the last name only differentiates us from one another. However, Segal (1974) asked members of a police academy to indicate who their best friends were. Trainees at this academy were assigned dormitory rooms and to seats in classes on the basis of the alphabetic order of their last names. And so, trainees whose last names were closer alphabetically were in closer physical proximity. Segal found that persons whose names were closer together in the alphabet were more likely to be friends. In this case the first letter of one's last name, which is ordinarily a nominal variable, became an ordinal variable.

In the section of this book dealing with descriptive statistics, Chapters 2 through 7 consider primarily variables measured at the interval level of measurement. Chapter 8 concerns the description of variables measured at the nominal and ordinal levels of measurement. In the inferential section, Chapters 10 through 18, the variable of prime importance is measured at the interval level in all chapters but 17 and 18. Inferential issues for nominal variables are considered in Chapter 17 and ordinal variables in Chapter 18.

It is common to consider a fourth level of measurement as well. At the *ratio level of measurement* it is permissible to compute the ratio between two measurements. For example, it might be said that John weighs twice as much as Sally. A key feature of the ratio level of measurement is that the value of zero is theoretically meaningful. However, the interpretation of no statistical technique discussed in this text requires that the variables be measured at the ratio level, and interval-level statistics are appropriate for ratio measurements.

# Mathematical Necessities

The purpose of this text is to help the student to understand better the meaning of numbers. To make sense out of numbers it is necessary to perform mathematical operations on them. Fortunately, most of the mathematics employed is at the high school level.

## Rounding

The data in statistics are usually stated as integer values—for instance, the number of bar presses, dollars earned annually, or the number correct on a test. Sometimes a number may have many decimal places and some of the trailing digits must be dropped to round the number. There are two decisions:

1. How many decimal places should be reported?
2. What is to be reported for the last digit?

These decisions must be made whenever computations are performed in data analysis. After dividing or taking a square root, the resulting number may have many trailing digits.

A good rule of thumb is to report a result of calculations to two more digits than the original data. So if the original data were integers or whole numbers, the result would be reported to two decimal places—that is, to the nearest hundredth. Because many statistics involve finding small differences between large numbers, one should use as many digits as possible during intermediate calculations. Minimally, during computations four more digits than the original data (twice as many as the number of digits to be reported) should be used. Any numerical result that is compared to a number in a table should be computed to at least as many digits as there are in the table.

After having decided on the number of digits to report, there is still the decision about what to do with the last digit. This decision concerns rounding. In rounding, one begins by examining the remaining quantity after the digit that is to be rounded. So if the number to be rounded is 4.3256 and the result is to be rounded to the nearest hundredth, then the remaining quantity is 56. If that quantity is greater than 50, one rounds up and less than 50 one rounds down. If the quantity number is exactly 50, then one rounds to the nearest *even* number. The numbers below are rounded to the nearest hundredth, as follows:

12.12123 is rounded to 12.12
12.12759 is rounded to 12.13
12.124 is rounded to 12.12
12.12507 is rounded to 12.13
12.12500 is rounded to 12.12
12.13500 is rounded to 12.14

When there is one number to be rounded that is exactly 5, it is not uncommon for there to be many such numbers. To avoid substantial rounding error, one should consider including another significant digit before rounding. For instance, instead of rounding to two decimal places, one rounds to three decimal places.

## Proportions, Percentages, and Odds

If you take a multiple-choice test and get 24 correct out of 30 questions, there are a number of ways to express how well you did on the test. You could compute the proportion of the items that you successfully completed. That would be 24 divided by 30 or .80. Because decimal points can be confusing, the proportion is often multiplied by 100 to obtain the percentage of correct items: $100\% \times .80 = 80\%$. Finally the odds of answering a question correctly can be computed. To compute this, the number correct, 24, is divided by the number incorrect: $24/6 = 4$. In brief, the proportion of correct answers is .80, the percentage is 80, and the odds of answering a question correctly are 4 to 1, or 4.

The formulas for proportion, percentage, and odds are as follows, where $n$ is the number correct and $m$ the number incorrect:

$$\text{proportion correct} = \frac{n}{n + m}$$

$$\text{percentage correct} = 100 \times \text{proportion}$$

$$\text{odds of being correct} = \frac{n}{m}$$

The odds can be derived from the proportion and from the percentage as follows:

$$\text{odds} = \frac{\text{proportion}}{1 - \text{proportion}} = \frac{\text{percentage}}{100 - \text{percentage}}$$

Alternatively, the proportion and the percentage can be derived from the odds; that is,

$$\text{proportion} = \frac{\text{odds}}{\text{odds} + 1}$$

$$\text{percentage} = 100\left[\frac{\text{odds}}{\text{odds} + 1}\right]$$

## Squares and Square Roots

In statistical work it is often necessary to square numbers and to compute square roots. Recall that a number squared equals the number times itself. The term $X$ squared is denoted by $X^2$. Thus, 6 squared is symbolized as $6^2$ and is 6 times 6, or 36. The square root of a number times the square root of the same number equals the number. Thus, the square root of 9 (3) times the square root of 9 (3) equals 9. The square root of a number $X$ is symbolized by the radical sign: $\sqrt{X}$. The square root of a negative number yields an imaginary solution, so one ordinarily does not attempt this computation. Also, the square root of $X^2$ equals either $+X$ or $-X$. In statistical work, only the positive square root is normally considered.

## Logarithms

A bit more complicated than squares and square roots are logarithms—or, as they are more commonly and simply called, *logs*. A logarithm is said to have a base. The logarithms that you probably learned about in high school are called common logarithms and their base is 10. The logarithm of $X$ for base 10 is defined as the number $Y$ that satisfies the equation

$$X = 10^Y$$

So if $X = 100$, then $Y = 2$ because

$$100 = 10^2$$

It is said that 2 is the log of 100 with a base of 10.

The antilog of a number is that quantity whose logarithm would produce the number. For instance, 100 is the antilog of the number 2 with a base of 10.

In scientific work it is more common not to use 10 as the base but to use a special number $e$. The number $e$, like the number $\pi(3.14\ldots)$, which is used to compute the diameters and areas of circles, has unique mathematical properties. Both $\pi$ and $e$ have an infinite number of trailing digits. The value of $e$ to three decimal places is 2.718. The number 2.718 can be used to approximate $e$. The number $e$ is very useful in accounting. Say I had $X$ dollars and I found a banker who would give me a 100% interest rate compounded instantaneously. At the end of the period I would have $X$ times $e$ dollars. The number $e$ is also very useful in demography in projecting population growth.

Logarithms to the base $e$ are called *natural logarithms* and they are usually symbolized as $\ln(X)$. If the common log of a number is known, one can convert the common log into the natural log. The formula for doing so is

$$\ln(X) \approx 2.303 \log_{10}(X)$$

In words, the natural logarithm of a number approximately equals 2.303 times the common logarithm of a number.

There are a number of facts about logarithms that hold regardless of base. First, it should be noted that the logs of zero or a negative number are not defined, regardless of the base. Second, the log of the product of two numbers equals the sum of the logs of the two numbers. This second fact explains how a calculator or a computer might multiply two numbers. It could multiply by adding the logs of two numbers and then taking the antilog.

Because logarithms are difficult to determine, they are usually tabled. Logarithms are available on many hand-held calculators and almost all computers. On calculators common logs are usually denoted as $\log(X)$ and natural (base $e$) logs are usually denoted as $\ln(X)$.

## Summation Sign

Many times in statistical work it is necessary to add a set of scores. For instance, suppose the sum of the following ten numbers is needed: 76, 83, 41, 96, 38, 71, 87, 39, 66, and 99. Their sum can be denoted as

$$76 + 83 + 41 + 96 + 38 + 71 + 87 + 39 + 66 + 99$$

but that takes up too much space, and so simplification is needed. First, let $X_1$ stand for 76, $X_2$ for 83, and so on. The sum of the ten numbers can be represented by

$$X_1 + X_2 + X_3 + X_4 + X_5 + X_6 + X_7 + X_8 + X_9 + X_{10}$$

This is more compactly represented by

$$\sum_{i=1}^{n} X_i$$

which is read as the sum of the $X$'s. The symbol $\Sigma$ is called a *summation sign*. The terms below ($i = 1$) and above ($n$) the summation sign mean that the $X$'s are summed from $X_1$ to $X_n$. The terms $i = 1$ and the $n$ that are below and above the summation sign are generally omitted and implicitly understood when the intent is to add *all* of the numbers in the set. The symbol $\Sigma$ is a Greek capital letter sigma which sounds like an "s." The first letter of sum is an "s," and so a Greek "s" is used to stand for sum. The symbol $\Sigma X$ is read as "sum all the $X$'s."

One basic theorem for summation signs is

$$\sum k = nk$$

where $n$ is the number of terms that are summed and $k$ is a constant, such as 2.0. In words, summing a constant $n$ times equals $n$ times the constant.

A second theorem is

$$\sum (X + Y) = \sum X + \sum Y$$

In words, sum of the sums of two sets of numbers equals the sum of the sums of the sets added separately. It is easy to verify the truth of this theorem. Consider the scores of four persons on the variables $X$ and $Y$:

| Person | X | Y | X + Y |
|--------|-----|-----|-------|
| 1 | 9 | 12 | 21 |
| 2 | 13 | 11 | 24 |
| 3 | 6 | 9 | 15 |
| 4 | 10 | 8 | 18 |
| Total | 38 | 40 | 78 |

The sum of the $X$'s or $\Sigma X$ equals 38, and the sum of the $Y$'s or $\Sigma Y$ equals 40. It is then true that $\Sigma X + \Sigma Y$ equals 78. The sum of $X + Y$ also equals 78.

It is important that $\Sigma X^2$ be clearly distinguished from $(\Sigma X)^2$. The term $\Sigma X^2$ is the sum of the squared numbers: square first, then sum. The term $(\Sigma X)^2$ is sum of all the numbers, which is then squared: sum first, then square. So if the set of numbers is 4, 7, and 9, $\Sigma X^2$ equals $16 + 49 + 81$ or 146, and $(\Sigma X)^2$ equals $(4 + 7 + 9)^2$ or $(20)^2$ or 400.

# *Using a Calculator in Statistical Work*

The analysis of data is aided by the use of a calculator, and it has become an indispensable tool in statistical work. Even with the increasing availability of computers, a calculator remains a very important tool in data analysis. A calculator eases the burden of tedious computation. Before purchasing a calculator you should consider the following facts.

1. The calculator should be able to handle large numbers, up to 99,999,999 at least. When balancing a checkbook, most people do not need eight digits, but in data analysis numbers that large are often encountered. Even though the data may have only a few digits, certain computations requiring summing and squaring can easily result in large numbers.
2. A calculator should have a square root key. Looking up square roots in tables creates too much rounding error, and computing them by hand is time-consuming.

For this text, it is helpful to have a calculator that has a memory key. This key can be used to store intermediate values to many significant digits. It is also desirable for the calculator to have a key for logarithms. Preferably the calculator has a natural logarithm key (ln). If there is only a key for common logs (log), then one can first compute the common log and then multiply by 2.303 to obtain the approximate natural log value.

It is possible to purchase, at greater cost, calculators that perform some of the statistical methods described in this book. However, for the beginner these calculators can create problems because it may be quite difficult to determine whether an error has been made. After completing this course, you might consider purchasing one of these more sophisticated calculators.

There are a number of helpful hints that can increase the accurracy in your computation:

1. Because most rounding error is introduced by square root, division, and logarithm operations, one should perform them as late as possible in a calculation.
2. Save preliminary computations in the calculator's memory or, if that is not possible, on a piece of paper. So if you make a mistake near the end of the calculation, you do not have to start all over.
3. All calculations should be repeated to check for errors.
4. On many calculators you must hit the equal sign to complete the calculations. Make certain that you have hit the equal sign when you have completed the calculation. If you fail to do so, your final result may be incorrect.
5. Before beginning the calculations, hit the clear key. If you fail to do so, prior calculations may carry over to the next set of calculations.

# Summary

Numbers are part of modern life. They help us in making decisions in our own lives and in decisions made by society. Persons create the numbers, not machines.

A number refers to an object or a person and is assigned to that object or person by a set of rules, called the *operational definition*. The *variable* is the construct that the researcher is attempting to measure. A set of numbers is called a *sample*, which in turn is part of a larger set called the *population*. Procedures that summarize the sample data are called *descriptive statistics*, and procedures used to draw conclusions about the population are called *inferential statistics*.

The number can be measured at one of three levels. At the *nominal* level the numbers only differentiate the objects. Examples of nominal variables are gender, ethnicity, and political party. At the *ordinal* level the numbers differentiate and rank the objects. An example of an ordinal variable is the order of finish in a race. At the *interval* level the numbers differentiate, rank, and quantify the objects. Examples of interval variables are weight, height, and age. For a statistic to be properly interpreted, it must be measured at the appropriate level of measurement. The level of measurement for a given variable is determined by theory and experience.

A percentage is 100 times a proportion. The odds are a proportion divided by one minus the proportion. The logarithm of a number is defined as that exponent for a base that equals the number. The base used in scientific work is $e$, which equals approximately 2.718. The summation sign $\Sigma$ is commonly used to denote the sum of a set of numbers.

# Problems

1. For each of the following, identify the number, object, and variable.

   a. a score of 76 for John on the midterm
   b. $6.98 for a Rolling Stones album
   c. 28 EPA estimated mileage for the 1986 Ford Tempo
   d. Mary, a brown-eyed person
   e. Sue, the third person to arrive at the party
   f. soft drink A, which has 40 calories
   g. Joe, whose telephone area code is 202
   h. the Conolly building with 44 floors

2. State the level of measurement for the following variables.

   a. heartbeats in a one-minute period
   b. blood type
   c. birth order (e.g., firstborn, second-born)

    d. rating a movie from one to ten

    e. age

    f. ethnicity (e.g., black, white, Hispanic)

    g. army rank (e.g., captain, lieutenant)

    h. eye color (e.g., blue, brown)

3. Compute the following and round to the nearest hundredth.

    a. $\ln(67)$          b. $\ln(.55)$

    c. $(\sqrt{15})(6.1)^2 \ln(15)$      d. $71/\ln(.01)$

4. One person's mood is measured for 20 straight days. The 20 numbers can be treated as a sample. For this example, what are the objects?

5. Round the following numbers to the nearest hundredth.

    a. .524     b. −.325     c. .835     d. .5251

    e. −.483     f. −.12563     g. −.130     h. .355

6. Harrison (1980) studied reactions to offers of aid. The eight subjects in the experimental condition were offered help on a boring task. They subsequently rated on an eleven-point scale how uncomfortable they expected to feel in future interactions with the person who had offered help. The results are given below. A higher score indicates more discomfort.

$$10 \quad 1 \quad 3 \quad 5 \quad 3 \quad 7 \quad 7 \quad 3$$

Find the following quantities.

    a. $\Sigma X$          b. $\Sigma X^2$          c. $(\Sigma X)^2$

    d. $\Sigma(X - 1)$     e. $\Sigma X - 1$     f. $\Sigma(X - 1)^2$

7. Cutrona (1982) asked 162 college freshmen what events or situations triggered loneliness. The percentage of responses in each category is listed below.

| Category | Percent |
| --- | --- |
| A. Leaving family and friends | 40 |
| B. Breakup of relationship | 15 |
| C. Problems with friends | 11 |
| D. Family problems | 9 |
| E. Academic difficulties | 11 |
| F. Living in isolation | 6 |
| G. Fraternity or sorority rejection | 3 |
| H. Medical problems | 2 |
| I. Birthday forgotten | 1 |

    a. Compute the proportion of responses in each category.

    b. Compute the odds of a response occurring in each category. Round the odds to two decimal places.

8. A researcher has 20 male subjects study a list of 40 words for five minutes. They are then asked to recall the 40 words. The researcher develops a memory score, which is the percentage of words correctly recalled. Identify

   a. the objects
   b. the data
   c. the variable
   d. the operational definition
   e. the units of measurement

9. Round the following numbers to the nearest tenth.

   a. .6666   b. −.333   c. .66   d. .55
   e. −.450   f. −.451   g. −.4501   h. .4999

10. Compute natural logarithms of the following numbers and round the result to the nearest thousandth.

   a. 15   b. 21   c. .33   d. .19
   e. 2.718   f. 1.000   g. 10.00   h. .3333

11. Consider a set of houses on one side of the street with the following addresses.

$$101, 121, 141, 181, 201, 221, 223$$

   At what level of measurement are these numbers?

12. For the following sample of numbers,

$$1, 6, 7, 4, 3, 6, 4$$

   compute

   a. $\Sigma X$   b. $\Sigma X^2$   c. $(\Sigma X)^2$
   d. $\Sigma(X - 1)$   e. $\Sigma X - 1$   f. $\Sigma(X - 1)^2$

PART 2

# Descriptive Statistics

# 2 | *The Distribution of Scores*

I think that we all enjoy going to a shopping mall and sitting and watching people go by. It is amazing how many sizes and shapes we observe in a very brief period of time. Some people are very tall and muscular with no necks. Some are absurdly overweight and do not seem to walk but to waddle. Some are so pencil thin that it is difficult to understand what keeps their pants on. Some have fat legs and skinny arms. People certainly come in different shapes and sizes.

Though not as fascinating as people, groups of numbers or samples also come in different shapes and sizes. The technical and more general name for the shape of a sample of numbers is *distribution*.

Numbers by themselves can overwhelm us. One purpose of data analysis is to simplify the presentation of the numbers so that their meaning becomes more apparent. Systematically rearranged numbers make much more sense than raw data.

To facilitate the comprehension of how to understand the distribution of a set of numbers, consider a report by Smith (1980). She reviewed 32 studies on the gender bias of therapists.[1] These studies investigated whether clinical psychologists, psychiatrists, and counselors were prejudiced toward one gender or the other. In each study, one or more therapists advised male and female clients. It was possible to measure whether the advice given to the *male* clients was more positive than the advice given to *female* clients. For instance, one study that Smith reviewed asked whether career counselors encouraged males to enter higher-prestige occupations than females. Smith's

---

[1]Smith included 34 studies in her report. For both pedagogical and scientific reasons, two studies that she felt were of low quality are dropped. For studies with more than one outcome, the outcomes are simply averaged. Some of the studies in Smith's review did not compare the reaction of therapists to males and females but rather compared their reaction to a gender role stereotypic person versus a nonstereotypical person. In these studies bias toward the stereotypic person was coded positively and bias toward the nonstereotypic person was coded negatively.

index works as follows: A positive score indicates a male bias, zero indicates neutrality, and a negative score indicates a female bias. A small amount of gender bias would be indicated by a score of $\pm.2$, a moderate gender bias by a score of $\pm.5$, and a large bias by a score of $\pm.8$. As an example, a score of .23 would indicate that therapists reacted more positively to males, but the difference is small. The scores for her 32 studies are contained in Table 2.1.

# The Frequency Table and Histogram

It is difficult to make any sense immediately out of the 32 numbers as they are presented in Table 2.1. With a little study, however, some patterns do appear in the data. There seem to be just about as many studies with negative scores as positive scores. Thus the therapists do not seem to be consistently biased one way or the other. But this is much too coarse a judgment. A way of rearranging the numbers is needed so that their meaning can be better understood.

The first thing that can be done with sample data is to rank order them from smallest to largest. Recall that a *larger* negative number, such as $-1.03$, is farther from zero than a *smaller* negative number, such as $-.56$. The rank ordering of Smith's scores is as follows:

|       |       |     |     |
|-------|-------|-----|-----|
| −1.03 | −.23  | .00 | .14 |
| −.56  | −.22  | .00 | .23 |
| −.40  | −.10  | .00 | .24 |
| −.36  | −.03  | .01 | .29 |
| −.31  | .00   | .01 | .35 |
| −.31  | .00   | .02 | .56 |
| −.23  | .00   | .05 | .56 |
| −.23  | .00   | .11 | .60 |

The picture is now slightly less confused. Overall there is one more study that shows a bias favoring males than ones favoring females, but a fairly large number of studies show little or no gender bias.

There is still too much detail that gets in the way of understanding the

*TABLE 2.1*   **Data for the Studies of Gender Bias**

|       |       |       |       |
|-------|-------|-------|-------|
| .29   | .01   | −.40  | .00   |
| .56   | −.31  | .00   | .35   |
| .00   | .14   | .02   | .11   |
| −.31  | −.22  | −.03  | −.23  |
| .56   | .01   | .00   | −.56  |
| −1.03 | .00   | −.23  | .23   |
| .00   | −.10  | .05   | .00   |
| .60   | −.23  | .24   | −.36  |

numbers. One way of reducing the confusion is to remove some of the detail. It is not that crucial to know that one study has a score of .24 and another a score of .29. What is needed is to gather those studies having similar scores into groups or classes. Therefore the measure of gender bias is divided into intervals of a certain width, and the number of scores that fall within each interval is tallied. Each interval has a *lower limit,* which is the lowest possible score that can fall in the interval, and an *upper limit,* which is the highest possible score that can fall in the interval.

The *class interval* then defines the range of possible scores that can be a member of a given class. For instance, for the class interval .10 to .29, the lower limit is .10 and the upper limit is .29. So any score that falls between .10 and .29 would fall in that class. The complete set of class intervals for the gender bias data is as follows:

| Class Interval | Frequency | Relative Frequency |
|---|---|---|
| −1.10 to −.91 | 1 | 3 |
| −.90 to −.71 | 0 | 0 |
| −.70 to −.51 | 1 | 3 |
| −.50 to −.31 | 4 | 12 |
| −.30 to −.11 | 4 | 12 |
| −.10 to .09 | 13 | 41 |
| .10 to .29 | 5 | 16 |
| .30 to .49 | 1 | 3 |
| .50 to .69 | 3 | 9 |
| Total | 32 | 99 |

In the first column are the classes, each with its lower and upper limit. The *class width* is defined as the difference between adjacent lower limits; in this case the class width is .20. The second column gives the *frequency* or number of cases in the class interval; for example, four studies have scores between −.30 and −.11. The final column gives the *relative frequency,* which is 100 times the frequency divided by sample size. For instance, for the interval −1.10 to −.91 the frequency is 1 and the sample size is 32 making the relative frequency 100 times 1 divided by 32 which equals 3, when rounded to the nearest whole number. Relative frequencies need not always be calculated, but they are especially informative when two different samples with different sample sizes are being compared.

The complete table of class intervals, frequencies, and relative frequencies is called a *frequency table.* The frequency table for Smith's scores shows that the scores cluster around zero, with about as many studies showing a male gender bias as a female gender bias. Scores between −.1 and +.1 can be considered as showing virtually no gender bias. The relative frequency of scores in the −.10 to .09 class is 41%, and therefore 41% of the studies show little or no gender bias.

Sometimes it is useful to compute the cumulative frequency. As the name implies, the *cumulative frequency* for a given class is the sum of all the frequencies below or equal to the upper limit of that class. So for the interval of −.70 to −.51, the cumulative frequency is $1 + 0 + 1 = 2$. It is also possible to compute the cumulative relative frequency by dividing the cumulative frequency by the sample size.

A graph of the frequency table is called a *histogram*. A graph has two lines that intersect at a right angle. These lines are called axes. The horizontal axis in a graph is called the $X$ axis. The vertical axis is called the $Y$ axis.[2] In a histogram the class intervals are on the $X$ axis. The frequency, raw or relative, is on the vertical or $Y$ axis. The resulting graph for the Smith data is presented in Figure 2.1. The histogram shows the shape of the 32 scores. The dominant feature of the graph is the peak in the middle of the numbers at about zero.

The basic steps to determine the shape of a sample of numbers are then

1. rank ordering the scores,
2. grouping the scores by class intervals, and
3. graphing the frequency table.

These steps are now discussed in more detail and generality.

## Rank Ordering

This step is fairly simple. The numbers are ordered from smallest to largest. Although this step is not absolutely necessary, it is generally advisable to do it for the following reasons. First, the mere rank ordering already begins to describe the shape of the scores. Second, it makes the steps of creating a frequency table and a histogram easier because the frequency of scores that

*FIGURE 2.1*    **Histogram for 32 studies of gender bias of therapists.**

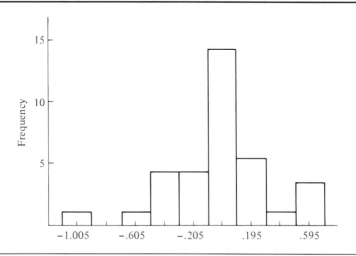

fall in a given interval can be quickly determined. Third, the computation of various statistical summaries that are discussed in the following chapters require a rank ordering of the numbers.

## *Grouping*

To group the data, class intervals must be created. The first question is that of how many classes there should be or, correspondingly, how wide the class intervals should be. One reasonable guideline is to have between 8 and 15 classes. To determine the width of the class interval, the smallest score in the sample is subtracted from the largest score. This quantity is divided by 8 to determine the maximum class width and by 15 to obtain the minimum class width. For the gender bias studies, the smallest score is −1.03 and the largest is .60. The maximum class width is 1.63/8 or .20 and the minimum class width is 1.63/15 or .11. So the class width should be somewhere around .20 and .11.

The second guideline in determining class width is the sample size. The number of scores (the sample size) divided by the number of classes should be at least three. Stated differently, the average number of scores per class should be at least three. Given 32 scores, dividing by 15 classes, there are about only two observations per class. If there were 8 classes, there would be four observations per class. Thus .20 as the class width with about 8 classes seems like a good choice. It is perfectly permissible to have more than 15 classes when the sample size is large (more than 60), and with small sample sizes (less than 20) there should be fewer than 8 classes.

Once the class interval has been chosen, the lower limit of the lowest class interval or the *lowest lower limit* must be determined. The lowest class limit is the smallest score in the sample, rounded down to a convenient number. For the gender bias study, the smallest score is −1.03, and rounding down yields −1.10.

It is often necessary to try out a number of alternative class widths and lowest lower limits and see which works out best. Also certain features of the data must be considered when choosing these values. For instance, for the gender bias data, I made sure that zero, which indicates no bias, was near the middle of a class interval. To have zero near the middle of the interval was accomplished by having −1.10 as the lowest lower limit and not some other value such as −1.05.

In determining the class width, attention should be paid to the unit of the distribution. The *unit of the distribution* is the smallest possible difference between a pair of scores. For the gender bias study, the unit of the distribution is hundredths or .01. The lower limit must be in the same unit and the class width should be an integer multiple of the unit. For example, consider the Milgram study (1963), which is presented in detail in the next chapter. Some of the data are

300, 315, 450, 345, 330, 375

All of these scores are in multiples of 15, so the unit of the distribution is 15. The class widths should be in multiples of 15. Thus 30 and 45 are acceptable class widths, whereas 10 and 40 are not.

## Graphing

Although the tally of the number of scores in an interval is helpful, a graph or histogram of the tally is even more informative. Though trite, it is still true that a picture is worth a thousand words.

In a histogram, the $X$ axis or horizontal axis is the variable of interest divided into classes. The usual convention is to demarcate the $X$ axis in a histogram by the class midpoints. The *midpoint of a class interval* is defined as half the sum of the upper limit of the class interval and the lower limit. So for the interval .10 to .29, the midpoint is $(.29 + .10)/2$, which equals .195. To have midpoints that do not have the additional trailing digit (the 5 in .195), it is advisable to have class intervals whose widths are odd. For the gender bias data, a class width of .15 or .25 might be a good alternative to .20.

The $Y$ axis or vertical axis in a histogram is the frequency for a class. Either the frequency or the relative frequency can serve as the $Y$ axis. Occasionally the cumulative frequency is used.

# Outliers

The procedures that have been described are especially helpful in identifying outliers. An *outlier* is a score in the sample that is considerably larger or smaller than most of the other scores. In Chapter 4 a quantitative definition of "considerably" larger or smaller will be given.

As an example of outliers consider a second data set. DePaulo and Rosenthal (1979) had a number of persons, called targets, describe someone they liked. Forty persons, called perceivers, subsequently viewed videotapes of the targets' descriptions. The perceiver judged on a nine-point scale how much the target liked the person that the target was describing. These ratings of liking made by each perceiver were then averaged across the targets that the perceiver viewed. Although none of the targets lied in their descriptions, the perceivers were led to believe that some of the targets may have been lying. High average liking scores for a perceiver indicate that the perceiver correctly judged the targets as liking the persons that they were describing. Low scores indicate inaccuracy. The rank-ordered scores for the 40 perceivers are as follows:

| | | | | |
|---|---|---|---|---|
| 1.37 | 5.70 | 6.16 | 6.55 | 7.05 |
| 2.21 | 5.70 | 6.16 | 6.63 | 7.05 |
| 3.21 | 5.80 | 6.26 | 6.63 | 7.26 |
| 4.65 | 5.95 | 6.35 | 6.65 | 7.30 |
| 5.05 | 6.05 | 6.40 | 6.73 | 7.35 |
| 5.55 | 6.05 | 6.42 | 6.75 | 7.35 |
| 5.65 | 6.10 | 6.45 | 6.89 | 7.89 |
| 5.70 | 6.15 | 6.52 | 7.00 | 7.94 |

The largest possible score is 9.00 and the smallest is 1.00. The numbers seem to cluster around 6, and there seem to be some rather small numbers.

Because the numbers are already rank ordered, the class width must now be determined. Because the largest score is 7.94 and the smallest is 1.37, the maximum class width is (7.94 − 1.37)/8 or .82 and the minimum class width is (7.94 − 1.37)/15 or .44. Two possible choices are .50 or .75. A .75 class width seems more reasonable than .50. First, .50 would result in 14 classes and an average of only 2.8 persons per class. Recall that at least three persons should be in each class. Second, because .75 is odd, the histogram would have class midpoints with two digits, not three as would happen if .50 were used. Because the lowest score is 1.37, rounding down to 1.00 yields the lowest lower limit. The resulting frequency table is

| Class Interval | Frequency | Relative Frequency |
|---|---|---|
| 1.00 to 1.74 | 1 | 2.5 |
| 1.75 to 2.49 | 1 | 2.5 |
| 2.50 to 3.24 | 1 | 2.5 |
| 3.25 to 3.99 | 0 | 0.0 |
| 4.00 to 4.74 | 1 | 2.5 |
| 4.75 to 5.49 | 1 | 2.5 |
| 5.50 to 6.24 | 13 | 32.5 |
| 6.25 to 6.99 | 13 | 32.5 |
| 7.00 to 7.74 | 7 | 17.5 |
| 7.75 to 8.49 | 2 | 5.0 |
| Total | 40 | 100.0 |

(In this case, for the relative frequency, it is sensible to round to the first decimal point because many of the numbers have a 5 at that point.) A histogram of the frequency table is presented in Figure 2.2.

An examination of both the frequency table and the histogram shows that the scores cluster near 6.25. Quite clearly the values 1.37, 2.21, and 3.21 are considerably smaller than the other 37 scores. They are all outliers.

After an outlier has been identified in the sample, it must be carefully considered why it is that the score is so atypical. Outliers are due to one of two reasons. First, they may be caused by a computational or data entry mistake. For instance, the recording of 6.42 as 642 would result in an outlier. Second, the outlier is not the result of a mistake, but rather it is generated by a different process than the other numbers. For instance, an abnormal physiological

*FIGURE 2.2*    **Histogram for the DePaulo and Rosenthal data.**

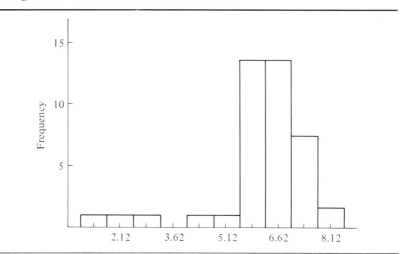

reading often indicates the presence of a disease. In industrial work, the presence of extremely large or small readings has often led to the discovery of a new manufacturing process. The outlier may be telling the researcher that the object is very different in some way from the others.

The outliers in the DePaulo and Rosenthal data are not the result of a computational mistake. Rather, they are attributable to different cognitive processes operating for the perceivers who obtained low scores. Before viewing the videotape the perceivers were led to believe in some of the descriptions the targets would be lying. That is, the target would be pretending to like someone they did not actually like. It is then plausible that the perceivers who have very low liking scores have these low scores because these perceivers felt that the targets were lying about liking the people that they were describing. Apparently most of the other perceivers took the targets at face value, but these three perceivers with low scores were very suspicious.

Although a frequency table and histogram were not necessary for the identification of the three outliers, they certainly facilitate that process. As will be seen in later chapters, the identification of outliers is an essential step in data analysis.

# Features of Distributions

People have certain characteristic shapes: fat, thin, muscular, and so on. When looking at distributions of numbers, there is a parallel set of descriptive categories.

Before detailing these categories, a description of the figures that will be used to illustrate the characteristics of distributions must be presented. A histogram of actual data, as in Figures 2.1 and 2.2, is quite jagged. However, if there were many scores and it was possible to have a very narrow class width, then the histogram would look quite smooth. When speaking about characteristics of distributions, it is helpful to consider these idealized distributions with large sample sizes and very narrow class widths. In practice, actual histograms only approximate the distributions that will be presented.

One of the first things examined in a histogram or frequency table are peaks in the distribution. A *peak* in a distribution is a frequency or set of adjacent frequencies that are larger than most of the other frequencies. So for the gender bias data, there is a peak at the class interval from −.10 to .09 because it has the highest frequency of 13.

Whereas most distributions have a single peak, some have more than one peak. Distributions that have two peaks are called *bimodal* distributions. For instance, Hammersla (1983) measured the duration that college students played a game. She was interested in whether paying someone to play a game they already found fun would decrease their desire to play the game. Subjects were observed for a period of five minutes, and Hammersla measured the duration of game playing in seconds. Her histogram had two peaks, one near 0 seconds and the other near 300 seconds. Evidently subjects either played or did not play the game during the entire time period. When two different types of persons—such as those who like a game and those who no longer do—are mixed together, a bimodal distribution can result.

Examples of distributions with different types of peaks are presented in Figure 2.3. The top distribution has a single peak in the center. The middle distribution is bimodal, and the bottom has a peak on the left side of the distribution.

Besides the peaks, the low frequencies are also informative to the data analyst. For most distributions the smallest frequencies occur for the very large and very small values of the variable. For example, usually in a test few students do very well or very poorly. Most students fall in the middle. Because this pattern of dwindling frequencies at the extremes looks like a tail, the frequencies for the very large and very small values of a variable are called *tails*. The tails of distributions can be either fat or skinny. In Figure 2.4 are examples of distributions whose tails are either fat or skinny, or both. The top distribution has skinny tails, the middle one fat tails, and the one on the bottom has a fat left tail and a skinny right tail.

The size of the tails and the peak are related. For a distribution that is peaked in the center, the higher the peak in the distribution the skinnier are the tails, and the lower the peak the fatter are the tails. Distributions with a very high peak in the center and skinny tails are said to be *leptokurtic*. Distributions with a low peak in the center and fat tails are said to be *platykurtic*. In Figure 2.4 the upper distribution is said to be leptokurtic and the middle one is said to be platykurtic.

*FIGURE 2.3*     **Examples of different types of peaks.**

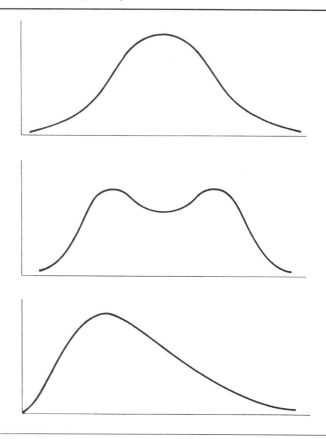

Some distributions have no peak at all. Distributions in which all scores are equally likely are called *flat* or *rectangular distributions*. The following set of scores have a flat distribution

$$3, 3, 4, 4, 5, 5, 6, 6, 7, 7$$

because each score occurs twice. The histogram for this flat distribution is contained in Figure 2.5. Perfectly flat distributions of naturally occurring variables are rarely encountered in the social and behavioral sciences. The distributions of most variables have a peak.

A distribution is said to be *symmetric* if its shape is such that if the data were regraphed, reversing the order of the class intervals, the shape would not change. Stated equivalently, a distribution is symmetric if its shape does not change when its mirror image is examined. If a perfectly symmetric distribution is plotted on a piece of paper and the paper is folded vertically, the two

*FIGURE 2.4*   **Examples of different types of tails.**

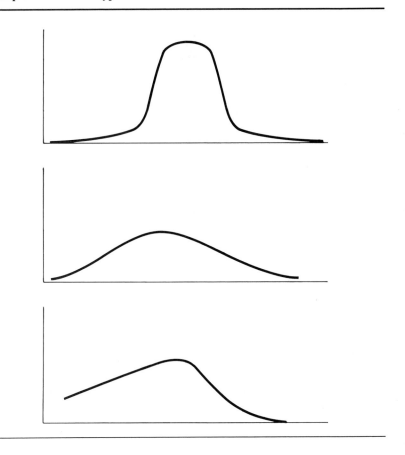

*FIGURE 2.5*   **Example of a flat distribution.**

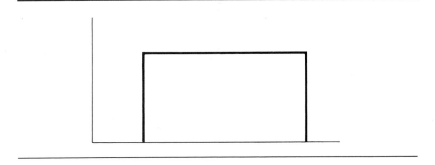

sides of the histogram should completely coincide. Figure 2.6 shows examples of symmetric and asymmetric distributions. The top two distributions are symmetric, whereas the bottom two are asymmetric.

Some asymmetric distributions are said to be skewed. A *skewed distribution* is one in which the frequencies for the class intervals trail off in one direction but not the other. The direction of skew is determined by the skinny

*FIGURE 2.6*    **Examples of symmetric and asymmetric distributions.**

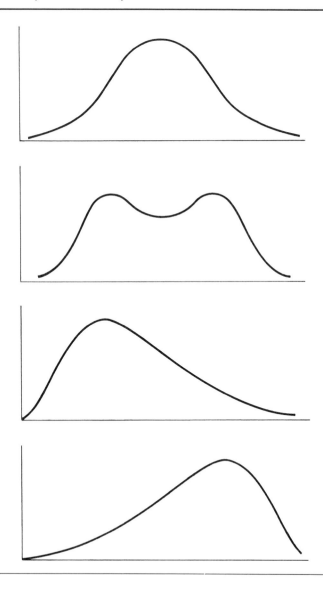

tail. The skinny tail can be on the left- or right-hand side along the $X$ axis. If the trailing values are to the left of the peak, *negative skew* is present. Trailing values to the right indicate *positive skew*. Because many distributions have lower limits of zero, distributions with a positive skew are quite common. Examples of variables with a positive skew are income, number of home runs, traffic accidents per week, number of bar presses by a laboratory rat, and the number of children per family. Scores on an easy test have a negative skew and scores on difficult test have a positive skew. The bottom distribution in Figure 2.6 has a negative skew and the one above it has a positive skew.

One type of distribution—commonly assumed in data analysis—is called the normal distribution. The *normal distribution* is a symmetric distribution with a single peak in the center. The resultant shape is often called bell-shaped. The normal distribution is discussed in detail in Chapter 10.

# Stem and Leaf Display

The frequency table and the histogram are traditional ways of arranging and displaying a sample of numbers. A newer, simpler, and more elegant procedure has been developed by John W. Tukey, called a *stem and leaf display*. In a stem and leaf display the classes are called *stems*. With a stem and leaf display there may be more stems than the 8 to 15 classes that are in a frequency table. The *stem* is essentially the lower limit of a class. So for instance, for the gender bias data, the stem could be the first two digits from the left of the score: −1.0, −0.9, −0.8 and so on. Then entered are the leaves, which are the trailing digit or the next digit to the right after the stem. So for the number 0.56 the stem is 0.5 and the leaf is 6. If there were any trailing digits to the right of the 6, they would be dropped.

The stems are arranged in a vertical order and to their right a vertical line is drawn. On the right of the line, the leaves are entered, one for each score in the sample. Each leaf is entered next to its stem. The stems can be separated by commas, but the common practice is not to do so. After the display has been completed, it is customary to redraw the display by rank ordering the leaves within each stem. The stem and leaf display for the gender bias data is presented in Table 2.2. Both the unranked (entering the leaves as they are presented in Table 2.1) and the ranked displays are presented. The stem and leaf display very clearly shows the peak at zero.

There are two features of the display that must be noted. First, zero has plus and minus stems of −.0 and +.0 stems. This looks odd but it is necessary to have categories for numbers from −.09 to −.00 and from +.00 to +.09. Second, in the ranked display on the right of Table 2.2, the leaves of the negative numbers appear to be ranked backward. They are not, because −.36 is less than −.31.

For the stem and leaf display of the DePaulo and Rosenthal data, the first

*TABLE 2.2*    **Unranked and Ranked Stem and Leaf Displays for the Gender Bias Data**

| | Unranked | | Ranked |
|---|---|---|---|
| −1.0 | 3 | −1.0 | 3 |
| −.9 | | −.9 | |
| −.8 | | −.8 | |
| −.7 | | −.7 | |
| −.6 | | −.6 | |
| −.5 | 6 | −.5 | 6 |
| −.4 | 0 | −.4 | 0 |
| −.3 | 116 | −.3 | 611 |
| −.2 | 2333 | −.2 | 3332 |
| −.1 | 0 | −.1 | 0 |
| −.0 | 3 | −.0 | 3 |
| .0 | 00110020500 | .0 | 00000001125 |
| .1 | 41 | .1 | 14 |
| .2 | 934 | .2 | 349 |
| .3 | 5 | .3 | 5 |
| .4 | | .4 | |
| .5 | 66 | .5 | 66 |
| .6 | 0 | .6 | 0 |

digit from the left in the number might be used as the stem and the second digit from the left as the leaf. It is common practice to just drop any other digit. And so, the ranked stem and leaf display for the DePaulo and Rosenthal data is

| | |
|---|---|
| 1 | 3 |
| 2 | 2 |
| 3 | 2 |
| 4 | 6 |
| 5 | 05677789 |
| 6 | 0011112344455666778 |
| 7 | 000233389 |

It should be noted that for each entry the third digit from the left was dropped. Thus, some information was lost, but that always happens in descriptive statistics. Why is the trailing digit dropped and why is the digit not rounded? For a stem and leaf display rounding really does not make much difference and so it is preferable to do the simpler thing by dropping the trailing digit.

Because for the stem and leaf display of the DePaulo and Rosenthal data, almost half the numbers pile up on 6, it is better to split the stems in half. Thus, for the stem 6.0, there are two stems of 6.0 and 6.5. The leaves are the second digit of the original scores. Here is the display with the stems split in half.

| 1.0 | 3 |
|-----|---|
| 1.5 |  |
| 2.0 | 2 |
| 2.5 |  |
| 3.0 | 2 |
| 3.5 |  |
| 4.0 |  |
| 4.5 | 6 |
| 5.0 | 0 |
| 5.5 | 5677789 |
| 6.0 | 00111123444 |
| 6.5 | 55666778 |
| 7.0 | 0002333 |
| 7.5 | 89 |

Having more stems provides a better view of the distribution and shows the outliers more clearly.

After one stem and leaf display has been constructed, it is very simple to construct another. The stems could be separated into fifths: 6.0, 6.2, 6.4, 6.6, and 6.8. To prevent the display from being too long, the three outliers are not displayed. The display can be reworked with the stems split in fifths, as follows:

| 4.6 | 6 |
|-----|---|
| 4.8 |  |
| 5.0 | 0 |
| 5.2 |  |
| 5.4 | 5 |
| 5.6 | 6777 |
| 5.8 | 89 |
| 6.0 | 001111 |
| 6.2 | 23 |
| 6.4 | 44455 |
| 6.6 | 66677 |
| 6.8 | 8 |
| 7.0 | 000 |
| 7.2 | 2333 |
| 7.4 |  |
| 7.6 |  |
| 7.8 | 89 |

The stem and leaf display has some important advantages over the more traditional frequency table and histogram. First, it is easier and faster to prepare than a frequency table or a histogram. Second, except for the dropped digits, the raw data can be recovered. Third, the display can be used to compute various statistical summaries.

# Smoothing the Frequencies

In creating the class intervals, the class width and the lowest lower limit must be chosen. For instance, for the gender bias data, the class width was set at .20 and the lowest lower limit at $-1.10$. These choices can affect the shape that is shown in the histogram. That is, a class width of .30 and a lowest lower limit of 1.20 might considerably alter the shape of the histogram. This section of the chapter describes a procedure called *smoothing,* which takes a histogram and yields a shape that would be essentially the same regardless of the choice of class width or lowest lower limit.

Smoothing is a way of mathematically adjusting the frequencies to remove the rough edges. The class frequencies can be denoted $f_1, f_2, f_3$, and so on, where $f_1$ is the frequency for the lowest scores, $f_2$ for second lowest class interval, and so on. The *smoothed frequency* for a class interval is one-half the frequency for that interval plus one-quarter the frequency of each adjacent frequency. So for the gender bias data, $f_4$ is 4 and its adjacent frequencies are 1 for $f_3$ and 4 for $f_5$. The smoothed frequency for the fourth class interval is $(.5)(4) + (.25)(1 + 4)$, which equals 3.25. In terms of a formula, the smoothed frequency for the class interval i is

$$.5f_i + .25(f_{i-1} + f_{i+1})$$

A problem arises when the first and last class frequencies are smoothed. They each have only one adjacent frequency. Two new classes must be added, one just before the smallest class interval and one just after the largest class interval. Before smoothing, these classes have zero frequencies. The frequencies of these two new classes can be smoothed by taking one-quarter of the adjacent frequency. The sum of the smoothed frequencies should always equal sample size.

The smoothed frequencies for the Smith study are as follows:

| Class Interval | Observed Frequency | Smoothed Frequency | Relative Smoothed Frequency |
|---|---|---|---|
| $-1.30$ to $-1.11$ | 0 | $.25 = .5(0) + .25(0 + 1)$ | 1 |
| $-1.10$ to $-.91$ | 1 | $.50 = .5(1) + .25(0 + 0)$ | 2 |
| $-.90$ to $-.71$ | 0 | $.50 = .5(0) + .25(1 + 1)$ | 2 |
| $-.70$ to $-.51$ | 1 | $1.50 = .5(1) + .25(0 + 4)$ | 5 |
| $-.50$ to $-.31$ | 4 | $3.25 = .5(4) + .25(1 + 4)$ | 10 |
| $-.30$ to $-.11$ | 4 | $6.25 = .5(4) + .25(4 + 13)$ | 20 |
| $-.10$ to $.09$ | 13 | $8.75 = .5(13) + .25(4 + 5)$ | 27 |
| $.10$ to $.29$ | 5 | $6.00 = .5(5) + .25(13 + 1)$ | 19 |
| $.30$ to $.49$ | 1 | $2.50 = .5(1) + .25(5 + 3)$ | 8 |
| $.50$ to $.69$ | 3 | $1.75 = .5(3) + .25(1 + 0)$ | 5 |
| $.70$ to $.99$ | 0 | $.75 = .5(0) + .25(3 + 0)$ | 2 |
| Total | | 32.00 | 101 |

(The total relative frequencies is 101 because of rounding error.) Note how much smoother and simpler the frequencies are after smoothing. The distri-

*FIGURE 2.7*     **Histogram of the smoothed frequencies for gender bias data.**

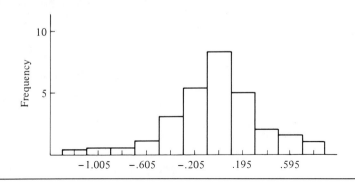

bution is clearly peaked near zero and is fairly symmetric. This can be seen best by graphing the smoothed frequencies in a histogram as in Figure 2.7 and comparing it with the unsmoothed histogram, which is reproduced in Figure 2.8.

Although smoothing does remove the rough edges in a frequency table, it also alters the actual frequencies. The effect of smoothing is that the peaks are lowered and the tails of the distribution are fattened. Moreover, smoothing can result in some anomalous results. For instance, if the DePaulo and Rosenthal data set is smoothed, it would happen that .25 person scored in the interval .25–.99. However, because 1.00 is the lowest possible score on the DePaulo-Rosenthal measure, the result makes no sense. Smoothing provides

*FIGURE 2.8*     **Histogram of the unsmoothed frequencies for gender bias data.**

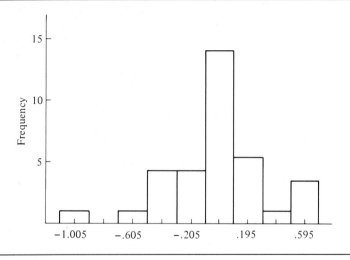

a clearer picture, but at the cost of removing important details in the data and producing an anomalous class.

# Summary

A sample of numbers can be summarized by grouping the numbers into a set of classes. Each class has a lower limit, which is the lowest possible score that can fall into the class and an upper limit. The *class width* is the difference between adjacent lower limits. The *class midpoint* is the average of the class's lower and upper limits. The number of scores that falls into a class is called the *class frequency*. The *relative frequency* of a class is the class frequency divided by the total number of scores. A table of class intervals, class frequencies, and relative frequencies is called a *frequency table*. A graph of the frequency table is called a *histogram*. In a histogram the *X* axis is the variable that is divided into classes, and the *Y* axis is frequency.

Highly deviant scores are called *outliers,* and they should be noted. The researcher should discover whether the outliers are due to a computational error or to a different process.

When a distribution is examined, the number and location of peaks should be noted. A distribution with two peaks is said to be *bimodal*. The *tails* of the distribution are the frequencies on the far left and right of the distribution. Distributions are characterized as having fat or skinny tails. A distribution with skinny tails is said to be *leptokurtic*. A distribution with fat tails is said to be *platykurtic*. A distribution with no peak at all is said to be *flat* or *rectangular*.

A distribution is said to be *symmetric* if, when the *X* axis is reversed, the shape does not change. A distribution with a peak on one side and a skinny tail on the other is said to be *skewed*. A *positive skew* has a skinny tail on the right, and a *negative skew* has a skinny tail on the left. A *normal distribution* is a unimodal and symmetric distribution that looks bell shaped.

A set of data can also be summarized by a *stem and leaf display,* which is a type of vertical histogram. The stems correspond to lower class limits and the leaves to the scores.

The shape of this histogram can be smoothed so that its true shape can be better revealed and so that chance fluctuation due to grouping is reduced.

# Problems

1. Prepare the frequency table for the following data using 46 as the lowest lower limit and 5 as the class width:

   68, 73, 81, 76, 83, 96, 76, 83, 95, 81, 48, 56,
   75, 79, 90, 73, 76, 77, 84, 63, 68, 65, 62, 70

2. Draw histograms for the following shapes.

   a. a single-peaked asymmetric distribution
   b. an asymmetric bimodal distribution
   c. a flat distribution
   d. a unimodal leptokurtic distribution
   e. a symmetric bimodal distribution

3. Why must a single-peaked, symmetric distribution be peaked in the middle?

4. Can a flat distribution be bimodal?

5. For the data in Table 2.1 prepare a frequency table using a class width of .20 and a lowest lower limit of −1.20. Smooth the frequency table.

6. The following sample of numbers consists of the rents of apartments listed for rent in a university town.

   | | | | | |
   |---|---|---|---|---|
   | 298 | 288 | 300 | 300 | 385 |
   | 310 | 230 | 385 | 325 | 375 |
   | 350 | 300 | 265 | 340 | 310 |
   | 285 | 260 | 425 | 275 | 300 |
   | 320 | 275 | 300 | 310 | 285 |
   | 260 | 375 | 295 | 250 | 275 |
   | 385 | 310 | 380 | 265 | 285 |
   | 310 | 300 | 310 | | |

   a. Discuss the choice of class width and lowest lower limit.
   b. Construct a frequency table showing the frequency of rents in each class interval. Use 25 as the class width and 226 as the lowest lower limit.
   c. Describe the shape of the distribution.

7. For the data in problem 6, construct a histogram of the frequencies.

8. Below is the life expectancy in years at birth of males and females in the 30 most populous countries.

   | *Male* | | *Female* | |
   |---|---|---|---|
   | 68.7 | 65.2 | 76.5 | 71.4 |
   | 45.8 | 57.6 | 46.6 | 61.0 |
   | 48.6 | 59.9 | 51.5 | 63.3 |
   | 59.2 | 51.6 | 62.7 | 53.8 |
   | 36.5 | 69.0 | 39.6 | 76.9 |
   | 68.3 | 41.9 | 74.8 | 40.6 |
   | 47.5 | 57.6 | 47.5 | 57.4 |
   | 69.0 | 72.2 | 74.9 | 77.4 |
   | 63.0 | 62.8 | 67.0 | 66.6 |
   | 37.2 | 53.7 | 36.7 | 48.8 |
   | 56.9 | 66.9 | 60.0 | 74.6 |
   | 49.8 | 69.7 | 53.3 | 75.0 |
   | 53.6 | 53.7 | 58.7 | 53.7 |
   | 64.0 | 67.8 | 74.0 | 73.0 |
   | 43.2 | 41.9 | 46.0 | 45.1 |

  a. Construct a frequency table for the males and another for the females using 35.0 as the lowest lower limit and 5.0 as the class width.
  b. Compare the two frequency tables. Do women live longer than men?

9. For the data in problem 8 construct a stem and leaf display for both males and females.

10. Smooth the frequencies for the data in problem 8.

11. Prepare a frequency table for the DePaulo and Rosenthal data using 1.25 as the lowest lower limit and .75 as the class width.

12. Near the end of the fall semester, Harrison (1984) gave the UCLA Loneliness Scale to freshman women living in dormitories. The possible scores on the scale range from 20 to 80, higher scores indicating more loneliness. Below are the results for the women who were assigned to their dormitories.

| Class Interval | Frequency |
|---|---|
| 23 to 25 | 2 |
| 26 to 28 | 9 |
| 29 to 31 | 11 |
| 32 to 34 | 6 |
| 35 to 37 | 2 |
| 38 to 40 | 3 |
| 41 to 43 | 1 |
| 44 to 46 | 5 |
| 47 to 49 | 2 |
| 50 to 52 | 3 |
| 53 to 55 | 2 |

  a. Compute the relative frequencies.
  b. Smooth the observed frequencies.
  c. Compute the relative smoothed frequencies.
  d. Compute the cumulative frequencies (of the unsmoothed data).

13. Below are the loneliness scores for the women who chose their dormitories.

| Class Interval | Frequency |
|---|---|
| 20 to 22 | 1 |
| 23 to 25 | 1 |
| 26 to 28 | 1 |
| 29 to 31 | 5 |
| 32 to 34 | 4 |
| 35 to 37 | 4 |
| 38 to 40 | 4 |
| 41 to 43 | 3 |
| 44 to 46 | 2 |

  a. Compute the relative frequencies.
  b. Compare this distribution with the distribution in problem 12. Where does each distribution peak?

    c. Are there relatively more residents with low loneliness scores (31 and below) in the assigned dorm group or in the group that chose their dorms? What about those with scores of 44 or higher? What about those with scores in the middle range, from 32 to 43?

14. For the following samples, state the unit of the distribution.

    a. 1.75, .25, 3.50, 4.50, 6.50, 7.00, 3.75
    b. 40, 120, 80, 160, 60, 100, 200
    c. 12, 18, 10, 26, 14, 18, 16, 14
    d. 1.33, 3.67, .67, 4.00, 3.33, 1.67

15. Prepare a stem and leaf display for data in problem 1 of this chapter.

16. Prepare a stem and leaf display for the data in problem 6 of this chapter.

17. Construct a histogram of the frequencies for the data in problem 12 of this chapter.

# 3 | *Central Tendency*

Numbers usually come in sets, also referred to as *samples*. A set of numbers by itself makes little or no sense if the numbers are not organized in some coherent manner. To understand the numbers and determine their meaning, they must be arranged in certain ways. In the previous chapter, methods for determining the shape, or distribution, of a set of numbers were presented. From the shape, the peak in the distribution and whether the distribution is symmetric can be determined.

This chapter discusses the typical or most representative value of a set of numbers. Of interest is the value around which the observations cluster. This value, the *central tendency* of a sample, estimates the typical value of an observation from the sample. A measure of central tendency represents all the numbers in the sample.

Central tendencies are a very common part of modern life. In sports, we hear about the average number of points a basketball player scores per game. In economics, we hear about the average cost of buying a home. In health, we hear of the average age at which children get a particular disease. Central tendencies are so common that they are not usually viewed as statistics.

There are two major reasons for knowing the central tendency: simplification and prediction. *Simplification* is needed because the whole sample of numbers often contains just too much information. Imagine that you are keeping a record of your monthly expenses for gasoline. Instead of remembering the dollar figures for each month over the past four years (48 numbers), it might be much more convenient to know only the average (one number). So for reasons of economy, it is often much more useful to record the mean, or the average of the numbers, instead of all the numbers. For the United States Census, it would be unthinkable not to compute a measure of central tendency. Imagine trying to report the incomes of 125 million households. There are too many numbers to keep track of. These numbers need to be boiled down to one measure of central tendency.

The second reason for computing the typical value is *prediction*. The knowledge of the average winter temperature in New York City for the past ten years can assist in predicting how cold it will be there this year. This knowledge will be useful in the determination of how much energy will be needed to keep the house warm. Thus, by knowing the average temperature, one can make a prediction and make choices that are consistent with that prediction. An average over ten years is probably a better predictor of next year's temperature than that of any one year. Let us consider a second example. Say a city is faced with a steady stream of immigrants from a foreign country. If the city is to plan for its future, it will need to know the age of these immigrants. Thus, by learning the average age of the immigrants, the city can predict future demand for schooling.

There are many ways of determining the typical value of a sample. First, consider one seemingly reasonable procedure. If it were possible to determine what person or observation is typical, that observation and that value would provide a measure of central tendency. This strategy is often employed by journalists. To predict an election, the journalist travels to a typical town and interviews the typical person in that town. There is one major advantage of using the response of the typical person as the typical response: It seems so sensible. The response of the typical person seems more valid than some statistical amalgamation of numbers. However, a more careful examination reveals two major drawbacks to the "typical person" strategy. First, in order to measure the typical value, where the typical town is and who in it is a typical person must be determined. Thus, there is a definitional problem. How can a typical value be determined if first it must be determined who is typical? The "typical person" definition of central tendency is circular, so it cannot be used. Second, the definition is not a very efficient way of determining the central tendency. To see this, it is likely that the typical value would be quite different if a different "typical" person was chosen.

The strategy of locating the typical person must be abandoned and a new strategy must be developed to determine the central tendency. One way is to determine what value is most frequent in the sample. A second way is to determine what value is in the middle of the distribution. Finally, a third way is to determine what value is the closest to all the others. These three definitions of typical values—most frequent, in the middle, and closest— correspond to the three standard ways of determining the typical value: the mode, the median, and the mean.

# Measures of Central Tendency

Figure 3.1 illustrates a hypothetical distribution that is somewhat positively skewed. The *mode* is the value that occurs most often. As shown in Figure 3.1, the mode is the highest point or peak in the distribution.

The *median* is the value in the distribution above which 50% of the scores

*FIGURE 3.1*    **Illustration of mode, median, and mean.**

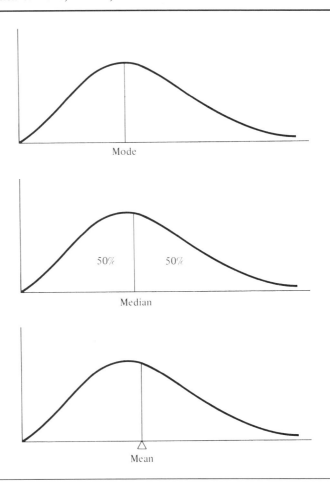

lie and below which 50% of the scores lie. The median divides the distribution in half in terms of frequency.

The *mean* is the arithmetic average of the set of numbers. It is the balance point in the distribution. If the distribution were a stack of toothpicks lying on a board, the mean would be the point of balance.

The mode requires only nominal data. It simply notes the most frequent occurrence and so the objects need only be differentiated. The median requires that the numbers be rank ordered and so only ordinal data are required. Finally, the mean requires interval data because the scores must be summed to calculate it. Note that the mode and median can be meaningfully calculated for interval data, but the mean should be used only with interval data.

# *Computation of Central Tendency*

Now that three measures of central tendency have been defined, their computational methods are presented.

The mode is easy to determine. *Count the number of times each observation occurs. The observation that occurs most frequently is the mode.* There may be a tie for the mode and in that case there would be two modes. If the distribution has two peaks of unequal height, it may be useful to report both peaks.

For some variables, measurements are fine-grained, with the consequence that no two scores have exactly the same value. For instance, for a sample of 100 persons whose weight is measured in grams, it is very unlikely that two or more persons would have exactly the same weight. In such cases, one may create a frequency table, smooth the frequencies, and report the mode of the smoothed distribution (smoothing was described in Chapter 2).

The median is determined by the following procedure.[1] *Rank order the scores; the median is the value of the score that falls in the middle.* The middle observation is determined as follows: The numbers are rank ordered from the smallest to the largest, just as was done in the previous chapter in order to make a frequency table. If $n$, the sample size, is odd then the middle observation is the $(n + 1)/2$th largest observation. So if there are eleven observations the median would be the sixth largest observation because there are five larger and five smaller scores. If $n$ is even, the median is defined as the average of $(n/2)$th and $(n/2 + 1)$th observations. In words, if the sample size is even, the median is one-half the sum of the two middle scores. So if $n$ is ten the median would be the average of the fifth, or $n/2$th, score and the sixth, or $(n/2 + 1)$th, score.

The mean is usually denoted in statistical work by the symbol $\overline{X}$ which is read as "*X*-bar." (Less frequently *M* is used to denote the mean.) *The mean or $\overline{X}$ is defined as the sum of the observations divided by the number of observations.* So, one simply adds up all the scores and divides by the total number of scores. The mean is the arithmetic average of the sample. In terms of a formula,

$$\overline{X} = \frac{\sum X}{n}$$

where $\sum X$ is the sum of the numbers in the sample and $n$ is the sample size. The mean is the most common measure of central tendency. It is as commonly used in statistical work as it is in everyday life.

---

[1]A more complicated formula is presented in some other texts. The formula presented here presumes that the numbers have not been rounded. If the scores have been rounded, a different formula must be used. The result using the more complicated formula differs only fractionally from the one presented in the text.

There are some procedures that can reduce the likelihood of computational errors in adding the scores. It may help to separate the positive and negative numbers and sum the two groups of numbers individually. Also if the numbers are very large, it may help to subtract a common number from all the scores and add that number back into the mean.

To illustrate the computation of measures of central tendency, three examples are provided. Given the following sample,

$$1, 1, 1, 1, 2, 2, 2, 3, 3$$

the mode is 1 because it is observed 4 times. The median is the fifth $[(9 + 1)/2\text{th}]$ largest observation, which is 2. The mean, rounded to two decimal places, is

$$\frac{1 + 1 + 1 + 1 + 2 + 2 + 2 + 3 + 3}{9} = 1.78$$

As another example consider Smith's study of bias of psychotherapists on the basis of the client's sex, which was presented in Chapter 2. A negative score indicates that therapists have a profemale bias, and a positive score indicates that they have a promale bias. The rank-ordered data are presented again here:

| | | | |
|------|------|-----|-----|
| −1.03 | −.23 | .00 | .14 |
| −.56 | −.22 | .00 | .23 |
| −.40 | −.10 | .00 | .24 |
| −.36 | −.03 | .01 | .29 |
| −.31 | .00 | .01 | .35 |
| −.31 | .00 | .02 | .56 |
| −.23 | .00 | .05 | .56 |
| −.23 | .00 | .11 | .60 |

The mode is .00, which indicates that therapists are most often neutral. The median is also .00. The mean is equal to the sum of the numbers divided by the number of studies. The sum of the 32 numbers equals −.84, and so the mean or $\overline{X}$ equals −.84 divided by 32, which equals −.02625. The mode, the median, and the mean virtually agree for this example and they all indicate that therapists show little or no bias on the average.

In 1960 Stanley Milgram conducted an experiment on obedience to authority. Residents of New Haven, Connecticut, were asked to shock someone who failed to learn material. Actually, the learner was a paid employee of Milgram and was not actually shocked. The "shocks" started at a low level but gradually escalated to a very high voltage. The person who was shocked begged the subject to stop shocking him and complained of a heart condition. The largest possible shock that could be administered, 450 volts, was labeled "danger—severe shock."

Psychiatrists had predicted that subjects would not shock the learner beyond 300 volts. What Milgram actually found is contained in Table 3.1.

TABLE 3.1   **Maximum Shock Level Administered by 40 Subjects in the Milgram Obedience Experiment**

| | | | |
|---|---|---|---|
| 300 | 330 | 450 | 450 |
| 300 | 345 | 450 | 450 |
| 300 | 360 | 450 | 450 |
| 300 | 375 | 450 | 450 |
| 300 | 450 | 450 | 450 |
| 315 | 450 | 450 | 450 |
| 315 | 450 | 450 | 450 |
| 315 | 450 | 450 | 450 |
| 315 | 450 | 450 | 450 |
| 330 | 450 | 450 | 450 |

Quite clearly subjects were very obedient. Of 40 subjects, 26 administered the maximum voltage.

The mode is 450, the maximum value possible. Because $n = 40$, the median is the average of the 20th and 21st observations. Both the 20th and 21st observations are 450, and so the median is also 450. The sum of the 40 observations is 16,200, and so the mean is $16,200/40 = 405$. Alternatively, the computations for the mean could be simplified by subtracting 300 from each number. The sum of the numbers would then be 4200. The mean would be $(4200/40) + 300$, which also equals 405, as it should. As will be seen later in this chapter, with distributions that are negatively skewed such as this one, the mean tends to be less than the median and the mode.

Three measures of central tendency have been defined: the mode, the median, and the mean. It should be noted that they do not always point to the same typical value. More often than not they disagree, if only by a small amount. The source of the disagreement is due to the shape of the distribution. Distributions with certain shapes yield a mode, median, and mean that differ from each other. When the distribution is exactly symmetric, the median and the mean are equal to each other. The earlier discussed gender bias data set has a fairly symmetric distribution, and the mean and the median are nearly equal to each other. Symmetry in the distribution is a sufficient but not necessary condition for the mean to equal the median. That is, in a symmetric distribution the mean must equal the median, but in a distribution that is *not* symmetric, the mean and median may also be equal. For instance, the following sample is not symmetric

$$1, 1, 2, 6, 7, 8, 17$$

but the mean and median are both equal to 6.

For symmetric distributions, the mode equals the median and the mean when the distribution is unimodal. A unimodal, symmetric distribution must be peaked in the center of the distribution. For bimodal, symmetric distributions, the peaks are not in the center and the mode does not equal the mean or the median.

For skewed distributions (see Chapter 2 for a definition of skew) the following generally holds ($<$ symbolizes less than)

Positive skew: mode $<$ median $<$ mean

Negative skew: mean $<$ median $<$ mode

which is shown graphically in Figure 3.2. So, for a positive skew, the mean is usually the largest measure of central tendency, whereas for a negative skew the mean is usually the smallest.

Because the value of central tendency depends on what measure of the central tendency is used, which one should be preferred? There is no universal answer to this question. The answer depends on what the *purpose* is in determining the central tendency. If the purpose is ease of computation, the mode is probably the best measure. However, for large data sets (sample size greater than 100), the mean is probably easier to determine than either the mode or the median. If the purpose is prediction, and given a symmetric distribution with observations bunched in the center, the mean tends to be a better predictor of future values than the median or the mode. For ease of interpretation the median is useful. For multimodal distributions reporting the mode is best. So, the determination of which measure of central tendency is best depends on the shape of the distribution, ease of computation, and simplicity of interpretation.

*FIGURE 3.2*   **Direction of skew and relationship between the mode, median, and mean.**

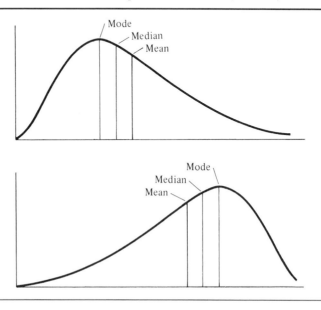

One guideline to determining which measure of central tendency to use is the level of measurement. For variables at the nominal level of measurement, the mode is the appropriate measure. For variables at the ordinal level of measurement, the median is the appropriate measure. Finally for variables at the interval level of measurement, the mean is the most appropriate measure of central tendency. But even if the level of measurement is at the interval level of measurement, the shape of the distribution may require using the median or the mode as the measure of central tendency.

# *Properties of the Mean*

The mean, or the arithmetic average, is the standard measure of central tendency. Given a sample of numbers, most researchers almost automatically compute the mean. Because it is an important and common statistic, special attention must be paid to its interpretation.

The mean is not necessarily equal to any observation in the sample. This point is abundantly clear for the variable of family size. Imagine that in a given area, the average size of a family is 4.2 persons. Of course, no family has 4.2 members, but 4.2 is a useful number nonetheless. Say, a housing development of 50 units is planned. Using the mean as a guide for forecasting, the expectation is for 50 times 4.2 or 210 to live in the development. The number 210 can be very useful in planning the need for social services in the community.

The mean can be quite misleading if there is an outlier in the sample. Say, the variable is income and the annual income of six college students is measured and the following numbers are obtained:

$2,700
$3,600
$2,800
$6,300
$1,800
$1,040,000

The millionaire student's income influences the mean so extremely as to result in a mean of $176,200, not at all typical of the income of the other five college students. Means can be grossly distorted by the presence of outliers. The median is much less affected. The median of the six incomes is $3,200.

When the sample contains a mixture of scores from two very different types of persons, the distribution is bimodal. In such cases the mean is not very informative.

The mean is the only measure of central tendency such that it can be subtracted from each observation and the sum of these differences is always zero. As a formula, $\Sigma(X - \bar{X})$ equals zero. One consequence of this fact is that the sum of observations above the mean is as far above the mean as the sum of the observations below the mean. It is this property that makes the mean the number that is closest to the other numbers in the sample.

A related property of the mean is that if a constant is subtracted from each score, and this difference is squared and summed across all the scores in the sample, this quantity is at its smallest value when the constant chosen is the mean. The mean is said to be a *least-squares* estimate of central tendency.

# Grouped Data

Sometimes a researcher has a frequency table but does not have access to original scores. The computation of the mode, median, and mean must be modified.

The mode of data in a frequency table is the midpoint of the class with the largest frequency. (As defined in Chapter 2, the midpoint is one-half the sum of the upper and lower limits for a class.) The median can be obtained by adding the relative frequencies starting with the lowest class. The median is the midpoint of the class interval whose cumulative relative frequency contains 50%.

When observations are grouped together into classes the mean can be computed as follows:

1. Multiply each score by its frequency.
2. Sum the score frequency products.
3. Divide this sum by the sample size.

For instance, in the Milgram study of obedience the shock level is first multiplied by the frequency.

| Score | | Frequency | | |
|---|---|---|---|---|
| 300 | × | 5 | = | 1,500 |
| 315 | × | 4 | = | 1,260 |
| 330 | × | 2 | = | 660 |
| 345 | × | 1 | = | 345 |
| 360 | × | 1 | = | 360 |
| 375 | × | 1 | = | 375 |
| 450 | × | 26 | = | 11,700 |

The sum of these products equals 16,200. The mean or $\bar{X}$ then equals 16,200 divided by 40, which is 405.

When numbers are grouped by class intervals and there is no access to the original data, one can use the above technique to approximate the mean. The midpoint of the class interval is used as the number to be multiplied by the frequency. For instance, in Chapter 2 the following class intervals were set up for Smith's review of 32 gender bias studies of therapists and counselors. Below are the class intervals and frequencies used in Chapter 2.

| Class Interval | Frequency |
|---|---|
| −1.10 to −.91 | 1 |
| −.90 to −.71 | 0 |
| −.70 to −.51 | 1 |
| −.50 to −.31 | 4 |
| −.30 to −.11 | 4 |
| −.10 to  .09 | 13 |
| .10 to  .29 | 5 |
| .30 to  .49 | 1 |
| .50 to  .69 | 3 |

The midpoint of the class intervals times the frequencies are

| Midpoint | | Frequency | | |
|---|---|---|---|---|
| −1.005 | × | 1 | = | −1.005 |
| −.805 | × | 0 | = | 0.000 |
| −.605 | × | 1 | = | −.605 |
| −.405 | × | 4 | = | −1.620 |
| −.205 | × | 4 | = | −.820 |
| −.005 | × | 13 | = | −.065 |
| .195 | × | 5 | = | .975 |
| .395 | × | 1 | = | .395 |
| .595 | × | 3 | = | 1.785 |

The sum of these numbers is −.960. The mean is then estimated by −.960 divided by 32, which equals −.03. This value closely approximates the actual mean of −.02625.

The mode for the gender bias data is −.005 because the interval −.10 to .09 has the largest frequency. The median is also −.005. Both of these values are close to the mode and median of the original data, each of which is .00.

Occasionally, there is a choice of what factor to use as the frequency when computing a measure of central tendency. What to use as the frequency is determined by what variable the researcher seeks to measure. Imagine a car manufacturer which produces five different kinds of cars. These cars have the following mean miles-per-gallon ratings.

$$18, \ 24, \ 27, \ 30, \ 45$$

What is the average miles per gallon for the car company's fleet? It all depends on what exactly the question is. If the mileage of the five cars is desired, the average the five numbers would suffice. However, if the mileage of cars sold is desired, the number of cars sold must be used as the frequency. Alternatively, if the interest is in the number of gallons of gas, the frequency to be used is the number of miles driven.

# Summary

A set of scores can be summarized by a measure of central tendency. The *central tendency* estimates a sample's typical value. Measures of central tendency are used to simplify a mass of data and to facilitate prediction. There are three common measures of central tendency: the mode, the median, and the mean. The *mode* is the most frequent number in the sample, the *median* is the middle of the distribution, and the *mean* is the arithmetic average.

The mode is directly determined by the most frequent score in the sample. The median is determined by first rank ordering the scores from smallest to largest. Given $n$ scores, for odd-sized samples, the median is the $(n + 1)/2$th largest score. For even-sized samples, the median is the average of the $n/2$th and the $(n/2 + 1)$th largest scores. The mean, which is symbolized by $\overline{X}$, is the sum of all the scores divided by the sample size.

For symmetric distributions the mean and the median are equal. For positively skewed distributions the mean tends to be greater than the median, whereas for negatively skewed distributions the mean tends to be less than the median. The mean is very sensitive to outliers, but the mode and median are less affected.

For data grouped into classes the mode, median, and mean of the original data can be closely approximated. The mode is the midpoint of the class interval that has the largest frequency. The median is the midpoint of the class interval that contains the cumulative relative frequency of 50. The mean is found by multiplying each class midpoint by its class frequency, summing the products, and then dividing by the sample size.

The measures of central tendency provide one way of summarizing a distribution. The next chapter presents various ways of determining how meaningful the central tendency is as a summary of the distribution. This can be done by computing a measure of variability.

# *Problems*

1. Compute the mode, median, and mean for the following samples.

   a. 6, 8, 3, 5, 6, 2, 7
   b. 4, 3, 2, 6, 2, 2, 5, 4
   c. 2, 8, 4, 2, 5, 8, 10, 1, 8, 2
   d. 2, 4, 3, 7, 5, 8, 3, 96

2. In which of the four samples in problem 1 is the mean less informative than the other measures of central tendency? Why?

3. a. Compute the mean, mode, and median of the following sample of numbers.

   2, 3, 3, 3, 3, 4, 4, 5, 6, 6

   b. Compute the mean, mode, and median for the same sample of numbers, but including the number 100.

2, 3, 3, 3, 3, 4, 4, 5, 6, 6, 100

Compare the results with the answers for part (a). Which measure of central tendency changed the most?

c. Again take the sample of numbers in part (a), include the number 10, and compute the mean, median, and mode.

2, 3, 3, 3, 3, 4, 4, 5, 6, 6, 10

Compare the results with the answers for part (a) and part (b). Which measure of central tendency was most affected by the size of the additional number?

4. A class, consisting of an even number of students, takes an exam, and 14 students score above the median. How many students are in the class?

5. Below is a table of the area of the New England states in square miles.

| State | Area |
|---|---|
| Connecticut | 5,009 |
| Maine | 33,215 |
| Massachusetts | 8,257 |
| New Hampshire | 9,304 |
| Rhode Island | 1,214 |
| Vermont | 9,609 |

a. Compute the mean and the median.
b. Which measure gives a better description of the area of a typical New England state?

6. Given below are test scores for a group of 20 subjects.

0, 3, 11, 24, 36, 47, 42, 53, 56, 59,
52, 58, 50, 64, 63, 61, 65, 78, 89, 91

a. Compute the mean and median. Which estimate, mean or median, best describes the central tendency of the data? Why?
b. Unfortunately, the values 0, 3, and 11 were incorrectly recorded. These three values should be 70, 73, and 81, respectively. Does this additional information change your answer to part (a)?

7. A survey of the ages of residents of nursing homes yielded the following measures of central tendency.

mean = 70     median = 78     mode = 83

In which direction is the distribution likely to be skewed?

8. Suppose that students in an introductory psychology class were tested on their knowledge of foreign policy. Suppose further that the following measures on central tendency were obtained.

mean = 80     median = 71     mode = 65

In which direction is the distribution likely to be skewed?

9. Draw histograms for samples with the following characteristics.

   a. The mean is greater than the mode.
   b. The mean is less than the median.
   c. The median is less than the mode.
   d. There are two modes.

10. Schwartz and Leonard (1984) studied the learning of words that refer to objects and words that refer to actions. Their subjects were children under 1.5 years old. The children repeatedly heard 16 nonsense syllables applied to unfamiliar objects or actions. The following shows how many object and action words were acquired by each child.

    | Child | Object Words | Action Words |
    |-------|--------------|--------------|
    | JR | 6 | 6 |
    | KP | 8 | 3 |
    | LT | 7 | 4 |
    | BG | 7 | 7 |
    | KS | 7 | 5 |
    | RC | 3 | 2 |
    | CB | 7 | 2 |
    | MK | 8 | 2 |
    | LF | 7 | 3 |
    | KB | 7 | 2 |
    | BP | 7 | 4 |
    | TH | 8 | 2 |

    a. Compute the mean number of object words the children acquired.
    b. Compute the mean number of action words.

11. From the data in problem 10, for each child subtract the action word *mean* from each child's action word score.

    a. How many children acquired more than the mean number of action words?
    b. How many acquired less than the mean number?
    c. Compute $\Sigma(X - \bar{X})$.
    d. If you did the same computations for the object words, what would you find for $\Sigma(X - \bar{X})$? Why?

12. a. From the data in problem 10, find the median number of action words acquired. Subtract the median from each child's action word score, square the result, and sum across children.
    b. For problem 11, you computed the deviation of each score from the mean. Square these deviations, and sum across children.
    c. Which is smaller, the sum of the squared deviations from the mean, or the sum of the squared deviations from the median?

13. Ballard and Crooks (1984) tried to increase the social involvement of preschool children who had low levels of interaction with others. They

observed the children after showing them a videotape in which a child modeled the activity of joining others in play. The subjects were scored on the degree of social involvement in their play and number of social interactions. Below are the mean scores per session each subject obtained after seeing the videotape.

| Subject | Interactions | Play Score |
|---------|--------------|------------|
| S1 | 8.17 | 3.29 |
| S2 | 8.50 | 2.69 |
| S3 | 8.38 | 3.25 |
| S4 | 7.92 | 2.73 |
| S5 | 14.43 | 2.45 |
| S6 | 14.36 | 3.33 |

a. Compute the mean number of interactions observed after the videotape, across all subjects.

b. Compute the mean play score obtained after the videotape, across all subjects.

14. A university bookstore stocks a range of calculators, from simple models for general use to scientific and business calculators. During one semester they sold the following number of calculators at the prices given.

| Model | Price (in dollars) | Number Sold |
|-------|--------------------|-------------|
| A | 8 | 12 |
| B | 15 | 7 |
| C | 22 | 5 |
| D | 45 | 12 |
| E | 70 | 2 |
| F | 120 | 1 |

a. Compute the measures of central tendency for the price of the calculators sold.

b. Which measure best describes how much was paid for a calculator?

15. The following frequency table of rent costs was compiled from a list of apartments for rent in a university town.

| Rent (in dollars) | Frequency |
|-------------------|-----------|
| 226–250 | 2 |
| 251–275 | 7 |
| 276–300 | 12 |
| 301–325 | 8 |
| 326–350 | 2 |
| 351–375 | 2 |
| 376–400 | 4 |
| 401–425 | 1 |

Compute the mode, median, and mean for the data.

# *4* | *Variability*

The average temperatures of San Francisco and Kansas City, Missouri, differ by only one degree Fahrenheit. The average temperature in San Francisco is 56 degrees and the average in Kansas City is 55 degrees. Does this mean that the two cities have the same weather? Not at all. In July, Kansas City tends to be 20 degrees warmer than San Francisco; in January, however, San Francisco tends to be 20 degrees warmer than Kansas City. Just because the averages are the same does not mean that the weather is the same. The weather in Kansas City is more variable than in San Francisco. The variability in the numbers is just as important as the average of the numbers.

A sample of numbers has a characteristic shape. Recall that the shape concerns the frequency of certain numbers relative to other numbers. Besides the shape, the central tendency of the numbers—or the typical value in the sample—can be determined. The mode, median, and mean can be used as measures of central tendency. A distribution of scores can also be characterized by the variability of the scores. Are the numbers bunched tightly together or do they vary over a wide range? For instance, the scores

$$3, 4, 5, 5, 5, 5, 5, 6, 6, 7$$

are bunched together near five, whereas the scores

$$1, 2, 3, 4, 4, 5, 6, 8, 11, 13$$

vary over a wider range. *Variability* refers to how much the scores differ from one another.

In everyday life, the term *consistency* is used to refer to variability. The weather in Honolulu is consistently around 75 degrees. The Boston Celtics basketball team is a consistent winner. Our favorite sports team is inconsistent. One person's weight may be very stable, whereas another's changes quite a bit from week to week. The concept of consistency implies that the person or unit has low variability, whereas inconsistency refers to high variability.

Variability is also the fundamental starting point in scientific explanation.

Because some people have higher GPAs, weigh more, live longer, and are more violent than others, there must be some reason why these people differ. Scientists observe variation and wonder what brings it about. To the extent that people vary, there must be forces that make them differ. Therefore, if it can be understood why people differ, it can be understood what causes human behavior. One goal of research is to explain what makes people vary, and one common way of stating a result from research is to give the percentage of the variability that is explained. For instance, Jencks and his colleagues (1972) claim that they are able to account for, or explain, about 30% of the variability in income earned by an individual in the United States.

The meaning of "explaining 30% of the variability" requires explanation. It does *not* mean that 3 out of 10 reasons that make persons vary are known. Nor does it mean that reasons are known why 30% of the people make money. Rather, it means that given what is known, existing variability could be reduced by 30%. If it were possible to equalize persons on education, motivation, intelligence, and other specified factors, the variability of the income distribution would be reduced by 30%.

There are three basic reasons for measuring the variability of a sample. They are as follows:

1. to determine how meaningful the measure of central tendency is,
2. to use it as a basis for determining whether a score is an outlier, and
3. to compare variability.

Consider each reason in turn.

The first reason for measuring variability is to judge how meaningful the measure of central tendency is. To the extent that the numbers vary widely, any measure of central tendency is somewhat less informative. Knowing that the average income of some group of persons is $13,432 is not very informative when some persons are earning nothing and others millions. A doctor would want to know a person's average blood pressure only if that person's blood pressure was not subject to huge swings. To the extent that the numbers are bunched together, the measure of central tendency is more descriptive of the numbers. Thus the variability tells how well the mean or any other measure of central tendency represents all the numbers.

Consider four persons who have dinner at a restaurant. The bill including tip comes to $60. One person suggests splitting the bill up evenly with everyone putting in $15. He or she has computed a measure of central tendency—the mean or arithmetic average. If one person's dinner cost $21 and the other three averaged $13, the mean would not be very representative of the cost of each person's meal. If two persons' meals were $16 and the other two $14, however, then the mean would be a reasonable approximation of the cost of each person's meal.

A measure of variability is also useful in assessing how deviant, or unusual, a given number is. To identify a genius or anyone who is exceptional some type of yardstick for assessment is needed. If it rained 15 inches more than the average this year, it is not known whether that is exceptional or typical. If it is known that the rainfall varies quite a bit from year to year, then being 15 inches above the average is not at all exceptional. But if the amount of rainfall hardly ever changes from year to year, then having 15 more inches is quite unusual.

A measure of variability is also useful for comparison purposes. Some faces are easier to read than others. That is, some persons clearly express their emotions; others are poker-faced and are not expressive. People vary in expressiveness—that is, their ability to send nonverbal messages. It also seems true that some people are better at reading or receiving other people's emotional reactions than others. Some people are sensitive and others are not. Research has shown that there is much more variability in expressiveness than in sensitivity. There is more variability among the senders of nonverbal messages than there is among the receivers (Kenny & La Voie, 1984).

Throughout this chapter, the computations will be illustrated with the same sample of numbers. During the fall of 1964, psychologists Robert Rosenthal and Lenore Jacobson entered a South San Francisco grade school and told teachers that a number of students would "bloom" during that year. The teachers were told that the students were expected to experience an unusual forward spurt in academic and intellectual performance during the year. Actually the students were no more likely to excel than their class-mates. Rosenthal and Jacobson were interested in studying whether merely suggesting to teachers that certain students would improve would create an actual change. As part of this research, Rosenthal and Jacobson measured the children's intelligence, or IQ. Table 4.1 lists the scores of 22 first-grade children. An IQ score is set so that a score of 100 is normal.

*TABLE 4.1*    **Intelligence Tests Scores for 22 First-Grade Children**

| | |
|---|---|
| 94 | 95 |
| 102 | 98 |
| 117 | 92 |
| 91 | 86 |
| 90 | 92 |
| 106 | 120 |
| 112 | 97 |
| 101 | 116 |
| 122 | 130 |
| 111 | 117 |
| 100 | 108 |

Data were taken from Rosenthal and Jacobson (1968).

# *Measures of Variability*

The three common measures of variability are the *range,* the *interquartile range,* and the *standard deviation.* The range looks at only the variability between the largest and smallest numbers. The interquartile range looks at the variability in the middle of the sample. The standard deviation examines the variability between all possible pairs of numbers.

## *The Range*

The range is the difference between the largest and smallest numbers in the sample. Thus, for the sample

$$3, 5, 6, 9, 11$$

the range is 11 minus 3, which is 8. For the sample

$$101, 190, 236, 436$$

the range is 436 minus 101, or 335. *The range is the largest minus the smallest number of the sample.* For the numbers in Table 4.1 the range is 130 − 86 = 44. The range concentrates on the extremes and ignores all the other numbers in the distribution. Because the range examines only the two most extreme numbers in the sample, it is influenced by outliers.

Although the range is a natural and simple measure of variability, it has an important limitation. The size of the range depends, in part, on how many scores there are in the sample. In general, the greater the sample size the greater the range. Consider the example of a flat distribution of numbers from 1 to 10. (Recall from Chapter 2 that a flat distribution is one in which each number is just as likely as any other number.) For the smallest possible sample size for which the range can be computed, 2, the range is, on the average, about 3.3. That is, if every combination of two numbers from 1 to 10 is taken, the range averages 3.3. As the sample size increases, the range increases. For the sample size of 100 the range is nearly 9.0, more than twice the expected range for a sample size of 2.

The reason for this situation is that extreme scores (i.e., very large or very small scores) are much more likely to appear when the sample size is large. Because extreme numbers are associated with large sample size, larger range tends to be as well. This is not to say that the range is a useless measure of variability; however, it would be misleading to compare the ranges of two samples with very different sizes.

One advantage that the range has is that it is a relatively easy measure to compute and it is simple to understand. For instance, when you receive a grade in a course, you may want to know the range of scores. The range is a natural measure of variability, but the following two measures of variability are more commonly used.

## The Interquartile Range

The interquartile range is another measure of variability, and it is almost as easy to compute as the range. The range measures the variability from the largest to the smallest number. The *interquartile range* measures the variability in the middle half of the distribution and is, therefore, less influenced by outliers. Basically, to compute the interquartile range, the sample is subdivided into two groups: those above the median and those below the median. The "median" is then computed for each of these groups. The median for the group above the overall median will be called the *upper median;* for the group below the overall median, it will be called the *lower median*. The interquartile range is the difference between the median of the group above the overall median and the median of the group below the overall median. The interquartile range is then the difference between the upper and the lower medians. So, for the following distribution

$$1, 2, 3, 3, 3, 4, 6, 9, 11, 20$$

the overall median is 3.5. The values below the median are

$$1, 2, 3, 3, 3$$

and their median is 3. The values above the overall median are

$$4, 6, 9, 11, 20$$

and their median is 9. The interquartile range is 9 minus 3 which equals 6. These computations are illustrated in Table 4.2.

*TABLE 4.2*   **Illustration of the Interquartile Range for Two Samples**

| | Sample 1 | | | Sample 2 | |
|---|---|---|---|---|---|
| Upper median | 20 11 9 6 4 | Scores of those above the median | 9 7 6 6 | Upper median |
| Overall median | 3.5 | | 6 | Overall median |
| Lower median | 3 3 3 2 1 | Scores of those below the median | 5 4 4 1 | Lower median |
| Sample 1 interquartile range: 9 − 3 = 6 | | | Sample 2 interquartile range: 6.5 − 4 = 2.5 | |

The rule for determining which scores are above and which are below the median is quite simple. If there is an even number of scores in the sample, the first half or $n/2$ scores are below the median, and the second half or $n/2$ scores are above. So, if there are 20 scores, $n/2$ is 20/2 or 10. The median is then computed of the 10 smaller scores and the median of the 10 larger scores. If there is an odd number of scores in the sample, the score at the median is excluded. Thus, for the sample

$$1, 4, 4, 5, 6, 6, 6, 7, 9$$

exclude the 6, the overall median, and use 1, 4, 4, and 5 as the scores below the median and 6, 6, 7, and 9 as the scores above the median. Again, these computations are illustrated in Table 4.2. Once it has been determined which scores are above and which are below, the medians are computed for these groups of scores by the simple procedure described in Chapter 3. Given $n$ scores, when $n$ is an odd number, the median is the $(n + 1)/2$th score. If $n$ is even, then the median is the average of the $n/2$th and $(n/2 + 1)$th scores.

For the IQ data in Table 4.1 there are 22 scores. The median is then the average of the eleventh and twelfth largest scores of the sample. It is

$$\frac{101 + 102}{2} = 101.5$$

The scores below the median are

$$86, 90, 91, 92, 92, 94, 95, 97, 98, 100, 101$$

and so the lower median is 94. The scores above the median are

$$102, 106, 108, 111, 112, 116, 117, 117, 120, 122, 130$$

and so the upper median is 116. The interquartile range is then

$$116 - 94 = 22$$

Therefore the middle half of the numbers varies over 22 IQ points.

This measure is called the interquartile range because it is based on the separation of the sample into four quartiles. The sample is separated into four groups with an equal number of scores per group: four quartiles. The first quartile contains the 25% of the scores that are the smallest. The second quartile contains the next 25%. The third and fourth quartiles are similarly defined. The boundary between the first and second quartile is the lower median. The boundary between the second and third quartile is the overall median of the sample. Finally, the boundary between the third and fourth quartile is the upper median.

## The Standard Deviation

Both the range and the interquartile range are sensible ways to measure variability in the sample. But they are limited because each of them only looks at part of the data. The range uses only the two most extreme numbers, and the interquartile range throws away all the extreme numbers. It would seem sensible to have a measure of variability that looks at all the numbers. The standard deviation is just such a measure.

To measure how different the numbers are from each other, one natural measure would be to subtract them from each other. So for the numbers 9, 10, 11 the difference between all possible pairs is computed as follows:

$$10 - 9 = 1$$
$$9 - 11 = -2$$
$$10 - 11 = -1$$

Because it is not important whether the difference is positive or negative, the differences are squared.

$$(10 - 9)^2 = 1$$
$$(9 - 11)^2 = 4$$
$$(10 - 11)^2 = 1$$

The mean of these three squared differences is $(1 + 4 + 1)/3 = 6/3 = 2$. This seems to be a very sensible measure of variability. It is the average of all possible squared differences, which will be called the *average squared difference*. Although the average squared difference is sensible, it presents a computational nightmare. If there are 50 numbers in the sample there are 1225 differences! It would take hours to compute even with a calculator. Fortunately, there is a computational shortcut. It is called the standard deviation.

The most common measure of the variability is the *standard deviation,* or as it is usually symbolized, $s$. Instead of first presenting the formula for the standard deviation, a rationale for it and its close relative, the variance, or $s^2$, is discussed. As was detailed in the previous chapter, the mean (the sum of the numbers divided by the sample size) is a measure of central tendency. One reasonable measure of variability would be to simply compute how far each score is from the mean. A deviation of each score from the mean is then computed. If the numbers are tightly bunched together, these deviations from the mean would be small and so the measure of variability would be small. Alternatively, if the numbers were spread over a wide range, the deviations from the mean would be large and so would the variability. Thus, a basic building block of a measure of variability can be the deviation of scores from the mean. For the sample

$$1, 6, 10, 12, 16$$

with a mean of 9, the deviations from the mean are

$$1 - 9 = -8$$
$$6 - 9 = -3$$
$$10 - 9 = 1$$
$$12 - 9 = 3$$
$$16 - 9 = 7$$

These deviations from the mean are each squared and then summed. Thus, the numbers −8, −3, 1, 3, and 7 are squared, to obtain 64, 9, 1, 9, 49, the sum of which equals 132. Then to adjust for sample size, the sum is divided by sample size less one. (The reasons for using sample size less one will be explained later.) So for the example above, the measure of variability is 132 divided by 4 (sample size less one) and it equals 33. This measure of variability is called the variance. The *variance,* symbolized by $s^2$, is the sum of squared deviations about the mean divided by sample size less one. The *standard deviation,* symbolized by $s$, is the positive square root of the variance.

The measure $s^2$, or variance, is closely related to the average squared difference. The average squared difference is determined by taking the difference between pairs of scores, squaring each of them, and then summing these squared differences and dividing that sum by the number of pairs. The variance is one-half the average squared difference. The variance and, therefore, the standard deviation are very closely related to the average squared difference. The variance is, however, much simpler to compute than the average squared difference, and hence it is generally preferred.

Consider, for instance, the sample 5, 11, 15, 17. The mean is 12 and so the deviations from the mean are −7, −1, 3, 5. To compute the variance, the deviations are squared, summed, and that sum is divided by the sample size less one. The sum of squared deviations is then $49 + 1 + 9 + 25 = 84$ and the variance is $84/3 = 28$. To compute the average squared difference, the difference between all possible pairs is computed:

$$
\begin{aligned}
11 - 5 &= 6 \\
15 - 5 &= 10 \\
17 - 5 &= 12 \\
15 - 11 &= 4 \\
17 - 11 &= 6 \\
17 - 15 &= 2
\end{aligned}
$$

The sum of the squares of these quantities is $36 + 100 + 144 + 16 + 36 + 4 = 336$. And 336 divided by 6 (the number of pairs) equals 56. Note that one-half of 56 is 28, the variance. So, one-half the average squared difference equals the variance, as it should.

Normally, it is advisable to use the standard deviation, not the variance, as a measure of variability. By computing the square root of the variance to obtain the standard deviation, the unit of measurement becomes interpretable. For the example in Table 4.1, the variable is intelligence or as it is usually called, IQ. When variance is computed, the deviations are squared. So for intelligence, the unit of measurement for the variance is IQ points squared. It is not too clear how to interpret intelligence squared. To return the unit of measurement to that of the original metric, the square root of the variance is computed.

The standard deviation is a measure of the "average" distance from the mean. It is not a straightforward arithmetic average because of the squaring, the summing, and the taking of the square root. The effect of the squaring is to make the standard deviation more heavily reflect scores that are farther from the mean. Consider the two following samples. Sample A contains the numbers

$$1, 3, 7, 9$$

and sample B contains the numbers

$$1, 3, 5, 11$$

Both samples have a mean of 5.0 and the sum of the absolute deviations from the mean are 12.0 for both samples. However, the standard deviation for sample A is 3.65 and the standard deviation for sample B is 4.32. Why are they different? Sample B has the relatively extreme score of 11, which is 6 units from the mean while for sample A the most deviant score is only 4 units from the mean. When the deviation of 6 is squared, it dominates the variance.

The variance and, therefore, the standard deviation are quite affected by the presence of outliers. Consider the following sample.

$$4, 5, 5, 6, 7, 7, 8$$

The standard deviation of this sample is 1.41. Consider now the same sample with the addition of a single outlier.

$$4, 5, 5, 6, 7, 7, 8, 44$$

The standard deviation has now exploded to 13.50. A single outlier can dramatically affect the size of the standard deviation.

*Why Squared Deviations?*    For the variance and the standard deviation, mean deviations are squared. An alternative is to just sum the deviations. For the sample whose numbers are 1, 6, 10, 12, and 16, the deviations are –8, –3, 1, 3, and 7. Their sum is zero. This is no accident. The sum of deviations about the mean is always zero. This is due to the definition of the mean. Recall that the mean is the balance point of a distribution; a balance point requires that the sum of deviations must be zero.

Because the sum of the deviations cannot be used as a measure of variability, an alternative to squaring them might be to compute the *absolute value* of deviations. An absolute value of a number is that number with the sign always positive. Therefore the deviations for the example above are –8, –3, 1, 3, 7, and the sum of their absolute values is 22. There is, however, no simple relationship between the absolute values of mean deviations and the absolute value of the deviation between all pairs of numbers. So, the average absolute deviation from the mean is not related to the average absolute deviation between all possible pairs.

***Why Sample Size Less One?***    There would not be much harm in dividing by sample size instead of sample size less one.[1] However, it is just a little better to divide by sample size less one. Why? Consider what the variance would be when the sample size is one. It is, of course, impossible to measure variability when there is only a single score in the sample. However, dividing by sample size (instead of sample size less one), the variance would always be zero when sample size is one. However, if the denominator of the variance is sample size less one, the variance is undefined when sample size is one. (Division by zero is not mathematically permissible.) So one reason for "less one" is to make the variance of the sample size of one to be undefined.

A related reason for dividing by sample size less one is because the mean is computed from the numbers that are used to compute the variance. If the mean were known without having to compute it from the data, it would be correct to divide by $n$ and not $n - 1$. In Chapter 9 the question of dividing by $n - 1$ instead of $n$ is discussed.

***Computation of the Standard Deviation.***    The definitional formula for the standard deviation is to

1. take each score and subtract the mean,
2. square each of the deviations from the mean,
3. sum the squared deviations,
4. divide the sum of squared deviations by sample size less one, and
5. take the square root.

In terms of a formula, the standard deviation is

$$s = \sqrt{\frac{\Sigma(X - \bar{X})^2}{n - 1}}$$

The numerator is the sum of squared deviations from the mean ($\bar{X}$) and the denominator is sample size less one.

The computations of $\Sigma(X - \bar{X})^2$ can be simplified by using the formula

$$\Sigma X^2 - \frac{(\Sigma X)^2}{n}$$

The first term of the formula, $\Sigma X^2$, is simply the sum of the squares of each number in the sample. The second term, $(\Sigma X)^2/n$, is the result of summing all the scores, squaring the result, and then dividing by $n$, the sample size. You might want to review the difference between $\Sigma X^2$ and $(\Sigma X)^2$, described in Chapter 1. The resulting complete computational formula for the standard deviation is

---

[1] Some texts recommend dividing by $n$ instead of $n - 1$. However, most statistical formulas presume that the denominator is $n - 1$.

$$s = \sqrt{\frac{\sum X^2 - \frac{(\sum X)^2}{n}}{n-1}}$$

Table 4.3 presents the computations required for the computational formula for the standard deviation using the IQ example in Table 4.1. The first column consists of the raw data. The second column gives the squares of each score. The totals or sums of the numbers of the two columns are then written beneath each column. The total for the first column in Table 4.3 is symbolized by $\sum X$ and the second column by $\sum X^2$. Taking the quantities in Table 4.3 and entering them into the computational formula yields

$$\sqrt{\frac{\sum X^2 - (\sum X)^2/n}{n-1}} = \sqrt{\frac{242967 - (2297)^2/22}{21}} = 12.23$$

*TABLE 4.3*    **Computations Necessary for the Standard Deviation**

| X (score) | $X^2$ (score squared) |
|---|---|
| 94 | 8836 |
| 102 | 10404 |
| 117 | 13689 |
| 91 | 8281 |
| 90 | 8100 |
| 106 | 11236 |
| 112 | 12544 |
| 101 | 10201 |
| 122 | 14884 |
| 111 | 12321 |
| 100 | 10000 |
| 95 | 9025 |
| 98 | 9604 |
| 92 | 8464 |
| 86 | 7396 |
| 92 | 8464 |
| 120 | 14400 |
| 97 | 9409 |
| 116 | 13456 |
| 130 | 16900 |
| 117 | 13689 |
| 108 | 11664 |
| Total   2297 | 242967 |

# Detection of Outliers

As discussed in Chapter 2, an outlier is a very large or small score in the sample. Now that quantitative measures of variability and central tendency have been presented, a quantitative rule for determining an outlier can be given.

Two different definitions of an outlier are presented. The first involves the interquartile range and the median. An outlier is any number that is more than two times the interquartile range away from the median. So one computes

$$\text{median} \pm 2 \times \text{interquartile range}$$

Scores larger than the median plus twice the interquartile range and smaller than the median minus twice the interquartile range are deemed outliers. For the IQ example, the median is 101.5 and the interquartile range is 22. So for a score to be an outlier it must exceed 101.5 + 44, or 145.5, or be less than 101.5 − 44, or 57.5. Certainly IQs of greater than 145.5 and less than 57.5 should be carefully examined. Using this definition of outliers, none of the scores in Table 4.1 can be considered outliers.

An outlier can be alternatively defined by using the standard deviation and the mean. The following quantity is computed:

$$\text{mean} \pm 2\frac{1}{2} \times \text{standard deviation}$$

The reason that it is two and one half times the standard deviation and only two times the interquartile range is that the standard deviation tends to be smaller than the interquartile range. For the IQ example the mean is 104.41 and the standard deviation is 12.23. So for a score to be an outlier it must exceed 104.41 + 30.57, or 134.98, or be less than 104.41 − 30.57, or 73.84. Using this definition of outliers, none of the scores in Table 4.1 can be considered outliers.

If one has a choice concerning the two measures of outliers, the interquartile range and the median measure is preferred to the standard deviation and the mean measure. The reason is that the mean and standard deviation are themselves influenced by the presence of an outlier. Say the sample in Table 2.1 had the three largest IQs changed to 175. This change would raise the mean and double the standard deviation, and the value of 175 would not be recognized as an outlier. However, the interquartile range and the median are less affected by the presence of outliers, and the 175 IQs would be deemed as outliers.

# Computational Errors and Checks

All measures of variability must be nonnegative. Because the least amount of variability is none, the lowest value of a measure of variability is zero. When variability is zero all the values of the sample are the same.

Because the range and the interquartile range basically involve only a single subtraction, it is unlikely that they would be mistakenly computed as negative. However, the variance involves much computation, and so errors can occur. Whenever the numerator of the variance is negative, one knows with certainty that there has been an error of computation. The usual reason for such an error is that one has incorrectly summed the squares of the scores. Also, it is possible that the sum of all the scores has been incorrectly computed.

A good way to locate a computational error is to study the range, the interquartile range, and the standard deviation. First, determine whether the estimated value of variability is possible. As was just stated, zero is the lowest possible value for variability. If the numbers themselves have a lower and an upper bound, then the measure of variability can be no larger than the upper bound minus the lower bound. (This is true of the standard deviation but not the variance.) So if a group of persons rate a movie on a scale from one to ten, then the measures of variability (except the variance) computed from these numbers must be nine or less. The standard deviation must be less than or equal to the range. If it is not, then an error has been made in computing the variability or there has been an error in the recording of the numbers.

One obvious, but commonly made error is to report the variance as the standard deviation. Recall that the standard deviation is the square root of the variance. For the standard deviation, always make certain that the square root of the variance has been calculated.

Recall from the previous chapter that the mode, the median, and the mean can be equal to the same value. However, one should not expect that the range, the interquartile range, and the standard deviation to be equal or nearly equal. Although all three measure the variability in the sample, they do so in different ways. For instance, the range must always be at least as large as the interquartile range. To see this, note that the largest value of a sample must be at least as large as the median of the upper half of the distribution. Similarly, the lowest value must be at least as small as the median of the lower half of the distribution, and so the range is at least as large as the interquartile range. Generally the standard deviation is smaller than the interquartile range, as it is for the IQ example used in this chapter. For unimodal symmetric distributions a good rule of thumb is that three-quarters of the interquartile range is about equal to the size of the standard deviation. For the IQ example the interquartile range is 22 and so three-quarters of 22 is 16.5, which is still larger than but nearer to the standard deviation of 12.23.

# *Summary*

The numbers in a sample vary. Some are larger than others and some are smaller. Measures of variability quantify how diverse the numbers are.

The simplest measure of variability is the *range*. The range is the largest number in the sample minus the smallest one. The range is not a very good measure of variability because it uses only two extreme scores and because it depends on the sample size. As the sample size increases, the range tends to increase.

The *interquartile range* is the difference between the median of the upper half of the numbers and the median of the lower half. The interquartile range measures how different the scores in the middle half of the distribution are.

The *standard deviation* is the most commonly used measure of variability. The standard deviation is the square root of the variance. The variance is one-half the average squared difference between all possible pairs of numbers. The variance is defined by the following steps.

1. Compute the deviation of each score from the mean.
2. Square each of these deviations.
3. Sum the squared deviations.
4. Divide the sum by sample size less one.

Computations can be simplified by these steps.

1. Square each score.
2. Sum the squares.
3. Subtract the square of the sum of all the scores divided by the sample size.
4. Divide by sample size less one.

Again, the square root of the variance is the standard deviation. The computational formula for the standard deviation is

$$s = \sqrt{\frac{\sum X^2 - \frac{(\sum X)^2}{n}}{n - 1}}$$

The variability of the numbers is a fundamental part of a distribution. The range, the interquartile range, and the standard deviation provide three ways of quantifying the variability of the numbers.

# *Problems*

1. Compute the range, interquartile range, and standard deviation for the following samples.

   a. 1, 3, 4, 6, 6, 9, 10, 11
   b. 2, 4, 6, 8, 9, 11, 13, 14, 15
   c. 3, 4, 4, 6, 7, 9, 10, 13, 15, 16

2. a. If the standard deviation of a sample is 6.5, what is the variance?
   b. What is the average squared difference?

3. a. Find the standard deviation, range, and interquartile range of the following sample.

$$6, 8, 10, 11, 12, 14, 16, 22, 150$$

   b. Compute the standard deviation, range, and interquartile range of the same sample, without the number 150.

$$6, 8, 10, 11, 12, 14, 16, 22$$

   c. Which measure of variability changed the least?

4. Two automobile manufacturers, Marvel Motors and Amazing Autos, have each produced six different models. Below is the number of years each model of car typically runs before requiring major repairs.

   Marvel Motors:  3  5  10  7  2  9

   Amazing Auto:  6  7  7  6  5  5

   The mean for both manufacturers is six years.

   a. Compute the range, interquartile range, and standard deviation for each manufacturer.
   b. Which manufacturer produces cars of more consistent quality?

5. The following table presents the number of milligrams of phosphorus, calcium, and vitamin C contained in one cup of various fruit juices.

   | Juice | Phosphorus | Calcium | Vitamin C |
   |-------|-----------|---------|-----------|
   | Apple | 23.0 | 15.0 | 2.5 |
   | Apricot | 30.0 | 23.0 | 7.5 |
   | Cranberry | 7.5 | 13.0 | 40.0 |
   | Grapefruit | 35.0 | 20.0 | 77.5 |
   | Orange | 45.0 | 25.0 | 100.0 |
   | Pineapple | 23.0 | 38.0 | 23.0 |

   a. Compute the variance and standard deviation of each nutrient.
   b. On which nutrient do these juices vary the most?

6. a. Given a standard deviation of 9.6, could the range equal 20?
   b. Given a range of 42, could the interquartile range equal 53.8?
   c. Given a range of 12.25, could the standard deviation equal –3.5?
   d. Given a range of 11.3, could the standard deviation equal 12?

7. Hughes and McNamara (1961) studied the effectiveness of programmed instruction versus traditional lecture-discussion for training computer service personnel. The experimental group, which received the programmed instruction, consisted of 70 individuals. There were 42 in the control group, which received traditional instruction. The table below gives the scores received by each group at the end of the course.

| Score | Control (n = 42) | Experimental (n = 70) |
|---|---|---|
| 65–69 | 1 | |
| 70–74 | 1 | |
| 75–79 | 5 | |
| 80–84 | 7 | 3 |
| 85–89 | 9 | 5 |
| 90–94 | 14 | 15 |
| 95–99 | 5 | 47 |

Use the class midpoint for the score for each group.

a. Compute the mean, variance, and standard deviation for each group.
b. Which group's scores are more variable?

8. Suppose two classes, each consisting of eight students, take an exam. Suppose further that the scores for class A are more variable than the scores for class B.

a. Create a sample of scores for class A and a sample for class B, each with a mean of 70.
b. Compute the standard deviation for each class.
c. In which class is the mean more representative of the students' performance?

9. For the following sample determine which observations are outliers. Use the median plus or minus twice the interquartile range as the definition of an outlier.

15, 19, 25, 22, 8, 19, 18, 15, 37

10. Below are the gender bias data discussed in the previous two chapters.

| | | | |
|---|---|---|---|
| .29 | .01 | −.40 | .00 |
| .56 | −.31 | .00 | .35 |
| .00 | .14 | .02 | .11 |
| −.31 | −.22 | −.03 | −.23 |
| .56 | .01 | .00 | −.56 |
| −1.03 | .00 | −.23 | .23 |
| .00 | −.10 | .05 | .00 |
| .60 | −.23 | .24 | −.36 |

Compute and interpret the range, interquartile range, and the standard deviation for the data.

11. Compute the standard deviation for the following sample.

6, 8, 12, 9, 13, 12, 10, 9

12. For the data in problem 11, add five to each score and compute the standard deviation. How does the standard deviation change?

13. The following data represent annual rainfall of two cities (A and B) over the past eight years.

    A:   24, 18, 19, 21, 28, 17, 32, 24

    B:   6, 19, 14, 7, 21, 9, 17, 4

    In which city is rainfall more variable?

14. You want to determine how variable the high temperature is in two exotic isles. You note that in San Luca the temperatures have been 87 and 93 for two days. But for Dolores you have more information. The temperatures are 84, 86, 87, 88, 93, 94, and 94 for seven days. Use the range and standard deviation to determine variability of weather of the two islands. Decide which island has more variable weather.

15. Using the median and the interquantile range, state which scores, if any, are outliers:

    14, 16, 13, 12, 16, 17, 14, 13, 17, 14, 21, 12, 14, 13, 6

# 5 | *Transformation*

The numbers in a sample are assigned to objects by a set of rules. For instance, in the United States the fuel efficiency of automobiles is commonly measured in miles per gallon. When the United States switches to the metric system, kilometers per liter will be used instead. Because one mile equals 1.61 kilometers and one gallon equals 3.79 liters, one mile per gallon equals 1.61/3.79 or .43 kilometer per liter. So if a car gets 22 miles per gallon, it gets 22 × .43 or 9.5 kilometers per liter. What has been done is take a number and transform it. *Transformation* is the process by which numbers in a sample are altered by some mathematical operation. So if a sample of numbers were measured in terms of miles per gallon, and there was a need to remeasure fuel efficiency in terms of kilometers per liter, the values of the numbers could be systematically changed by multiplying them by .43. Numbers are not immutable. The rules that are used to assign numbers to objects can be changed, and possible transformation of the numbers should be considered.

Transformations of data are more common than might be realized. Most of us received a gross Scholastic Aptitude Test (SAT) score, say 580 in verbal and 560 in math. But attached to each is a percentile rank, say 76% and 68%. A percentile rank is a transformation of the raw SAT score. Also, when a person compares his or her test score to a friend's test score, the person might subtract his or her own score from the friend's score. The person then knows how much better or worse is the friend's score. Another commonly used transformation is to rank order the sample of numbers. Typically the times from a running race are rank ordered from fastest to slowest.

A *transformation* takes the original numbers and performs a mathematical operation on them. A transformation can be represented by

some mathematical operation on the original sample of numbers
= transformed sample of numbers

The mathematical operation can be as simple as addition or as complex as some trigonometric function, but the principle remains the same.

Transformations have three major purposes: to increase *interpretability*, *comparability,* and *symmetry.* Numbers can be confusing, and so if a transformation makes the meaning of the numbers any clearer, they should be transformed. For instance, if it is known that a consultant made $15,000 in 20 weeks, it is more informative to know the consultant's rate of pay per hour. It is $15,000 divided by the number of hours worked. The number of hours is 20 × 5 × 8, or 800, and so the hourly rate is 15,000/800, which is $18.75. By transforming the number from dollars per 20 weeks to dollars per hour, it is clearer how much the person made. Transformations are also used to make samples of numbers more comparable than they would be otherwise. One reason for converting to the metric system is to make numbers in the United States comparable with those in the rest of the world.

Finally, transformations are used to make distributions more symmetric than they would be otherwise. Symmetry in a distribution is desired because it is assumed by many inferential statistical techniques. Moreover, experience shows that asymmetric distributions sometimes are not at the interval level of measurement. Transformations that promote symmetry tend to increase the likelihood that the level of measurement is interval in nature.

Data transformations are a basic part of data analysis. There are four major types of transformations: no-stretch, one-stretch, two-stretch, and flat transformations. This classification system focuses on the effect of each type of transformation on the shape of the distribution. Each is considered in turn.

One example will be used to illustrate the various transformations. It is taken from Duncan and Fiske's (1977) study of two-person interactions. Two strangers interacted for a period of seven minutes, the last five of which were videotaped. In Table 5.1 are the number of smiles during the five minutes of the interaction. The numbers in the table were based on measurements from 22 women, graduate students at the University of Chicago, who interacted with 22 men. The mean number of smiles for the women is 9.55 and the standard deviation is 4.77.

*TABLE 5.1*   **Number of Smiles of 22 Females During Five Minutes of Interaction with a Male**

| | |
|---|---|
| 8 | 13 |
| 4 | 10 |
| 18 | 4 |
| 11 | 7 |
| 5 | 11 |
| 7 | 9 |
| 5 | 9 |
| 21 | 7 |
| 14 | 14 |
| 1 | 9 |
| 9 | 14 |

Data were gathered by Duncan and Fiske (1977).

# No-Stretch Transformations

A no-stretch transformation is the simplest and most common type. Most of the time that this transformation is used, the researcher is not even aware of it. When quarts are converted to gallons or feet are converted to miles, a no-stretch transformation (sometimes called a linear transformation) is performed. A *no-stretch transformation* does not alter the basic shape of the distribution. Although the distribution is stretched, it is uniformly stretched.

This transformation is called a no-stretch transformation for the following reason: The effect of this type of transformation is not to alter the basic shape of the distribution. So, if the distribution is symmetric it remains symmetric. If there is a positive skew it remains. If the distribution is flat it remains flat. Many of the statistics that are discussed in later chapters do not change after a no-stretch transformation. For instance, the correlation coefficient and $t$ test do not change.

One type of no-stretch transformation is one in which all the numbers in the sample are multiplied or divided by the same number. If the score is denoted as $X$ and the number that the score is multiplied by is denoted as $k$, the transformed score is $kX$. For example, if $X$ is a score whose unit of measurement is gallons and the researcher wishes to convert to quarts, then $k$ is set to four because four quarts equal one gallon. Alternatively each score can be divided by a constant: $X$ divided by $k$. For instance, if $X$ is a score whose unit of measurement is feet and the researcher wishes to convert to miles, then $k$ is 1/5280 or .0001894 because one mile equals 5280 feet. As another example, the data in Table 5.1 could be divided by five. Because the numbers in the table refer to the number of smiles in five minutes, dividing by five would yield smiles per minute.

Another type of no-stretch transformation is the adding of the same number to all the scores. If the numbers are denoted as $X$ and the number added to all the scores is denoted as $m$, the transformed score is $X + m$. For instance, if $X$ is a student's score on a statistics test and $m$ is a five-point bonus that the teacher gives each student, then the transformed score is the previous score plus five bonus points. Alternatively, a constant may be subtracted from each score: $X - m$. For instance, if $X$ is annual income earned by workers at a factory and $m$ is $-10,000$, then the transformed score is a worker's earnings over $10,000. As another example, the value of 9.55 could be subtracted from the numbers in Table 5.1. Because 9.55 is the mean, a score minus the mean measures a given person's number of smiles relative to the mean. So a negative score would indicate that a person smiles less than average and a positive score indicates more smiling.

These two types of no-stretch transformations can be combined. That is, each score can be multiplied (or divided) by a number and then have another number added (or subtracted) to the score. If the score is denoted as $X$, the multiplier as $k$, and the score to be added as $m$, the transformed score is $m +$

$kX$. The temperature conversion from Celsius to Fahrenheit is of this form. The conversion for 20 degrees Celsius into Fahrenheit is $(1.8)(20) + 32$, which equals 68 degrees Fahrenheit. Thus, $k = 1.8$ and $m = 32$ for the conversion from Celsius to Fahrenheit.

Measures of central tendency and variability are altered in a no-stretch transformation. The three measures of central tendency are changed in the following way. If $M$ is used to denote the measure of central tendency, then the new measure of location is $kM + m$, where $k$ is the term that is multiplied and $m$ is added. So if the mean is 10.4 and $k$ and $m$ are 2.0 and 1.0, respectively, then the transformed mean is $(2.0)(10.4) + 1.0 = 21.8$.

The variability is not affected by adding or subtracting a constant to the scores. For example, if ten is added to all the scores, the variability does not change. If scores are multiplied by $k$, however, the range, interquartile range, and the standard deviation of the transformed scores equal the measure of variability of the untransformed scores multiplied by $k$. The variance of the transformed scores equals the variance of the original scores multiplied by $k^2$.

There are four basic reasons for employing a no-stretch transformation: change in unit of measurement, change in scale limits, reversal of scale, and standardization.

## Change in the Unit of Measurement

This purpose has already been discussed. Converting from inches to feet or from pounds to grams are examples of transformation to change the unit of measurement. This transformation is so common that is not even viewed as a transformation.

## Change of Scale Limits

Imagine a set of scores with a possible range of from 100 to 500. It may be desirable to change the upper limit of 500 to 10 and the lower limit from 100 to 1. Thus, the transformed measure would range from 1 to 10. A little notation can help:

  UL    upper limit of the original sample
  LL    lower limit of the original sample
  TUL   transformed upper limit of the transformed sample
  TLL   transformed lower limit of the transformed sample

So, UL after transformation the becomes TUL and LL becomes TLL. To change the limits, one computes for each score $X$

$$(X - \text{LL}) \frac{\text{TUL} - \text{TLL}}{\text{UL} - \text{LL}} + \text{TLL}$$

for the above example

$$UL = 500$$
$$LL = 100$$
$$TUL = 10$$
$$TLL = 1$$

The appropriate transformation is then

$$(X - 100)\frac{10 - 1}{500 - 100} + 1$$

or

$$(X - 100)\frac{9}{400} + 1$$

For example, if $X$ is 300, then transformed $X$ equals 5.5.

## Reversal

Sometimes scales are oriented in the "wrong direction." Generally larger numbers indicate more of some quantity. If this is not the case, the scale needs to be reversed. For instance, a variable may be the rating of a political leader on a ten-point scale—that is, a scale from one to ten. On some questions a response of ten is a favorable response toward the political leader and on others a one is a favorable response. It may be desirable to reverse the questions in which one is a favorable response. That is, make a response of one a ten and a ten a one. This transformation of the response $X$ is accomplished in this example by $11 - X$. In general, the transformation for reversal is $LL + UL - X$, where LL is the lower limit, UL is the upper limit and $X$ is the score to be transformed. So, the score to be transformed is subtracted from the sum of the lower and upper limits.

## Standard Scores

The most often used no-stretch transformation in statistical work is one that is said to standardize the scores. To *standardize* a set of numbers, the mean is subtracted from each score, and this difference is divided by the standard deviation. The transformed score for person $i$ is denoted by $Z_i$ and is given by

$$Z_i = \frac{X_i - \overline{X}}{s}$$

where $\overline{X}$ is the mean of the $X$ variable and $s$ is the standard deviation. This score is called a *Z score* or a *standard score*. The effect of this transformation is to make the mean of the transformed scores equal zero and to make the standard deviation equal one. Researchers employ this transformation because a standard score (or Z score) tells them how far each person is from the mean in standard deviation units. For instance, a $Z$ score of $-2$ indicates that the person scores two standard deviations below the mean. When the $Z$ score transformation is used, the scores are said to be *standardized*.

For the smile numbers in Table 5.1, the mean is 9.55 and the standard deviation is 4.77. The formula for the $Z$ score of variable $X$ for person $i$ is

$$Z_i = \frac{X_i - 9.55}{4.77}$$

So if $X = 14$, then $Z = (14 - 9.55)/4.77 = .933$, which is about one standard deviation above the mean. A positive $Z$ score indicates that the score is above the mean, and a negative $Z$ score indicates a score below the mean.

# One-Stretch Transformations

A no-stretch transformation does not alter the basic character or the shape of the distribution. So if the shape of the distribution needs to be changed, a no-stretch transformation does not do the job. For instance, if there is a large positive skew (the scores trail off in the positive direction), it may be desirable to transform the numbers to remove that skew. One way to remove it would be to stretch the lower numbers. A method is needed to stretch one end of the distribution, and so the transformation is called *one-stretch*. The presence of positive skew is quite common when scores have a lower bound of zero and no upper bound. Examples of such variables are income, reaction time, age, and number of cars on a freeway. The three major one-stretch transformations are square root, logarithm, and reciprocal.

## Square Root

The square root transformation, or $\sqrt{X}$, is relatively simple. A square root transformation is simply the square root of every number in the sample. One should employ this transformation only if all the numbers are positive. It is generally appropriate when $X$ is a count. Examples of counts are the number of bar presses by laboratory rats or the number of cars passing through an intersection. This transformation has become easier to perform now because most hand-held calculators have a square root key.

## Logarithm

The logarithm, or log, is the most commonly used one-stretch transformation, but it is the most complicated numerically. (See Chapter 1 for a review of logarithms.) For instance, income is regularly subjected to a logarithmic transformation. The logarithm of $X$ is the number that satisfies the equation $X = 10^Y$. The term $Y$ is said to be a common logarithm, base 10, of $X$. It is much more common in scientific work to use natural logarithms, which have *e* (approximately equal to 2.718) as the base. So *Y* is determined from

$$X = e^Y$$

To convert from base 10 to base *e*, multiply the base 10 logarithm by 2.303. It should be noted that log(1) = 0 regardless of the base. Also the logarithm of zero and negative numbers is undefined regardless of the base. Thus, this transformation is only feasible when the numbers are positive. If zero is a possible value, it is necessary to compute $X + .5$ or $X + 1.0$ and then compute the logarithm. That is, .5 or 1.0 is added to all the numbers and then the logarithms are computed.

## *Reciprocal*

The least commonly employed one-stretch transformation is the reciprocal transformation. The reciprocal of *X* is defined as $1/X$. In words, one is divided by the score. It is particularly useful when *X* is time. Say the number of minutes it takes to run a mile is measured. The reciprocal would measure how many miles or fractions thereof were run in a minute, or the rate at which one runs. Thus, the reciprocal of time equals rate, and the reciprocal of rate equals time. Unlike the square root and logarithm transformations, the reciprocal transformation reverses the order of scores, and thus the largest score becomes the smallest.

Of the one-stretch transformations, the square root stretches the least and the reciprocal stretches the most. To measure how much stretch there is in a one-stretch transformation, the amount of stretch in the lower numbers is compared with the amount of stretch in the higher numbers. To do this, the lower numbers are 1 and 5 and the higher numbers are 11 and 15. The stretch index is

$$\frac{\text{transformation}(5) - \text{transformation}(1)}{\text{transformation}(15) - \text{transformation}(11)}$$

so for instance, for reciprocal, the value is

$$\frac{1/5 - 1/1}{1/15 - 1/11}$$

Using this stretch index, the following values are obtained.

| No transformation | 1.00 |
|---|---|
| Square root | 2.22 |
| Logarithm | 5.19 |
| Reciprocal | 33.00 |

Although the square root exhibits some stretch, it stretches the least. The reciprocal transformation dramatically stretches the numbers.

One-stretch transformations are used primarily to remove positive skew.[1] To simplify the presentation, positive skew is indicated by the median being smaller than the mean. In practice, to determine the presence of positive skew a more detailed analysis is required. In particular, the histogram would have to be created and examined. One clue that a positively skewed distribution will be aided by a one-stretch transformation is the *coefficient of variation,* which equals $s/\bar{X}$, the ratio of the standard deviation to the mean. If the coefficient of variation is greater than .25, a one-stretch transformation is probably needed to remove skew.

To see how one-stretch transformations affect the shape of a distribution consider the following sample.

$$1, 4, 4, 9, 9, 9, 16, 16, 25.$$

This sample is positively skewed because the median is 9 and the mean is 10.33. However, after a square root transformation, the distribution becomes perfectly symmetric, with the mean and the median equal; that is,

$$1, 2, 2, 3, 3, 3, 4, 4, 5.$$

However, the logarithm transformation of the original numbers results in the following numbers:

$$0, 1.39, 1.39, 2.20, 2.20, 2.20, 2.77, 2.77, 3.22$$

and the skew has been overcorrected, as is indicated by the fact that the median (2.20) is now greater than the mean (2.02). There is no longer a positive skew but a slightly negative one. Thus researchers must be careful to avoid overcorrecting the skew.

In Table 5.2 are the numbers from Table 5.1 and their one-stretch transformed scores. There is a slight positive skew in the original scores, making the mean larger than the median. Applying the square root transformation makes the mean and median nearly equal. However, the log transformation overcorrects for skew and the reciprocal even more so. (Recall that the reciprocal transformation reverses the scores and so the relative size of the mean and median is reversed.) The square root transformation seems the most appropriate here.

The mode and median have an advantage over the mean when a one-stretch transformation is used. Assume that the numbers are each logged. The mean of the logs does not ordinarily equal the log of the mean of the untransformed

---

[1]One-stretch transformations have two other purposes. These transformations tend to remove heterogeneity of variance and nonlinearity. Heterogeneity of variance means that the variance changes for different samples. It is not uncommon for the standard deviation of a sample to be related to the mean of the sample. Typically, samples with larger means will have larger standard deviations. This is an example of heterogeneity of variance. One-stretch transformations generally bring about equal standard deviations. This topic is discussed in Chapter 13. Finally, in Chapter 7 the effect of one-stretch transformations on making relationships linear is considered.

*TABLE 5.2*  **One-Stretch Transformation**

|  | Raw Score | Square Root | Logarithm | Reciprocal |
|---|---|---|---|---|
|  | 8 | 2.828 | 2.079 | .1250 |
|  | 4 | 2.000 | 1.386 | .2500 |
|  | 18 | 4.243 | 2.890 | .0556 |
|  | 11 | 3.317 | 2.398 | .0909 |
|  | 5 | 2.236 | 1.609 | .2000 |
|  | 7 | 2.646 | 1.946 | .1429 |
|  | 5 | 2.236 | 1.609 | .2000 |
|  | 21 | 4.583 | 3.045 | .0476 |
|  | 14 | 3.742 | 2.639 | .0714 |
|  | 1 | 1.000 | 0.000 | 1.0000 |
|  | 9 | 3.000 | 2.197 | .1111 |
|  | 13 | 3.606 | 2.565 | .0769 |
|  | 10 | 3.162 | 2.303 | .1000 |
|  | 4 | 2.000 | 1.386 | .2500 |
|  | 7 | 2.646 | 1.946 | .1429 |
|  | 11 | 3.317 | 2.398 | .0909 |
|  | 9 | 3.000 | 2.197 | .1111 |
|  | 9 | 3.000 | 2.197 | .1111 |
|  | 7 | 2.646 | 1.946 | .1429 |
|  | 14 | 3.742 | 2.639 | .0714 |
|  | 9 | 3.000 | 2.197 | .1111 |
|  | 14 | 3.742 | 2.639 | .0714 |
| Mean | 9.55 | 2.986 | 2.100 | .1625 |
| Median | 9.00 | 3.000 | 2.197 | .1111 |

scores. However, the log of the mode equals the mode of the logs. The same holds for the median when sample size is odd and is closely approximated when sample size is even. So if the data are transformed by a one-stretch transformation, the mean must be recomputed, but the mode and median can be computed from the mode or median of the original data.

Before discussing two-stretch transformations, there is one other one-stretch transformation. Occasionally it happens that scores have a negative skew. For instance, the scores

$$0, 6, 6, 8, 8, 10$$

show a slight negative skew (the median, 7, being greater than the mean, 6.33), which is removed by squaring the numbers, as follows:

$$0, 36, 36, 64, 64, 100$$

Thus, squaring the numbers can remove a negative skew. What squaring the data does is to stretch the larger numbers so that the sample has less of a negative skew.

# Two-Stretch Transformations

A one-stretch transformation is generally useful when the numbers have a lower bound (usually zero) but no upper bound. With a lower bound, numbers tend to pile up near it, creating a positive skew. What a one-stretch transformation does is to remove this lower bound.

A two-stretch transformation is useful for samples that have both a lower and an upper bound. There is one major type of data that has both an upper and lower bound: a proportion. A proportion has a lower bound of zero and an upper bound of one. (The student might review the discussion of proportions, percentages, and odds in Chapter 1.) One can score no higher than one and no lower than zero. The purpose of a two-stretch transformation is to stretch both the numbers near one and those near zero. It thus stretches twice.

Many times a researcher has proportions and does not realize it. For instance, if the variable for a test is the number correct out of 40, the numbers 16, 25, 38 do not look like proportions. But when the number of correct items on the test is divided by 40, the result is a proportion: $16/40 = .40$.

The numbers near one and zero need stretching because a small change is more difficult when a proportion is near zero or one. For instance, reducing the risk of surgical procedure from .02 to .01 is more impressive than reducing it from .55 to .54. Although in both cases the risk has been reduced by 1%, the odds of dying have been cut in half in the .02 to .01 case and have hardly changed at all in the .55 to .54 case. Small differences between proportions near zero and one can be viewed as larger than small differences near .5.

The three types of two-stretch transformations are arcsin, probit, and logit. All three of these transformations are rather mathematically complicated; tables are given in Appendix A.

## Arcsin

The arcsin is the most commonly used transformation for proportions in psychology. The sine is a trigonometric function and the arcsin is its inverse. Fortunately, one need know nothing about trigonometry to apply this transformation. The effect of this transformation is to stretch the distribution's tails.

## Probit

The probit transformation is relatively uncommon in psychology but is very common in other sciences. For instance, political scientists routinely transform the proportion of persons voting for a candidate to a probit. To understand this transformation fully requires knowledge of the normal distribution, which is explained in detail in Chapter 10. In that chapter it is explained how the values for the probit transformation are determined. For the moment, probit can be viewed as one of three possible transformations for percentage data.

## Logit

The logit transformation is less commonly employed than either arcsin or probit. It is based on a very commonsense approach to probability. For instance, at the racetrack a horse is said to have odds of winning of two to one. These odds mean that for every three races, the horse would win twice. So a two to one odds mean a two out of three probability or a 67% chance of winning the race. For the logit, one first converts a proportion into an odds. The formula for an odds given a proportion $p$ is

$$\frac{p}{1.0 - p}$$

The odds range from zero to positive infinity. The *logit* is the natural logarithm of the odds. In equation form the logit is

$$\ln\left[\frac{p}{1.0 - p}\right]$$

where ln is the natural logarithm (base $e$) and $p$ is a proportion. Recall that $e$ approximately equals 2.718. By convention a proportion of 1.00 is set to .9975 and .00 is set to .0025. (This same change is done for probit but is not necessary for arcsin.) It might be noted that a logit of zero corresponds to a proportion of .50.

The effect of the two-stretch transformation is to stretch numbers more as they move away from .50. That is, the smallest and the largest scores are stretched the most. The most stretching transformation is logit and the least is arcsin. If most of the values are between .25 and .75, the effect of these transformations is so small as to make them unnecessary.

# Flat Transformations

A flat transformation turns the sample's distribution into a flat distribution regardless of the shape of the original distribution. There are two types of major flat transformations: rank order and percentile rank.

## Rank Order

This transformation has already been implicitly employed when medians and interquartile ranges are computed. For this transformation the numbers are rank ordered from the smallest to the largest. (Occasionally, numbers are ranked from largest to smallest, as in the case of class rank.) For instance, the numbers

$$7, 8, 3, 9, 1, 11$$

are rank ordered

$$1, 3, 7, 8, 9, 11$$

Then the smallest value is assigned a one, the next smallest a two, and so on. Thus, for the above sample,

$$
\begin{array}{cccccc}
1 & 3 & 7 & 8 & 9 & 11 \\
\downarrow & \downarrow & \downarrow & \downarrow & \downarrow & \downarrow \\
1 & 2 & 3 & 4 & 5 & 6
\end{array}
$$

The distribution is now flat. In the case of ties, the rule is to assign the average rank. So for the sample

$$1, 2, 2, 2, 2, 7, 8$$

there are four observations equal to 2. They have ranks 2 through 5 and so the average rank equals $(2 + 3 + 4 + 5)/4$ which equals 3.5. The ranks would then be

$$1, 3.5, 3.5, 3.5, 3.5, 6, 7$$

When there are ties, the rank-order transformation does not produce a perfectly flat distribution.

## Percentile Rank

Scholastic Aptitude Test (SAT) scores as well as many personality tests are presented in terms of percentile rank. In this section a method is presented to convert a score from a sample into its percentile rank. A percentile rank is measured by the following formula:

$$100 \left[ \frac{R - .5}{n} \right]$$

where $n$ is the sample size and $R$ is the rank order of the score. So if $n = 20$ and an observation has a rank order of 9, the percentile rank is

$$100 \frac{(9 - .5)}{20} = 42.5$$

There are two major purposes for employing flat transformations. First, rank orders and percentile ranks are often easier to interpret than raw scores. Second, some of the statistics to be discussed later in this book are based on the assumption that the level of measurement is at the interval level. The effect of the flat transformation is to remove distance information between observations and make the data ordinal. Once the data are at the ordinal level of measurement, statistical procedures are available that make fewer assumptions about the data. These procedures are discussed in Chapter 18.

# *Two-Variable Transformations*

All the transformations that have been considered involve a single sample of numbers. It is very common to have two samples of numbers from the same people. For instance, there could be scores on two quizzes for each person in a class. There are two major ways of combining information from two samples: the sum and the difference.

The simplest procedure is to sum the scores. This is sensible when the numbers measure the same trait. For instance, one can add the scores on the two quizzes. The resultant score should be more reliable than either score by itself.

A second alternative is to create a difference score. This is common for measuring change. If the earlier quiz score is subtracted from the more recent one, the resulting difference would be a measure of improvement. As a second example, studies on the effect of psychotherapy on adjustment typically have a baseline measure before psychotherapy. The outcome is the improvement from the baseline.

When working with two-variable transformations, the mean and variance have an advantage over other measures. The overall mean equals the mean of its components. For example, population change can be defined as the number of births minus the number of deaths plus the number of immigrants minus the number of emigrants. In equation form,

population change = births – deaths + immigrants – emigrants

The mean is the only measure of central tendency for which the mean number of births minus the mean number of deaths plus the mean number of immigrants minus the mean number of emigrants exactly equals the mean population change. The mean then equals the mean of its components.

The variance has a similar advantage over other measures of variability. The variance of the sum of two variables equals the sum of each variable's variance.[2] This fact is not true of either the interquartile range or the simple range.

---

[2]This fact holds only if the two variables are uncorrelated. When variables are correlated, the fact must be modified.

# *Summary*

Numbers are not immutable. Researchers implicitly decide how to measure the objects of interest. There should be explicit consideration concerning how to use these numbers. In some cases it may be desirable, or even necessary, to transform the numbers.

Table 5.3 outlines the transformations that have been discussed in this chapter. The simplest type of transformation is a no-stretch transformation. This type of transformation involves adding or multiplying a number by each score in the sample. A *no-stretch transformation* preserves the basic shape of the distribution. It is used to change the units of measurement—for example, to switch from inches to centimeters. It is also used to express the score in terms of standard deviation units or Z scores. A Z or *standard score* is the individual score minus the mean, divided by the standard deviation.

The second type of transformation is called a *one-stretch transformation.* Its primary purpose is to remove positive skew from a distribution and so make the distribution symmetric. These transformations are the *square root, logarithm,* and *reciprocal.* The square root stretches the least and the reciprocal stretches the most.

The third type of transformation is a *two-stretch transformation.* This

**TABLE 5.3    Summary of Transformations**

| No-Stretch | Two-Stretch |
|---|---|
| Types | Types (see Appendix A) |
|    Adding a number to all scores |    Arcsin |
|    Multiplying scores by a number |    Probit |
| Purposes |    Logit |
|    Change unit of measurement | Purpose |
|    Change limits |    Remove floor and ceiling |
|    Reverse scale direction | Requirements |
|    Standardization |    Scores between 0 and 1 |
| Requirements | |
|    None | |
| **One-Stretch** | **Flat Transformation** |
| Types | Types |
|    Square root |    Rank order |
|    Logarithm |    Percentile rank |
|    Reciprocal | Purposes |
| Purpose |    Facilitate interpretation |
|    Remove skew |    Make data ordinal |
| Requirements | Requirements |
|    Positive numbers |    Few ties |

transformation is appropriate for samples with a lower and an upper bound and is particularly appropriate for percentage data. These transformations are *arcsin, probit,* and *logit,* with arcsin making the smallest change and logit the largest.

The fourth type of transformation is a *flat transformation,* which turns the numbers into a flat distribution. The two transformations of this type are *rank order* and *percentile rank.* These transformations generally increase interpretability and allow for a relaxation of some assumptions for statistical tests.

At first reading, the topic of transformations may seem to be bewildering because some of the mathematical operations seem complex. However, because the one-stretch transformations can be performed with a calculator and two-stretch transformations are tabled in Appendix A, all the messy mathematical complications are circumvented. The other transformations involve only the simple operations of addition, multiplication, and rank ordering.

The second aspect of transformation that leads to confusion is the decision of which transformation to use. Say a researcher wishes to employ a one-stretch transformation. How does he or she know whether to use square root, logarithm, or reciprocal? Which one is best? The researcher generally does not know this in advance. It may be necessary to try each out and determine which one makes the distribution more symmetric. Eventually, it should be clear which transformation, if any, is best.

# *Problems*

1. Compute to three decimal points each of the following.

   a. ln 10
   b. square root of 39
   c. arcsin of .75
   d. 1/48
   e. probit of .66
   f. logit of .54
   g. ln 63
   h. square root of 75
   i. arcsin of .44
   j. logit of .17
   k. probit of .23
   l. logit of .50

2. Using the median and the mean for the following sample of numbers, which one-stretch transformation makes the distribution most symmetric?

|   |    |    |   |
|---|----|----|---|
| 1 | 4  | 5  | 3 |
| 9 | 3  | 1  | 4 |
| 3 | 15 | 2  | 7 |
| 4 | 8  | 18 | 6 |
| 8 | 7  | 9  | 3 |

3. Compute $Z$ scores for the numbers in problem 2. The mean equals 6.0 and the standard deviation is 4.40.

4. For a sample with limits of zero and 100, state how to change the limits to one and ten.

5. For the following transformations give the new mean and standard deviation for $Z$.

   a. $Z = 10X + 3$, where $\bar{X} = 5$ and $s_x = 3$.
   b. $Z = Y/5 + 15$, where $\bar{Y} = 9$ and $s_y = 1$.
   c. $(X - 10)/15 = Z$, where $\bar{X} = 10$ and $s_x = 15$.
   d. $Z = 6X + 4$, where $\bar{X} = 5$ and $s_x = 2$.

6. For the following sample, compute percentile ranks for the scores 8, 12, and 17.

|    |    |    |    |
|----|----|----|----|
| 7  | 13 | 14 | 14 |
| 9  | 12 | 6  | 9  |
| 15 | 18 | 17 | 3  |
| 6  | 6  | 21 | 15 |
| 8  | 11 | 9  | 20 |

7. Suppose that a sample of numbers has a mean of 25 and a standard deviation of 3.20. Compute the means, standard deviations, and variances that would result from the following transformations.

   a. Add five to all scores.
   b. Multiply scores by three.
   c. Subtract 20 and then multiply scores by five.
   d. Add 100 and then multiply scores by two.

8. Chapter 1, problem 6, gives ratings of discomfort by subjects who were offered aid. The data are repeated below. The scale runs from one to ten, with higher scores indicating greater discomfort. Transform the eight numbers by reversing the scale so that the scale still runs from one to ten, and a higher score indicates more comfort.

$$10, 1, 3, 5, 3, 7, 7, 3$$

9. a. For the data in the previous problem, transform the eight scores into $Z$ scores.
   b. Transform the eight scores into percentile ranks.

10. Suppose that the possible points on a final exam range from five to 75.

Change the lower limit to zero and the upper limit to 100 and transform the following scores.

a. 55    b. 72    c. 68    d. 40

11. Vinsel, Brown, Altman, and Foss (1980) conducted a study in which one of the dependent measures was the type of wall decorations used by dormitory freshmen. The following table shows the average area covered by each type of decoration by male and female subjects.

| | *Area* | |
|---|---|---|
| *Category* | *Males* | *Females* |
| A. Personal relationships | 8.5 | 20.1 |
| B. Abstract | 66.1 | 80.7 |
| C. Music/theater | 24.4 | 20.3 |
| D. Sports | 77.4 | 21.1 |
| E. Values | 9.4 | 16.8 |
| F. Reference items | 18.7 | 5.8 |
| G. Idiosyncratic | 46.4 | 17.8 |
| H. Entertainment/equipment | 15.4 | 10.0 |

a. For the males, rank the categories from least area used to most area used.
b. Do the same for the females.
c. For which category does the ranking for the females and males differ the most?

12. Chapter 4, problem 7 gave the scores from an experiment on programmed instruction. Some of the data are repeated below.

| Class Midpoint | Frequency |
|---|---|
| 67 | 1 |
| 72 | 1 |
| 77 | 5 |
| 82 | 7 |
| 87 | 9 |
| 92 | 14 |
| 97 | 5 |

a. Apply a square root transform to the scores.
b. Perform a natural log transform on the scores.
c. Perform a reciprocal transform on the scores.
d. Compute the mean and mode of the original data. (Make sure to weight by frequency.) In which direction are the data skewed? Do the same for the three sets of transformed scores. How do the transformations affect the skew?

# 6 | *Measuring Association: The Regression Coefficient*

A sample has a distribution, a central tendency, and a variability. Each of these characteristics helps the researcher make sense out of the numbers. Although each reveals some important aspect of the data, none of them measures the relationship or association between two sets of numbers.

Examples of questions of association are: If you receive an A on the midterm test, will you tend to get a high mark on the final examination? If you buy a car that is supposed to get 32 miles per gallon, what kind of mileage will your car actually get? If your roommate likes you, will you like your roommate? These are all questions of association or relationship.

To speak of relationship or association there are two separate samples of numbers that are linked together. For instance, consider the two samples of midterm grades and final grades. The two samples are linked together by persons to whom the grades refer. This linkage is illustrated in Table 6.1. For instance, John R. obtained a 36 on the midterm and a 48 on the final. The unit that links together the two samples need not necessarily be a person. For instance, to relate husband's height to wife's height, the unit that links together the scores is married couple and not person. One potential source of confusion in interpreting a measure of association is determining the object that links the pair of scores together. For instance, consider the statement that more discipline leads to more academic achievement. The relationship between these two variables can be measured for students: Do students who receive more discipline achieve more? For classrooms: In classrooms where there is more discipline, do the students achieve more? And for schools: In schools where there is more discipline, do the students achieve more? So the object can be the student, the classroom, or the school.

*TABLE 6.1*     **Linking Together of Two Samples**

| Midterm | Object | Final |
|---------|--------|-------|
| 36 ⟷ | John R. ⟷ | 48 |
| 89 ⟷ | Mary P. ⟷ | 78 |
| 93 ⟷ | Paul T. ⟷ | 81 |
| 78 ⟷ | Jane A. ⟷ | 95 |
| 90 ⟷ | James S. ⟷ | 82 |
| 81 ⟷ | Jean M. ⟷ | 89 |

If two variables are associated, then as the numbers in one sample vary, their partner numbers in the second sample vary in some related fashion. So association implies that the numbers vary together or, as stated in data analysis, the numbers *covary*.

The simplest way in which the numbers can covary is in a linear fashion. A relationship is said to be *linear* if a difference in one variable of a fixed amount results in a constant difference in the second variable. The term linear is used because when you plot a linear association on a graph, the result is a straight line.

*Linearity* requires that a one-unit increase in variable $X$ produces the same change in variable $Y$ regardless of where that one-unit increase in $X$ comes. Say, for instance, it is known that for every year of schooling a person earns on average $3000 more per year. If the relationship is linear, then an extra year of schooling provides the same amount of money ($3000) regardless of whether the extra year of schooling is one more year of college or one more year of high school.

The effect of a variable on another may be more complex than a linear one. Changes in a drug's dosage may be less potent at smaller concentrations than at larger ones. In this case the strength of a relationship between the variables increases as one variable increases. Any pattern of association between two variables that is not linear is referred to as *nonlinear* association. The important issue of nonlinear relationships is addressed in the next chapter.

There are two directions of linear association. The first type is positive association. Most often the expectation is that higher numbers in one sample are associated with higher numbers in the other sample. A student who does well on the midterm tends to do well on the final. The expectation is that the high numbers on the midterm are paired with the high numbers on the final and low numbers on the midterm go with low numbers on the final. Such a pattern of relationship is called a *positive* association. A *positive association* implies that as the numbers increase for one variable, they tend to increase for the other variable.

Sometimes as the numbers go up in one sample they go down in the other.

*FIGURE 6.1*   **Illustrations of positive (on the left) and negative (on the right) linear relationships.**

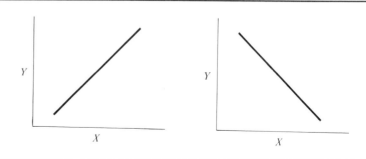

For instance, the more an adult weighs presumably the slower the person can run. This is an example of negative association: more weight, less speed. High numbers in one sample are associated with low numbers in the other. *Negative association* between two variables means that as the numbers increase in one sample, they decrease in the other.

The difference between a positive and a negative relationship is shown graphically in Figure 6.1. The positive relationship on the left of Figure 6.1 shows an ascending relationship, whereas on the right, the negative relationship is descending.

In Table 6.2 is a data set that will be used throughout this chapter. The data are memory scores from 16 men of various ages in the Boston area. It is then "person" that links the age and memory scores together. All men were given a test of short-term memory, which will be referred to as STM. Higher scores on STM indicate better short-term memory. The lowest possible score is zero and the maximum possible score is 24. All 16 men were being treated for alcoholism. The question considered in this chapter is the extent to which age and STM covary.

*TABLE 6.2*   **Ages and Short-Term Memory Scores (STM) of 16 Alcoholic Men**

| Person | Age | STM | Person | Age | STM |
|--------|-----|-----|--------|-----|-----|
| 1 | 48 | 14 | 9 | 56 | 2 |
| 2 | 46 | 7 | 10 | 54 | 12 |
| 3 | 44 | 12 | 11 | 65 | 12 |
| 4 | 52 | 10 | 12 | 35 | 18 |
| 5 | 22 | 24 | 13 | 63 | 5 |
| 6 | 43 | 11 | 14 | 39 | 18 |
| 7 | 51 | 9 | 15 | 30 | 14 |
| 8 | 54 | 19 | 16 | 47 | 8 |

Data were gathered by Dennis Ilchisin.

To describe the relationship between two variables the scores can be plotted on a graph. One variable is represented on the $X$ axis and the other on the $Y$ axis. Then each person's score is placed on the graph. A diagram in which the axes are the variables and the points are the data is called a *scatterplot*. Figure 6.2 is a plot of the score from person 1 in Table 6.2. The $X$ axis is age and the $Y$ axis is memory score. The person's age (48) and memory score (14) are located on the $X$ and $Y$ axes, respectively. Lines are drawn that are perpendicular to each axis from each score. The point at which the two lines intersect is a point in the scatterplot. When constructing a scatterplot, the perpendicular lines are not actually drawn. The dashed lines were drawn in Figure 6.2 only to show how to determine a point in a scatterplot.

Figure 6.3 illustrates the complete scatterplot for the age and memory data. The scatterplot itself tends to reveal whether there is any relationship between the two variables. Here, a negative relationship is suggested. Older persons tend to have lower memory scores.

## *The Regression Coefficient*

Although a scatterplot describes relationship, most relationships are too weak to be clearly discerned from an examination of the scatterplot. Some method of capturing the strength of a linear relationship in a scatterplot is needed. One strategy for measuring association is to draw a straight line through the set of points in the scatterplot.

The slope of a line is the standard measure of linear association. For the slope, the variable on the $X$ axis is called the predictor and the other is the

**FIGURE 6.2    How to determine a point in a scatterplot.**

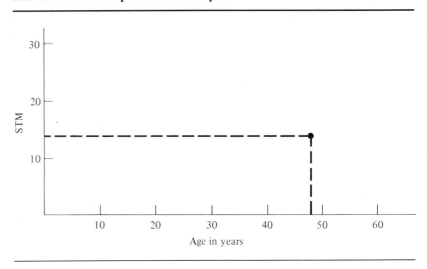

*FIGURE 6.3*     **Scatterplot for the age-memory study.**

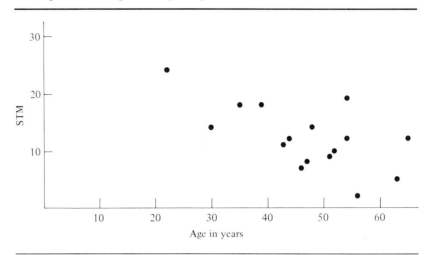

criterion. The *slope* measures the effect of a change of one unit in the predictor on the criterion. Consider the following examples.

An industrial psychologist is interested in predicting who is more productive at a given factory. She devises a test that she thinks can predict productivity. The test would be the predictor and productivity would be the criterion.

A sociologist believes that the number of dollars that a community spends on schools per pupil will depend on the percentage of persons over 65 in the community. He believes that the relationship is negative: the larger the percentage of elderly, the less the amount spent on education. The predictor is the percentage over age 65, and the criterion is money spent per pupil.

A clinical psychologist believes that depression is related to diet. She believes that sugar in the diet leads to depression. The predictor is sugar consumption, and the criterion is depression.

The major measure of slope between a predictor and a criterion is a measure called the *regression coefficient*. Two different rationales for the regression coefficient are developed in this chapter. One is based on the notion that the regression coefficient is an average slope. The other is based on the notion of the regression coefficient as the best fitting line.

## The Average Slope

Anyone who has decided to devote four years to obtaining a college degree must have wondered about the relationship between years of education and dollars earned. Does education predict income? So years of education is the predictor and income is the criterion. Imagine twin brothers Bob and Ray.

Ray graduated from college at State U. (16 years of education), and at age 30 he earns $35,000 a year. Bob, after finishing high school, elected not to go to college (twelve years of education), and at age 30 he earns $25,000 a year. Ray has four more years of education than Bob and he earns $10,000 more. Each year of education has brought Ray another $2,500 in income. The number 2,500 is a slope. Implicitly the following expression has been used.

$$\frac{\text{Ray's income} - \text{Bob's income}}{\text{Ray's education} - \text{Bob's education}}$$

The numerator is the difference between the brothers' incomes, which is $10,000. The denominator is the difference between the number of years of education. The *slope* states the amount of change in the criterion variable as a function of a one unit change in the predictor variable.

Suppose that there is a third brother, Mike, who went to college for only one year and so has 13 years of education. Mike's annual income is $26,000. There are now three different slopes: using Bob and Ray, Bob and Mike, and Ray and Mike. These three slopes are:

Bob and Ray:    $2,500

Ray and Mike:    $3,000

Bob and Mike:    $1,000

These slopes measure how much money is earned for every year of education. In Figure 6.4 are the scores of Bob, Ray, and Mike in a scatterplot. The

*FIGURE 6.4*    **Scatterplot for the education and income example.**

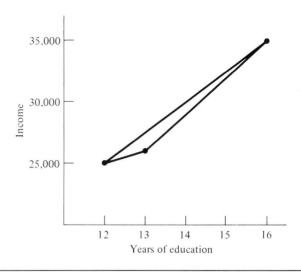

horizontal or $X$ axis is the predictor variable: number of years of education. On the vertical or $Y$ axis is the criterion: dollars earned per year. There are three points in the graph for the three persons. These three points can be connected to form three different measures of slope.

A little notation can help. Education is denoted as $X$ and income as $Y$. With three data points the scores are denoted $X_1$, $X_2$, and $X_3$, and $Y_1$, $Y_2$, and $Y_3$. The three measures of slope between the three pairs of lines are

$$\frac{Y_1 - Y_2}{X_1 - X_2}$$

$$\frac{Y_1 - Y_3}{X_1 - X_3}$$

$$\frac{Y_2 - Y_3}{X_2 - X_3}$$

These three measures of slope will not be equal unless the three points fall on a single straight line. In the example in Figure 6.4, the three pairs of points create three different measures of slope. To arrive at a single measure of slope, the three measures need to be averaged. The three measures could simply be averaged: $(2500 + 3000 + 1000)/3 = 2167$ for the example. Alternatively, the numerators and denominators of the three estimates of slope could be separately summed; that is,

$$\frac{(Y_1 - Y_2) + (Y_1 - Y_3) + (Y_2 - Y_3)}{(X_1 - X_2) + (X_1 - X_3) + (X_2 - X_3)}$$

However, this seemingly sensible solution results in the loss of the $X_2$, $Y_2$ data pair and yields, as an estimate of slope,

$$\frac{Y_1 - Y_3}{X_1 - X_3}$$

and so this average results in throwing away the measures of slope that involve $X_2$ and $Y_2$. To remedy this problem, the estimates of slope must be weighted in some fashion. The estimate of slope using two persons whose education differs markedly should be more reliable than using two persons whose education is quite similar. Therefore, one strategy is to weight by the difference between the scores on the predictor variable. Thus the more different two persons are on the predictor variable, the more their estimate of slope should be weighted.

Weighting by differences in the predictor variable makes intuitive sense. When persons do not differ at all on the predictor, the slope becomes impossible to measure. Weighting by differences in the predictor, the estimate of the slope becomes

$$\frac{(Y_1 - Y_2)(X_1 - X_2) + (Y_1 - Y_3)(X_1 - X_3) + (Y_2 - Y_3)(X_2 - X_3)}{(X_1 - X_2)^2 + (X_1 - X_3)^2 + (X_2 - X_3)^2}$$

or, for the example,

$$\frac{(10000)(4) + (9000)(3) + (1000)(1)}{16 + 9 + 1} = 2615$$

This is the measure of slope that is commonly used in social research and it is called the *regression coefficient*. The regression coefficient can be viewed as the average slope across all possible pairs of observations and weighted by difference on the predictor variable. Researchers never actually compute the slope for all possible pairs, but the regression coefficient does equal an average slope and can be interpreted as such.

## The Best-Fitting Line

There is a second and more commonly known rationale for the regression coefficient. It is the least-squares line, which is now described.

To measure the relationship between education and income, a line is fitted in the scatterplot between education and income. Because there are many such possible lines, some way is needed to determine what line is the "best" line. The best-fitting line is one that minimizes the errors.

Before proceeding any further, an error in prediction must be defined. In Figure 6.5 is the scatterplot of the three brothers' education and income. In the scatterplot there is a prediction line drawn. Also drawn are vertical lines from the line to the points in the scatterplot. The vertical length of the line can be viewed as an error in prediction. An *error* is defined as the vertical distance from the line to the score.

*FIGURE 6.5*    **Errors in a regression equation.**

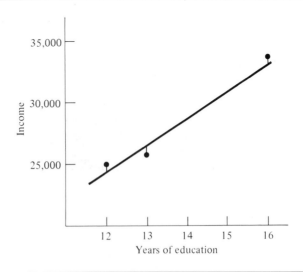

These errors in prediction can be squared. The regression coefficient is defined as the slope of the line that minimizes the sum of squared errors. It is for this reason that the regression line is called a *least-squares* estimate; that is, the line that is chosen has the least sum of squared errors. For instance, for the education and income example, a line with a slope of 2615 has the lowest sum of squared errors. A line with any other slope has a greater sum of squared errors.

There are then two rationales for the regression coefficient. One is that it is a weighted average of all possible slope measures. The other is that the regression coefficient is the slope of the best-fitting line.

# The Regression Equation

The regression line can be represented graphically as in Figure 6.5 or it can be represented by an equation. The equation is

$$Y = a + bX + e$$

The term $a$ is the intercept, $b$ is the slope or regression coefficient, and $e$ is the error. The *intercept* is the predicted score for $Y$ given that $X$ is zero. The intercept is the point at which the regression line intersects the $Y$ axis.

The predicted value $Y$ given $X$, or $\hat{Y}$, equals

$$\hat{Y} = a + bX$$

The term $\hat{Y}$ is the predicted $Y$ given a particular value of $X$. The error in prediction is then defined by $Y - \hat{Y}$. The error in prediction is the vertical distance of the regression line from the point in the scatterplot.

The regression coefficient has two important properties. First, the line always passes through the point $\bar{X}, \bar{Y}$. If the line did not pass through this point, it would no longer be the least-squares line. Second, the mean of the errors always equals zero; that is $\Sigma(Y - \hat{Y})/n = 0$.

The value of a regression coefficient depends on the unit of measurement. Adding or subtracting a constant to either the predictor or the criterion does not affect the regression coefficient; however, if the scores are multiplied or divided by a constant, the regression coefficient does change. If the predictor is multiplied by a constant, the regression coefficient is divided by the constant. If the criterion is multiplied by a constant, the regression coefficient is also multiplied by the constant.

# Computation

The standard formula for the regression coefficient $b$, where $X$ is the predictor and $Y$ is the criterion, is

$$b = \frac{\sum(X - \bar{X})(Y - \bar{Y})}{\sum(X - \bar{X})^2}$$

The denominator of the slope formula is the numerator of the formula for the variance of the predictor. The numerator of the formula for slope is called the *sum of cross-products*. The denominator of the formula is called the *sum of squares* of the predictor variable. In general, the regression coefficient equals the sum of cross-products between the predictor and the criterion divided by the sum of squares of the predictor.

There is a computationally more efficient formula for the regression coefficient:

$$b = \frac{\sum XY - (\sum X)(\sum Y)/n}{\sum X^2 - (\sum X)^2/n}$$

This is the formula that is generally used to compute the regression coefficient.

The formula for the intercept, which is symbolized by $a$, can be obtained by solving for the predicted value of $Y$ when $X$ is at the mean. Because $\bar{X}$, $\bar{Y}$ is a point on the regression line, the predicted value of $Y$ for $\bar{X}$ is $\bar{Y}$. The resulting prediction equation is $\bar{Y} = a + b\bar{X}$. Solving for $a$, the solution is

$$a = \bar{Y} - b\bar{X}$$

The intercept equals the mean of the criterion minus the product of the mean of the predictor and the regression coefficient.

The variance of the errors is symbolized by $s_{Y \cdot X}^2$ and is sometimes referred to as the mean square error or MSE. The formula for the variance of the errors is

$$s_{Y \cdot X}^2 = \frac{\sum(Y - \hat{Y})^2}{n - 2}$$

This formula does not look like a variance but it is. The numerator is the sum of squared errors. There is no need to subtract the mean of the errors because that mean is always zero. It is correct to divide by $n - 2$ instead of $n - 1$ because both the regression coefficient and the intercept have been estimated.

Actually the errors need not be individually computed to determine their variance. The following formula is often much simpler to compute:

$$s_{Y \cdot X}^2 = \frac{n - 1}{n - 2}(s_Y^2 - b^2 s_X^2)$$

To illustrate the use of the formulas consider the data previously presented in Table 6.2. To compute the regression coefficient, the intercept, and the variance of the errors, the following quantities are computed: $\sum X$, $\sum Y$, $\sum XY$, $\sum X^2$, and $\sum Y^2$. Because age is the predictor, it is denoted as $X$. And because STM is the criterion, it is denoted as $Y$. In Table 6.3, these computations for the age-memory study are illustrated. Laying out the numbers, as in Table 6.3, can simplify the computations. The slope for the example is

*TABLE 6.3*   **Computational Table for Age and STM Study**

| Person | Age | STM | Age$^2$ | STM$^2$ | Age $\times$ STM | $\hat{\text{STM}}$ | STM − $\hat{\text{STM}}$ |
|---|---|---|---|---|---|---|---|
| 1 | 48 | 14 | 2304 | 196 | 672 | 11.81 | 2.19 |
| 2 | 46 | 7 | 2116 | 49 | 322 | 12.44 | −5.44 |
| 3 | 44 | 12 | 193 | 144 | 528 | 13.08 | −1.08 |
| 4 | 52 | 10 | 2704 | 100 | 520 | 10.54 | −.54 |
| 5 | 22 | 24 | 484 | 576 | 528 | 20.05 | 3.95 |
| 6 | 43 | 11 | 1849 | 121 | 473 | 13.40 | −2.40 |
| 7 | 51 | 9 | 2601 | 81 | 459 | 10.86 | −1.86 |
| 8 | 54 | 19 | 2916 | 361 | 1026 | 9.91 | 9.09 |
| 9 | 56 | 2 | 3136 | 4 | 112 | 9.28 | −7.28 |
| 10 | 54 | 12 | 2916 | 144 | 648 | 9.91 | 2.09 |
| 11 | 65 | 12 | 4225 | 144 | 780 | 6.42 | 5.58 |
| 12 | 35 | 18 | 1225 | 324 | 630 | 15.93 | 2.07 |
| 13 | 63 | 5 | 3969 | 25 | 315 | 7.06 | −2.06 |
| 14 | 39 | 18 | 1521 | 324 | 702 | 14.66 | 3.34 |
| 15 | 30 | 14 | 900 | 196 | 420 | 17.52 | −3.52 |
| 16 | 47 | 8 | 2209 | 64 | 376 | 12.13 | −4.13 |
| Total | 749 | 195 | 37011 | 2853 | 8511 | 195.00 | 0.00 |

$$b = \frac{8511 - (749)(195)/16}{37011 - 749^2/16} = -.3169$$

As the scatterplot shows, the slope is negative. As these men age a year, their short-term memory declines by about three-tenths of a unit, or for a decade the men lose about three points of memory score. Because the mean of the predictor is 749/16 or 46.8125 and the mean of the criterion is 195/16 or 12.1875, the intercept is

$$a = 12.1875 - (-.3169)(46.8125) = 27.0224$$

This is the predicted score for a person whose age is zero—that is, newborns. The resulting regression equation is

$$\text{STM} = 27.0224 - .3169(\text{Age}) + e$$

The variance of the errors requires the computation of the variances for age and STM. For age the variance is

$$\frac{37011 - 749^2/16}{15} = 129.8958$$

And for STM the variance is

$$\frac{2853 - 195^2/16}{15} = 31.7625$$

The error variance is then

**FIGURE 6.6**  **Scatterplot and regression line for age-memory study.**

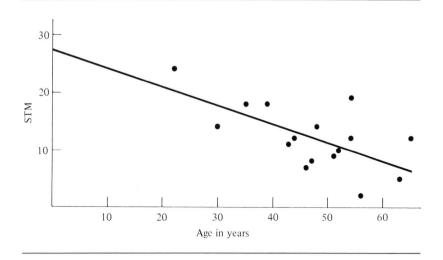

$$s_{Y \cdot X}^2 = \frac{15}{14}[31.7625 - (-.3169)^2(129.8958)] = 20.0546$$

Note that the error variation is considerably smaller than the variance of the criterion variable, which is 31.8292.

The predicted scores are also presented in Table 6.3. For instance, for person 1 whose age is 48, the predicted score is

$$27.0224 + (-.3169)(48) = 11.81$$

The error in prediction for person 1 equals the actual memory score of 14 minus the predicted score of 11.81, which is 2.19. The fact that the sum of the errors is zero is a mathematical necessity.

Finally, Figure 6.6 shows the scatterplot with the regression line plotted. To plot a line two points are needed. The two points that are used to plot the regression line are $X = 0$, $Y = 27.0224$ (the intercept) and $\bar{X} = 46.8125$, $\bar{Y} = 12.1875$ (the means). The line very clearly shows the declining memory scores with increasing age.

# Interpretation of a Regression Coefficient

To compute a regression line, one variable must be treated as the predictor and the other as the criterion. The choice of which variable to designate as the predictor and which to use as a criterion should not be arbitrary. That is, one should have either a practical or conceptual basis for making the designation. If one is not certain which variable to treat as the predictor, it may be more appropriate to use another measure of association such as the correlation coefficient, which is described in the next chapter.

If $X$ is used to predict $Y$, the line obtained is different from the one obtained when $Y$ is used to predict $X$. Hence there are two regression lines. To distinguish the two lines, the regression coefficient is often subscripted first by the criterion and then by the predictor. So $b_{YX}$ implies that $X$ is the predictor and $Y$ is the criterion, and $b_{XY}$ implies that $Y$ is the predictor and $X$ is the criterion. If $Y$ is used to predict $X$, the formula for $b_{XY}$ is as follows:

$$b_{XY} = \frac{\sum XY - (\sum X)(\sum Y)/n}{\sum Y^2 - (\sum Y)^2/n}$$

The relationship between $b_{YX}$ and $b_{XY}$ is straightforward. To convert $b_{YX}$ into $b_{XY}$, the following formula is used.

$$b_{XY} = b_{YX} \left( \frac{s_x^2}{s_Y^2} \right)$$

Conversely,

$$b_{YX} = b_{XY} \left( \frac{s_Y^2}{s_X^2} \right)$$

There are two major purposes for the regression coefficient: prediction and explanation. In prediction, the following question is asked: If one knew someone's standing on a variable, how well would one be able to predict the person's standing on another variable? The purpose in prediction is not to change or alter reality, but merely to make good guesses about the future. The predictive use of a regression equation is valid within only the range of the predictor variable. Using a regression equation to predict scores of persons who do not score within the range of the predictor variable can be quite misleading. For the age-memory example the youngest person is 22 the oldest is 65. So any prediction for subjects younger than 22 or older than 65 involves an extending or extrapolating of the regression line beyond the sample used to estimate it. For the age-memory example, the intercept can be viewed as an extrapolation because no subjects are zero years of age. The intercept is 27.0224 and it predicts newborns would remember more than 27 items. Because 24 is the maximum possible score, it is logically impossible for anyone to score so high. This illustrates the dangers of extrapolation.

The second major use of a regression equation is for explanation. Here the researcher wants to claim that the regression equation indicates a causal effect. A causal interpretation states what would happen when reality is changed, whereas a predictive relation describes reality as it is. For instance, attitudes and behavior are generally strongly related. As an example, individual attitudes toward the use of seat belts predicts fairly well who will use seat belts. Given this association, one can also attempt to change persons' attitudes to increase the use of seat belts. But just because attitude and behavior are correlated does not mean that attitude causes behavior. Using regression coefficients causally is even more dangerous than using them for prediction. More will be said about the causal interpretation of measures of relationship in the next chapter.

# Regression Toward the Mean

The variance in the predicted scores is never larger than the variance in the criterion. This is one of the first statistical facts ever discovered. Galton in 1890 was surprised to learn that tall fathers tended to have sons shorter than themselves. If father's height is used to predict child's height, the predicted child's height is closer to the mean height than is the father's. This also worked when Galton used the son's height to predict the father's. That is, if child's height is used to predict father's height, the predicted height for the father is closer to the mean height than the son's. Galton labeled this phenomenon as *regression toward the mean.*

When the slope is zero, the predicted scores take on one value: the mean of the criterion. In this case the predicted scores have no variance. When there are no errors in prediction, the predicted score equals the criterion score and hence the two have equal variance.

# Summary

Two variables are said to covary if differences in one variable are related in a systematic fashion to differences in a second variable. The most common form of a relationship is a linear one. In a *linear relationship,* the strength of the relationship does not depend on the values of the variables. Linear relationships can be positive or negative. In a *positive relationship* as one variable increases, the other variable also increases. In a *negative relationship* as one variable increases, the other decreases.

The *slope* is a measure of linear association. It measures the effect of a change in one variable as a function of a one-unit change in another variable. When measuring the slope, one variable is denoted as the predictor and the other as the criterion. The slope is given by

$$b = \frac{\sum XY - (\sum X)(\sum Y)/n}{\sum X^2 - (\sum X)^2/n}$$

where $X$ is the predictor and $Y$ the criterion. The *intercept* measures the predicted value for the criterion when the predictor is zero. In other words, it is the point at which the regression line intersects the $Y$ axis and is given as follows:

$$a = \bar{Y} - b\bar{X}$$

The *variance in errors* is given by the following:

$$s_{Y\cdot X}^2 = \frac{n-1}{n-2}(s_Y^2 - b^2 s_X^2)$$

The regression line can be used for either prediction or explanation. In prediction one variable is used to predict the other. In explanation, one variable is assumed to produce changes in the other.

# Problems

1. There is a country saying that one can determine the temperature (Fahrenheit) by counting the number of cricket chirps in 14 seconds and adding 40. Consider cricket chirps in 14 seconds as the predictor variable and temperature as the criterion. What, according to the saying, is the slope and intercept of this regression equation?

2. An industrial psychologist uses number of cigarettes smoked per day ($S$) to predict the number days absent during the year ($A$). Her regression equation is:

$$A = 2.23 + .081S + e$$

   a. How many days absent does the equation predict for someone who does not smoke? Someone who smokes 20 cigarettes a day? Someone who smokes 40 cigarettes a day?
   b. If the company were able to lower the number of cigarettes that each of its employees smoked by 15, and the company employs 94 people, how many fewer lost days per year would it be predicted to have?

3. For the following pairs of variables, which should be treated as the predictor and which as the criterion?

   a. marital satisfaction and similarity
   b. effort and performance
   c. sleep and efficiency
   d. health and mood

4. For the following pairs of variables what is the likely direction of the relationship, positive or negative?

   a. religious belief and church attendance
   b. vocabulary and intelligence
   c. rainfall and outdoor activity
   d. criminal behavior and alcoholism

  e. hours studied and midterm grade
  f. odometer mileage and repair costs
  g. grade point average and number of parties attended
  h. price of beer and enjoyable taste

5. The following scores are the height and weight of five persons.

| Height (inches) | Weight (pounds) |
|---|---|
| 60 | 140 |
| 64 | 170 |
| 72 | 210 |
| 68 | 180 |
| 70 | 150 |

Treat height as the predictor variable and weight as the criterion.

  a. What is the slope, intercept, and variance of the errors? Interpret each statistic.
  b. Compute the errors for the five scores and verify that their mean is zero.
  c. Draw a scatterplot and plot the regression line.

6. For the following pairs of variables what would be the object across which the measure of association would be computed?

  a. literacy rate and gross national product
  b. population and crime rate
  c. sense of control and happiness
  d. leadership and productivity

7. Describe each of the following relationships as either predictive or causal.

  a. predictor: cigarette smoking; criterion: lung cancer
  b. predictor: beer consumption; criterion: wine consumption
  c. predictor: child's height; criterion: child's reading skill
  d. predictor: physical attractiveness; criterion: popularity
  e. predictor: presence of smoke; criterion: fire

8. Harrison (1984) conducted a questionnaire study of crowding, privacy, and loneliness among female dormitory residents. Satisfaction with privacy in the dormitory $(X)$ was measured on a scale from one to seven, with higher scores indicating greater satisfaction. Subjects were also asked how often they avoided people other than friends in the dormitory $(Y)$. Again, responses were given on a seven-point scale, with higher scores indicating more frequent avoidance behavior. The data of 20 of the subjects, randomly chosen, are given below, along with some calculations.

| Subject | Satisfaction (X) | Avoidance (Y) | $X^2$ | $Y^2$ | XY |
|---|---|---|---|---|---|
| 1 | 4 | 2 | 16 | 4 | 8 |
| 2 | 6 | 1 | 36 | 1 | 6 |
| 3 | 6 | 1 | 36 | 1 | 6 |
| 4 | 5 | 2 | 25 | 4 | 10 |
| 5 | 2 | 6 | 4 | 36 | 12 |
| 6 | 6 | 2 | 36 | 4 | 12 |
| 7 | 6 | 1 | 36 | 1 | 6 |
| 8 | 5 | 2 | 25 | 4 | 10 |
| 9 | 6 | 5 | 36 | 25 | 30 |
| 10 | 4 | 1 | 16 | 1 | 4 |
| 11 | 6 | 1 | 36 | 1 | 6 |
| 12 | 4 | 4 | 16 | 16 | 16 |
| 13 | 4 | 1 | 16 | 1 | 4 |
| 14 | 5 | 4 | 25 | 16 | 20 |
| 15 | 7 | 2 | 49 | 4 | 14 |
| 16 | 3 | 2 | 9 | 4 | 6 |
| 17 | 4 | 3 | 16 | 9 | 12 |
| 18 | 2 | 7 | 4 | 49 | 14 |
| 18 | 6 | 1 | 36 | 1 | 6 |
| 20 | 6 | 1 | 36 | 1 | 6 |
| Total | 97 | 49 | 509 | 183 | 208 |

a. Construct a scatterplot. Does the plot tend to show a positive or negative relationship?

b. Compute the slope of the regression equation, with satisfaction as the predictor and avoidance as the criterion. Interpret the slope. Compute the intercept. What does it mean?

c. Compute the variance of X and of Y. Compute the variance of the errors.

9. From the results of problem 8, state the regression equation. Compute the predicted avoidance scores for each of the observed satisfaction scores. Compute the variance of the predicted scores. How does it compare with the variance of the observed avoidance scores? What is this change in variance called?

10. For the regression equation

$$Y = 10.3 + .6X + e$$

find predicted scores for the following values of X: 10, 12, 15, and 31.

11. If $s_Y^2 = 15$, $s_X^2 = 10$, $n = 25$, and $b_{YX} = .5$ find the following:

a. $b_{XY}$　　b. $s_{Y \cdot X}^2$　　c. $s_{X \cdot Y}^2$

12. Below is the temperature in Hartford, Connecticut, and the expected number of cars that will have starting difficulties.

| Temperature (Fahrenheit) | Number of Disabled Cars |
|:---:|:---:|
| 40 | 1159 |
| 32 | 1288 |
| 25 | 1519 |
| 20 | 2276 |
| 15 | 2941 |
| 10 | 3296 |
| 5 | 4481 |
| 0 | 5665 |
| −5 | 7210 |
| −10 | 8858 |

Treat temperature as the predictor and number of disabled cars as the criterion and estimate the slope and intercept. Interpret each. Compute the errors for each observation. Using the equation, how many cars will have starting difficulties when the temperature is −8 degrees? When it is 70 degrees?

# 7 | *Relationship: The Correlation Coefficient*

In the preceding chapter the regression coefficient was presented as a measure of association between two variables. The regression coefficient as a measure of association is asymmetric and is expressed in the units of measurement of the variables. It is asymmetric in the sense that its value depends on which variable is considered as the criterion and which is the predictor. It is expressed in the units of the variables in that it measures the amount of change in the criterion as a function of a one-unit change in the predictor.

Sometimes it is not possible to specify which variable is the predictor and which variable is the criterion. For instance, in measuring the degree of relationship between reading comprehension and vocabulary skill in school-children, one variable is not clearly the predictor and the other the criterion. Also, because the regression coefficient is expressed in the units of measurement of the variables, the strength of association is not very clear. It would be desirable to obtain a measure of association that was symmetric and expressed the degree of association between the variables. The correlation coefficient meets both requirements.

The *correlation coefficient* is symbolized by the letter $r$. Because it is a symmetric measure of association, it follows that $r_{xy} = r_{yx}$. The correlation coefficient is by far the most common measure of association used in the social and behavioral sciences. Only economists use the regression coefficient more frequently than the correlation coefficient. This is no doubt due to the fact that the unit of measurement in economics (the dollar) is readily interpretable.

As an example for this chapter, the variable of laughter in conversations will be considered. Duncan and Fiske (1977) coded the nonverbal behavior of 22 pairs of men and women for five minutes. The couples were instructed to get acquainted with each other. During these conversations, there was occasional laughter. Table 7.1 lists the number of laughs of each person for the 22 couples over the five-minute period.

# Rationale for the Correlation Coefficient

The correlation coefficient is a special regression coefficient. Consider the case in which there are two variables, $X$ and $Y$. First, the scores for the $X$ and $Y$ variables are separately standardized. Thus, $Z$ scores are created for each variable; that is, the mean for the variable is subtracted from each score and then this difference is divided by the variable's standard deviation. To compute the regression coefficient, one $Z$-scored variable is the predictor and the other is the criterion. The *correlation coefficient,* symbolized by the letter $r$, is the regression coefficient between two variables whose scores have been standardized.

The correlation coefficient is a symmetric measure of association and so $r_{XY}$ equals $r_{YX}$. Unlike the regression coefficient, a correlation coefficient has

**TABLE 7.1**   **Number of Laughs in 22 Conversations**

| Couple | Number of Laughs (Women) | Number of Laughs (Men) |
|:------:|:------------------------:|:----------------------:|
| 1 | 0 | 0 |
| 2 | 4 | 1 |
| 3 | 17 | 9 |
| 4 | 4 | 4 |
| 5 | 2 | 0 |
| 6 | 1 | 0 |
| 7 | 3 | 1 |
| 8 | 9 | 5 |
| 9 | 5 | 1 |
| 10 | 1 | 0 |
| 11 | 4 | 5 |
| 12 | 8 | 2 |
| 13 | 4 | 2 |
| 14 | 0 | 2 |
| 15 | 6 | 0 |
| 16 | 12 | 3 |
| 17 | 8 | 1 |
| 18 | 3 | 2 |
| 19 | 5 | 2 |
| 20 | 7 | 0 |
| 21 | 5 | 3 |
| 22 | 8 | 3 |

Data were taken from Duncan and Fiske (1977).

an upper limit of $+1$ and a lower limit of $-1$. A $+1$ correlation indicates a perfect positive correlation and a $-1$ correlation indicates a perfect negative correlation. In a perfect correlation, all the points fall on the regression line. The line is ascending if the correlation is $+1$ and descending if it is $-1$. Like the regression coefficient, a zero value indicates no *linear* association between the variables. Any nonlinear association may not be reflected by the correlation coefficient.

# *Computation*

As mentioned above, unlike the regression coefficient, the correlation coefficient is a symmetric measure of association: $r_{XY} = r_{YX}$. The relation of $r_{XY}$ to $b_{XY}$ and $b_{YX}$ is straight forward. (Recall from the previous chapter that for $b_{XY}$ the variable $Y$ is the predictor and $X$ the criterion and for $b_{YX}$ the variable $X$ is the predictor and $Y$ the criterion.) The formulas for turning $b$ into $r$ are

$$r_{XY} = b_{XY}\left(\frac{s_Y}{s_X}\right)$$

$$r_{XY} = b_{YX}\left(\frac{s_X}{s_Y}\right)$$

In words, the correlation coefficient equals the regression coefficient times the standard deviation of the predictor divided by the standard deviation of the criterion. It is also true that

$$r_{XY}^2 = b_{XY}b_{YX}$$

To convert from $r$ to $b$ the formulas are

$$b_{XY} = r_{XY}\left(\frac{s_X}{s_Y}\right)$$

$$b_{YX} = r_{XY}\left(\frac{s_Y}{s_X}\right)$$

In words, a regression coefficient equals the correlation times the standard deviation of the criterion divided by the standard deviation of the predictor.

More typically, the correlation is computed directly without computing the regression coefficient. There is also no need to standardize or compute $Z$ scores for each person. The correlation coefficient can be computed by the following formula.

$$r_{XY} = \frac{\sum(X - \bar{X})(Y - \bar{Y})}{\sqrt{\sum(X - \bar{X})^2 \sum(Y - \bar{Y})^2}}$$

The top term of the above formula is, as referred to in the previous chapter, the sum of cross-products. The denominator is the square root of the product of the sums of squares of both $X$ and $Y$. A simpler and more practical computational formula is

$$r_{XY} = \frac{\sum XY - (\sum X)(\sum Y)/n}{\sqrt{[\sum X^2 - (\sum X)^2/n][\sum Y^2 - (\sum Y)^2/n]}}$$

The computational formula for correlation has the following ingredients: the sum of the scores for all the subjects on both variables, $\sum X$ and $\sum Y$; the squared sum of the scores, $(\sum X)^2$ and $(\sum Y)^2$; the sum of each squared score for each variable; $\sum X^2$ and $\sum Y^2$; and the sum of the product of scores, $\sum XY$. One common computational error in computing a correlation coefficient is to forget to take the square root of the denominator.

Conventionally correlations are computed to two digits. This is a sensible strategy in that the third digit is not ordinarily interpretable. So if a correlation is to be computed and interpreted, rounding to the second digit should suffice. However, correlation coefficients are often used to compute other statistics, some of which are presented in Chapter 15. If a correlation is to be used to compute other statistics, it should be computed to three or possibly four digits. In this chapter, correlations will be given to three digits.

Table 7.2 displays the computations for the laughing in male-female conversations. The female laughs are denoted as $X$ and male laughs as $Y$. The sum of cross-products of $X$ and $Y$ is as follows:

$$369 - (116)(46)/22 = 126.4545$$

The sum of squares for $X$ is

$$954 - (116)^2/22 = 342.3636$$

and the sum of squares for $Y$ is

$$198 - (46)^2/22 = 101.8181$$

The correlation then equals

$$r_{XY} = \frac{126.4545}{\sqrt{(342.3636)(101.8181)}} = .677$$

Not surprisingly, there is a very large correlation in the amount of laughter between two persons in a conversation. Laughter is indeed contagious.

# Interpretation of r

One way to understand what a correlation of a given size means is to examine various correlations between variables. In Table 7.3 are a set of correlations taken from research. It contains correlations that are small (.1), moderate (.3),

*TABLE 7.2*     **Computations for Laughing Example**

|       | $X$ | $Y$ | $X^2$ | $Y^2$ | $XY$ |
|-------|-----|-----|-------|-------|------|
|       | 0   | 0   | 0     | 0     | 0    |
|       | 4   | 1   | 16    | 1     | 4    |
|       | 17  | 9   | 289   | 81    | 153  |
|       | 4   | 4   | 16    | 16    | 16   |
|       | 2   | 0   | 4     | 0     | 0    |
|       | 1   | 0   | 1     | 0     | 0    |
|       | 3   | 1   | 9     | 1     | 3    |
|       | 9   | 5   | 81    | 25    | 45   |
|       | 5   | 1   | 25    | 1     | 5    |
|       | 1   | 0   | 1     | 0     | 0    |
|       | 4   | 5   | 16    | 25    | 20   |
|       | 8   | 2   | 64    | 4     | 16   |
|       | 4   | 2   | 16    | 4     | 8    |
|       | 0   | 2   | 0     | 4     | 0    |
|       | 6   | 0   | 36    | 0     | 0    |
|       | 12  | 3   | 144   | 9     | 36   |
|       | 8   | 1   | 64    | 1     | 8    |
|       | 3   | 2   | 9     | 4     | 6    |
|       | 5   | 2   | 25    | 4     | 10   |
|       | 7   | 0   | 49    | 0     | 0    |
|       | 5   | 3   | 25    | 9     | 15   |
|       | 8   | 3   | 64    | 9     | 24   |
| Total | 116 | 46  | 954   | 198   | 369  |

and large (.5). Small correlations are the most common correlations in the social and behavioral sciences. The reason for so many small correlations is that most variables are caused by numerous factors, and so any one factor's correlation with a variable that it causes must be small. The relation between stress and physical disease, such as heart trouble, and the relation between intelligence and a grade in a course are in the .10 range. A moderate correlation is large enough for laypersons to recognize. An example of moderate correlation is general sense of self-worth and grade point average. Large correlations represent very strong correlations, such as the correlation between intelligence and overall GPA.

It is important to note that a large correlation is not a correlation of .90. Correlations of this size are often between two different measures of the same variable. For instance, the correlation of two measures of intelligence taken a year apart is about .90 once persons are age six or more. Also such large correlations often indicate not a meaningful relationship between variables, but an artificial one. For instance, the .677 correlation between male laughter and female laughter will be seen to be artificially high.

The differences between small, moderate, and large correlations can also be seen in their scatterplots. As explained in Chapter 6, a scatterplot is a graph

*TABLE 7.3*     **Illustration of Correlations of Various Sizes**

*Small: .10*

   Viewing television violence — Aggressive behavior
   Stress — Physical illness
   Intelligence — Grade in a particular course

*Moderate: .30*

   Psychotherapy — Adjustment
   Self-esteem — Grades in school
   Value similarity — Interpersonal attraction

*Large: .50*

   Intelligence — Grade point average
   Wife's satisfaction — Husband's satisfaction
   Father's occupation — Son's occupation
   Belief in God — Church attendance

in which the two variables form the $X$ and $Y$ axes. The pairs of scores for each person are plotted in a scatterplot. The scatterplots for .1, .3, and .5 correlations are presented in Figure 7.1. For a .1 correlation, the correlation is not even visible to the naked eye. For .3 to the trained eye there is the hint of association. For the .5 correlation the linear relationship is clearly visible.

A correlation coefficient is a regression coefficient between standardized scores. It can be directly interpreted then as a regression coefficient between standard scores. If $r_{XY}$ equals .5, then someone who is one standard deviation above the mean on $X$ would tend to be .5 standard deviation units above the mean on $Y$. So a correlation between $X$ and $Y$ measures how many standard deviation units above or below the mean a person's score is on $Y$ when the person is one standard deviation above the mean on $X$. Because it is a symmetric measure, it can also be interpreted as the predicted value for $X$ for someone who is one standard deviation above the mean on $Y$.

The most common way to interpret a correlation coefficient is by squaring the correlation and interpreting the result as the proportion of variance that the two variables share in common. The proportion can be multiplied by 100 to obtain the percent of shared variance. So, for instance, if high school grades and college grades correlate .6, then $.6^2$ or .36 of their variance is shared in common. Besides shared variance, the squared correlation can also be interpreted as the proportion of variance explained. So a .6 correlation between high school grades and college grades implies that high school grades can explain .36 of the total variance in college grades. The squared correlation for shared or explained variance is often used to trivialize small correlations. For instance, a .1 correlation represents only .01 shared variance. It should be noted that the squared correlation represents shared or explained *variance* and

*FIGURE 7.1*     **Scatterplots of .1, .3, and .5 correlation coefficients.**

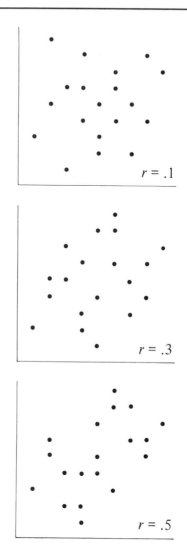

not standard deviation. Because variance is in squared units, the meaning of explained variance may be difficult to appreciate. For instance, if intelligence explains 25% of the variance in high school grades, it means that intelligence explains 25% of squared grade points.

A correlation can be viewed in terms of a probability. Consider two persons, one, called *A*, who is one standard deviation *above* the mean on *X* and the other, called *B*, who is one standard deviation *below* the mean on *X*.

Person A has a two-standard-deviation advantage on $X$ over person B. If the correlation between $X$ and $Y$ is known, then the probability that person A scores higher than B on $Y$ can be determined. For instance, assume that variable $X$ is education and variable $Y$ is income. Let one standard deviation above the mean on education be a master's degree and one standard deviation below the mean be a high school education. The issue is the probability of someone who has a master's degree earning more money than someone who has only graduated from high school. This probability will be referred to as the *two-standard-deviation advantage* and abbreviated as the 2sd advantage.

To determine the probability, it is assumed that both variables are normally distributed. The normal distribution is discussed in Chapter 10. Rosenthal and Rubin (1979) make radically different distributional assumptions, yet for $r$ between 0 and .5 they obtain virtually the same result. (They assume that $X$ and $Y$ are dichotomies as opposed to normally distributed variables measured at the interval level of measurement.)

In Table 7.4 are the 2sd advantage probabilities for correlations of various sizes.[1] So for instance, if $r$ is .45, then the probability that someone who is one standard deviation above the mean on $X$ will score on $Y$ above the person who is one standard deviation below the mean on $X$ is .762. If the correlation is negative, the probabilities in the table can be read as the probability of someone one standard deviation above the mean on $X$ scoring *below* someone one standard deviation below the mean on $X$.

The table is read as follows. First, find the correlation to be interpreted in the $r$ column. Second, the value in the probability column states the probability that a person who is one standard deviation above the mean will outscore

*TABLE 7.4*  **Correlation in Terms of the Two-Standard-Deviation Advantage**

| $r$ | Probability[a] | $r$ | Probability |
|-----|------------|-----|-------------|
| .00 | .500 | .50 | .793 |
| .05 | .528 | .55 | .824 |
| .10 | .557 | .60 | .856 |
| .15 | .585 | .65 | .887 |
| .20 | .614 | .70 | .917 |
| .25 | .642 | .75 | .945 |
| .30 | .672 | .80 | .970 |
| .35 | .701 | .85 | .989 |
| .40 | .731 | .90 | .998 |
| .45 | .762 | .95 | 1.000 |

[a] The probability of a person who is one standard deviation above the mean on $X$ scoring higher on $Y$ than someone who is one standard deviation below the mean on $X$.

[1]The 2sd advantage can be shown to equal the probability that $Z$ is less than $\sqrt{2}\rho/\sqrt{1 - \rho^2}$ where $\rho$ is the population correlation and $Z$ is a standard normal variable.

someone else on $Y$ who is one standard deviation below the mean on $X$. If the correlation is zero, the 2sd (two-standard-deviation) advantage is .5. That is, a person with a 2sd advantage on $X$ over another person has only a 50/50 chance of outscoring the other person. Even a seemingly low correlation like .2 carries with it an impressive probability of .614. Thus, for a correlation of .2, over 60% of the time person A (who is one standard deviation above the mean on $X$) will outscore person B (who is one standard deviation below the mean of $X$) on $Y$.

The 2sd advantages for small, medium, and large correlations are:

Small ($r = .1$): .557
Medium ($r = .3$): .672
Large ($r = .5$): .793

# Factors Affecting the Size of r

Special care must be taken in interpreting correlation and regression coefficients. At times, a coefficient can be artificially too small or too large. Various factors are discussed below that must be considered when interpreting measures of association, especially correlation coefficients.

## Nonlinearity

The fundamental definition of a regression coefficient is that of a slope of the straight line fitted to a set of points. A correlation coefficient is the slope of the line when the two samples have been converted into $Z$ scores. Both measures of association assume that the line to be fitted is *straight* and not curved. The association between variables may be systematic, but it need not be linear. There are two major types of nonlinear associations. They are nonlinear association in which the function changes direction and nonlinear association in which the function does not change direction.

In Figure 7.2 are examples of changes of direction. In the top diagram of the figure, the relationship starts as positive and then turns negative. In the bottom diagram, the relationship starts negative and then turns positive. Both of these patterns are called *curvilinear* association. More precisely, a relationship that begins as positive and turns negative (the upper half of the figure) is called a *convex curvilinear* or an *inverted U* relationship. And a relationship that begins as negative and turns positive (the bottom half of the figure) is called a *concave curvilinear* or *U-shaped* relationship. For either type of curvilinear association both the correlation and regression coefficient can be quite misleading measures of association. These measures may well be zero even when there is a strong curvilinear association. As an example of a

*FIGURE 7.2*    **Examples of curvilinear relationships.**

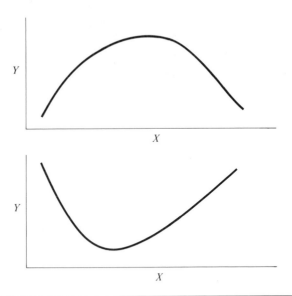

concave curvilinear association, amount of leisure time is curvilinearly related to age, with older and younger persons having more leisure time than middle-aged persons.

If the researcher expects that two variables are curvilinearly associated, the scatterplot should be carefully examined. If the point at which the relationship changes direction can be determined, a linear measure of association can be computed before and after that point. If the relationship is truly curvilinear, then one relationship should be positive and the other negative. For instance, if a researcher expects that the amount of leisure time begins to increase at age 45, then the correlation between age and leisure time should be negative for those under the age of 45 and positive for those who are 45 and older.

For the second type of nonlinear association, the relationship does not change in direction. This pattern is illustrated in Figure 7.3. For these relationships the direction of the relationship does not change but the strength does. For the three examples in Figure 7.3 the relationship is positive; that is, as $X$ increases, $Y$ increases. For the top diagram in the figure as $X$ increases, the relationship between $X$ and $Y$ increases. This is an accelerating function. For the middle diagram in the figure, as $X$ increases, the relationship between $X$ and $Y$ decreases. This is a decelerating function. For the bottom diagram in the figure, as $X$ increases, the relationship first increases and then it decreases. This type of relationship is called *S-shaped*. Correlation and regression coefficients are less affected by this form of nonlinearity than the form in

*FIGURE 7.3*    **Examples of nonlinear relationships that do not change direction.**

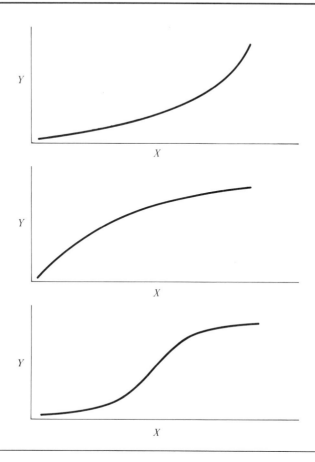

which the relationship changes direction. Nonetheless, it is important to attempt to straighten out the relationship.

Nonlinear relationships that do not change direction can be turned into linear relationships by transformation. For instance, for the pattern in the top of Figure 7.3, the relationship can be made more linear by applying a one-stretch transformation (square root, logarithm, or reciprocal) to the $Y$ variable. For the pattern in the middle of Figure 7.3, the relationship can be made more linear by applying a one-stretch transformation (square root, logarithm, or reciprocal) to the $X$ variable. For the pattern in the bottom of the figure, the relationship can be made more linear by applying a two-stretch transformation (arcsin, logit, or probit) to the $Y$ variable. If the researcher cannot specify the exact type of transformation, then Spearman's rank-order correlation (discussed in the next chapter) may be a more appropriate measure of association for any nonlinear relationship that does not change direction.

## *Unreliability*

Measurement in the social and the behavioral sciences is imperfect. Although every effort is made to measure persons as accurately as possible, unintentional errors of measurement are inevitable. Measurement involves not only what the researcher hopes to be measuring but noise or error as well. The first component is called the *true score* and the second is called *error of measurement*. The percent of variance of a measure that is due to the true score is called *reliability*. Constructs that social and behavioral scientists measure hardly ever have perfect reliability. Even a variable such as age has error due to distortion (people lying) and rounding. It is not at all unusual for a personality test to have a reliability of .80. A reliability of .80 means that 20% of the variance in the test is attributable to error.

Less than perfect reliability in a measure affects the size of the correlation and regression coefficients. The effect is one of attenuation. That is, the estimated size of the coefficient is nearer to zero than it ought to be. A regression coefficient is lowered only when the predictor variable is unreliable. Correlations are attenuated when either variable has less than perfect reliability.

## *Aggregation*

Sometimes researchers average the scores of a group of persons and use these averages as the basic data. For instance, students in the classroom are averaged and the basic analysis is on the classroom averages. When scores are averaged across persons, the data are called *aggregate data*. Generally, correlations computed from aggregate data are larger than what they would be if the individual scores were used. This increase is in part due to increased reliability, because aggregate data are generally more reliable than individual data. Though less likely, aggregation can reduce the size of a correlation.

Because a correlation computed from scores aggregated across persons can be quite different from a correlation of individual scores, one should never interpret the aggregated correlation as if it were the correlation from individuals. To do so would be what is called the *ecological fallacy*. An example of the ecological fallacy would be to correlate precinct voting data to make inferences about individual voting patterns. Correlations computed using aggregates (precincts) may not resemble correlations based on individuals (voters).

## *Part-Whole Correlation*

A correlation involves two variables. Sometimes one of the variables is derived from the other variable. When one variable is derived from a second variable, there can be a built-in correlation between the two. The variables must share variance because one is part of the other. In Table 7.5 the variable $X$ is used to derive a second variable, and the direction of bias is indicated. If the direction of bias is indicated in the table as positive, it does not mean that the correlation is necessarily positive, but that the correlation is larger than it should be.

In the first case in Table 7.5, the variable $X$ is used to derive the measure $X + Y$. Because $X$ is present in both measures, there is a built-in positive correlation. In the second case $X$ is subtracted from $Y$. In this case the correlation is negative. One should avoid computing correlations between variables that have common components.

## Restriction in Range

Correlations computed from scores that have low variability generally tend to be small. This phenomenon is called *restriction of range*. It can be illustrated graphically as in Figure 7.4. The data in the figure show that the $X$ variable has been split at the mean and the correlation has been recomputed for those scoring above and below the mean. Overall the correlation is .533, but for those who score below the mean (as is indicated by the dashed line in Figure 7.4) the correlation is .341 and for those who score above the mean, the correlation is also .341. When a variable has a narrow range of scores, correlations tend to be small.

Interestingly, restriction in range does not influence the regression coefficient nearly as much as it does the correlation coefficient. So if the range of a variable may be restricted, the regression coefficient is the preferred measure of association.

## Outliers

Extreme values or outliers in the sample can distort the size of a correlation. For instance, for the following set of data

**TABLE 7.5    Part-Whole Correlation**

| Variable 1 | Variable 2 | Bias in $r$ |
|---|---|---|
| $X$ | $X + Y$ | Positive |
| $X$ | $Y - X$ | Negative |
| $X + Z$ | $X + Y$ | Positive |
| $X + Z$ | $Y - X$ | Negative |
| $Y/X$ | $W/X$ | Positive |

*FIGURE 7.4*  **Effect of restriction of range.**

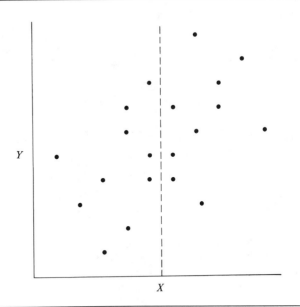

| Person | X | Y |
|--------|----|---|
| 1 | 3 | 3 |
| 2 | 2 | 2 |
| 3 | 1 | 1 |
| 4 | 4 | 4 |
| 5 | 10 | 1 |

the correlation is negative even though four of the five persons have the same score on $X$ and $Y$. The extreme value of 10 for person 5 on variable $X$ distorts the size of the correlation. Outliers can also cause a correlation that is truly zero to appear to be very large. A careful analysis of each variable should be done to identify outliers. In Chapter 4 an outlier is defined as a value that is away from the median by more than twice the interquartile range.

It is an outlier that brings about the very large correlation of .677 for the laughing data in Table 7.1. Note in Table 7.1 that couple 3 has the largest number of laughs for both males and females. In fact, each can be considered an outlier given the definition given in Chapter 4. What happens to the correlation coefficient when the data from this one couple is discarded? The resulting correlation is

$$r_{XY} = \frac{216 - (99)(37)/21}{\sqrt{(665 - 99^2/21)(117 - 37^2/21)}} = .410$$

Dropping this one observation changes what was an unreasonably large correlation into a moderate-to-large correlation. Besides dropping the one observation, an alternative would be to transform the observations. An examination of the histograms for the observations reveals that both variables are positively skewed. If the observations are square rooted, the scores for couple 3 are no longer outliers. The resulting correlation of the square rooted data is .551.

# Correlation and Causality

Correlations by their very nature seem to give rise to causal statements. If a newspaper publishes a report that persons who eat carrots live longer, it is a certainty that more carrots will be sold the following day. Finding out that carrot eating and longevity are associated inclines persons to jump to the conclusion that carrot consumption causes longer life. But correlation does not imply a particular causal relation. Just knowing that carrot eating and long life are associated does not mean that carrot eating causes longer life. There are other equally plausible explanations of the relationship. For instance, it may be that persons with more income tend both to live longer and also to eat carrots. And so the relationship between carrot eating and longevity may be due to the third variable of income.

Most of the time correlation *does* imply causality, but the exact form of the causality is uncertain. Consider another example. There is a small-to-moderate positive correlation between preference for violent television programs and the tendency to be physically and verbally aggressive among preadolescent males. Thus, boys who get into fights prefer to watch Kojak and the Three Stooges. The reason for this correlation is not clear. It could be that the violence on television makes the children more aggressive. Or it could be that being aggressive makes boys seek out more violent television shows. Or it may be that neither causes the other but both are caused by some other variable. For instance, parental socialization may affect both television viewing and aggressive behavior. It might be that authoritarian parental rearing leads to aggressive boys who watch violent television shows. Thus, knowing that there is a correlation between two variables does not tell us what brought about the correlation. As is often stated, "correlation does not imply causality." It is better to restate the maxim as "correlation does not imply one particular form of causality."

Sometimes the source of correlation is not a causal process but is just an accident. For instance, there is some indication that the economic climate is negatively correlated with the length of women's skirts. Good economic times have been associated with shorter skirts and bad times with longer skirts. Surely skirt length does not cause the financial climate. Nor is it likely that the financial climate causes the length of skirts. Most likely this correlation is an accident, a statistical freak. One way to determine whether the correlation is just an accident is to check its continuation into the future. If it disappears, then it is likely an accident.

One should not take all that has been said to mean that correlations tell us nothing about causation. It is true that from correlations it is not possible to determine the particular causal connections. But if there is reason to believe that one variable causes the other, then the two variables should be correlated. Thus, a correlation can be used to *verify* a causal linkage, but it is indeed perilous to infer a particular causal linkage from a correlation. Thus, causation implies a correlation but correlation does not specify the exact form of causality.

# Summary

The correlation between two variables is defined as the regression coefficient computed from two variables' $Z$ scores. The correlation coefficient, symbolized by $r$, is a directionless measure of association that varies between $-1$ and $+1$.

The formula for a correlation coefficient is

$$r_{XY} = \frac{\sum (X - \bar{X})(Y - \bar{Y})}{\sqrt{\sum (\bar{X} - \bar{X})^2 \sum (Y - \bar{Y})^2}}$$

The computational formula for a correlation coefficient is

$$r_{XY} = \frac{\sum XY - (\sum X)(\sum Y)/n}{\sqrt{[\sum X^2 - (\sum X)^2/n][\sum Y^2 - (\sum Y)^2/n]}}$$

A small correlation is .1, medium is .3, and large is .5. A correlaton can be interpreted as a regression coefficient. A squared correlation indicates the proportion of variance explained or variance shared. A correlation can also be interpreted as a probability of someone with a two-standard-deviation advantage over another person on one variable outscoring that person on a second variable.

Correlations are affected by nonlinearity, unreliability, aggregation, restriction in range, the part-whole problem, and outliers.

Just because two variables are correlated does not mean that one causes the other. It may be that the two variables are both caused by a third variable. A correlation indicates some type of causal connection but does not identify the particular type.

# *Problems*

1. For the data

| X | Y |
|---|---|
| 1 | 7 |
| 2 | 9 |
| 4 | 13 |
| 3 | 11 |

   compute $r_{XY}$.

2. Smith finds that the correlation between motivation and performance equals .391. How would you help her interpret her result?

3. If

| X | Y |
|----|----|
| −3 | 1 |
| 9 | 9 |
| 1 | 15 |

   compute $r_{XY}$.

4. Compute the following.

   a. $b_{XY}$ given that $r_{XY} = .4$, $s_X = 2$, $s_Y = 4$
   b. $b_{XY}$ given that $r_{XY} = -.3$, $s_X = 10$, $s_Y = 3$
   c. $r_{XY}$ given that $b_{XY} = .1$, $s_X = s_Y$

5. Baxter (1972) used an adaptation of traditional methods to teach clerical skills to mildly retarded adults. At the end of training, the skill of these adults in each task was rated by the same standards. The scale ranged from zero to ten, with zero the lowest possible rating and ten the highest. Some of Baxter's results are given below.

| Subject | Typing | Stencils |
|---------|--------|----------|
| 1 | 3 | 4 |
| 2 | 8 | 7 |
| 3 | 6 | 7 |
| 4 | 5 | 3 |
| 5 | 6 | 7 |
| 6 | 2 | 7 |
| 7 | 6 | 9 |
| 8 | 6 | 5 |
| 9 | 9 | 9 |
| 10 | 6 | 9 |

   Compute the correlation between typing and stencil preparation. Interpret the result.

6. Compute $r$ if $n = 10$, $\Sigma X = 20$, $\Sigma Y = 40$, $\Sigma X^2 = 60$, $\Sigma Y^2 = 180$, and $\Sigma XY = 90$.

7. For the age and memory data presented in Table 6.3 of the preceding chapter, compute and interpret the correlation coefficient.

8. Below are the life expectancies at birth for males ($X$) and females ($Y$) in six of the less developed countries of the world. Also given are $\Sigma XY$, $\Sigma X$, $\Sigma Y$, $\Sigma X^2$, and $\Sigma Y^2$. Compute the correlation between the life expectancies of males and that of females.

| | Life Expectancy (years) | |
|---|---|---|
| *Country* | *Males (X)* | *Females (Y)* |
| India | 51 | 50 |
| Indonesia | 45 | 48 |
| Brazil | 58 | 63 |
| Bangladesh | 50 | 47 |
| Pakistan | 49 | 47 |
| Nigeria | 40 | 43 |
| Total | 293 | 298 |

$\Sigma XY = 14{,}737$; $\Sigma X^2 = 14{,}491$; $\Sigma Y^2 = 15{,}040$

9. Below are the life expectancies at birth for males ($X$) and females ($Y$) in six of the more developed countries of the world. Also given are $\Sigma XY$, $\Sigma X$, $\Sigma Y$, $\Sigma X^2$, and $\Sigma Y^2$. Compute the correlation of the life expectancies of males and females.

| | Life Expectancy (years) | |
|---|---|---|
| *Country* | *Males (X)* | *Females (Y)* |
| France | 70 | 78 |
| U.S.A. | 70 | 78 |
| Japan | 73 | 79 |
| W. Germany | 70 | 76 |
| Italy | 70 | 76 |
| United Kingdom | 70 | 76 |
| Total | 423 | 463 |

$\Sigma XY = 32{,}647$; $\Sigma X^2 = 29{,}829$; $\Sigma Y^2 = 35{,}737$

10. Below are the life expectancies for males and females in the twelve countries given in problems 8 and 9.

|                  | Life Expectancy (years) | |
|------------------|-------------|-------------|
| Country          | Males (X)   | Females (Y) |
| India            | 51          | 50          |
| Indonesia        | 45          | 48          |
| Brazil           | 58          | 63          |
| Bangladesh       | 50          | 47          |
| Pakistan         | 49          | 47          |
| Nigeria          | 40          | 43          |
| France           | 70          | 78          |
| U.S.A.           | 70          | 78          |
| Japan            | 73          | 79          |
| W. Germany       | 70          | 76          |
| Italy            | 70          | 76          |
| United Kingdom   | 70          | 76          |
| Total            | 716         | 761         |

$\Sigma XY = 47,384$; $\Sigma X^2 = 44,320$; $\Sigma Y^2 = 50,777$

a. Compute the correlation of life expectancies for males and females.

b. Compare this correlation to the correlation obtained in problems 8 and 9. What caused the correlation to change?

11. Draw a scatterplot for the data in problem 12 in Chapter 6. Describe the nonlinearity in the relationship and suggest a transformation to remove it.

12. For the following studies, state what might affect the size of the correlation, and explain how the correlation would be affected.

a. using the average score of children in 500 schools, the correlation between vocabulary and reading comprehension

b. the correlation between a child's height at birth and growth in the first year of life

c. the correlation of stress in the workplace with physical ailments among air traffic controllers

d. the correlation between number of calories ingested during the day and happiness

e. the correlation between intelligence, as measured by one item of an IQ test, and a student's grade-point average

# 8 | *Measures of Association: Ordinal and Nominal Variables*

Researchers in the social and the behavioral sciences generally study variables measured at the interval level of measurement. It is also possible for a variable to be measured at either the nominal or ordinal level of measurement. (The topic of level of measurement was presented in Chapter 1.) For instance, the variable of political party is at the nominal level of measurement. In this case each person does not have a score or number but rather each person is a member of a category. A variable, such as political party, that is categorical and not numeric is called a *nominal variable*. For a nominal variable each person is a member of one discrete category as opposed to each person receiving a numeric score. A nominal variable with only two categories is called a *dichotomy*.

Ordinal variables are also encountered in the social and behavioral sciences. Variables measured at the ordinal level of measurement permit a rank ordering of the objects. One can make a case that many variables in the social and behavioral sciences are measured at the ordinal level.

Although not as common as interval variables, nominal and ordinal variables are hardly unusual. Many of the standard demographic variables are nominal. For instance, gender, ethnicity, religion, and geographic region of residence are all nominal variables. Many medical and physical variables are also nominal: eye color, blood type, left- versus righthandedness, and diagnostic category. Many of the responses that are studied in research are nominal in nature. Whether a person agrees or disagrees on a survey, whom the voter prefers in an election, what product a customer purchases, whether a surgical procedure results in life or death, whether a subject recalls a nonsense syllable, and whether a nerve cell fires when stimulated are all examples of nominal variables. Nominal variables are quite common in research.

Ordinal variables can be established if the measurements are a rank ordering. Some variables, such as birth order and military rank, are clearly at the ordinal level of measurement. The boundary between the ordinal and the interval level of measurement is quite cloudy. The determination is often a matter of preference.

The previous six chapters focused on the various descriptive statistics for variables measured at the interval level of measurement. In this chapter various descriptive statistics for nominal and ordinal variables are presented. Most of the discussion concerns measures of association for these variables.

# Shape, Location, Variation, and Transformation

This section concerns the issues of shape, location, and variation for both nominal and ordinal variables. First considered are nominal variables.

## Nominal Variables

The distribution of a nominal variable is simply a set of frequencies. For instance, of 132 people polled, 28 people intend to vote Democratic, 36 Republican, and 68 intend not to vote at all. The numbers 28, 36, and 68 are frequencies. Or a particular surgical procedure resulted in 42 deaths and 778 lives saved. Or 91 nerve cells fired and only 3 did not. The distribution of a nominal variable is the number of observations in each category.

The categories of a nominal variable are not numeric. So when the numbers are graphed in a histogram, the bottom axis or $X$ axis is not numeric. The resulting distribution is called a *bar graph*. In a bar graph, unlike a histogram, the bars are separated by a space. Because the categories are not numeric, the ordering of the categories on the $X$ axis is arbitrary. One still should compute the relative frequencies for each category. For instance, if 58 respondents say yes and 21 say no, one would compute $58/(58 + 21) = .73$ for yes and $21/(58 + 21) = .27$ for no.

What then can be said about the shape of the distribution? With a nominal variable, one can describe how flat the distribution is. A flat distribution is one in which each of the categories is equally likely. A dichotomy in which the categories are quite uneven (say 75% to 25%) is said to be skewed.

Because the categories of a nominal scale are not ordered, a median makes no sense; and because they are not quantitatively ordered, a mean likewise does not represent a meaningful measure of central tendency. However, the mode can be determined. The modal category has the largest frequency. Hence for the nominal scale of political party, the party most frequently chosen would be the mode.

The formulas for variability presented in Chapter 4 are not appropriate for categorical variables. However, there is relatively more variation if the distribution is flat. Nearly equal numbers of persons in the categories results in greater variation. If the number of persons in one category is relatively large, then there is relatively little variation.

All of the data transformations discussed in Chapter 5 involve a quantitative operation on the data. Because with nominal variables there are only categories and not numbers, all of those transformations are inappropriate. Nonetheless, nominal variables are in a sense transformed when categories are collapsed. For instance, in surveys it is common practice to treat "don't know" and "no opinion" as the same response. There are two helpful rules for determining which categories to collapse. First, one should consider as good candidates for collapsing those categories whose occurrences are infrequent, say less than 5%. The second rule is to combine categories that are conceptually similar. For instance, if the categories are white, black, and Hispanic, it may be sensible to collapse black and Hispanic to form a minority group category.

Recall that the definition of a nominal variable is one in which each observation is placed in one and only one category. The categories of a nominal variable are said to be mutually exclusive (no person may be in two categories) and exhaustive (each person must be in at least one category). Occasionally, a given nominal variable may violate these rules, but the violations can be easily remedied. For instance, there may be some persons who are neither Democrat, Republican, nor Independent, and, less likely, there could be a person who claims to be a member of two parties. Those persons who do not fall into any category can be put into a residual category of "other." Alternatively, if there are few persons who fit in no category (less than 5%), they can be dropped from the sample. For those who are members of two categories, a new category of "both" could be created or again if they are few in number they could be dropped from the sample.

Occasionally, it is useful to treat a nominal variable *as if* it were a numeric variable. In this case, a researcher arbitrarily assigns numbers to the various categories of the nominal variable. For instance, for the variable of gender, men may be given a score of zero and women a score of one. This is said to be an arbitrary assignment because the researcher could have just as easily given men a score of one and women a score of zero, or men a score of 50 and women a score of −38. The arbitrary assigning of numbers to levels of a nominal variable is called creating a *dummy variable*. Another example of dummy variables is as follows:

assign a 1 for Catholics,
assign a 2 for Protestants,
assign a 3 for Jews, and
assign a 4 for others.

Dummy variables are used in computing the phi coefficient, which is a measure of relationship between two dichotomies that is presented later in this chapter.

For a dichotomy, the usual convention is to assign a one to one category and a zero to the other. The mean of such a dummy variable is the proportion of persons assigned a one. The standard deviation[1] is the square root of $n/(n - 1)$ times the product of the proportion assigned a one and one minus that proportion. So if $p$ is the proportion of people assigned a one, then the mean of the dummy variable is $p$ and the standard deviation is the square root of $np(1 - p)/(n - 1)$.

## *Ordinal Variables*

If a variable is truly measured at the ordinal level of measurement, then its shape contains nothing theoretically interesting. The numbers reveal only the relative position of the persons in the sample and nothing about the distance between two scores. Shape then is virtually meaningless for a variable that is at the ordinal level of measurement. However, if an ordinal variable has relatively few levels and there are many observations, there are many tied observations. Two observations are tied if the researcher is unable to determine which observation has more of the quantity that is measured. When an ordinal variable has many tied observations and few levels, a bar graph can be drawn. The ordinal variable would be on the $X$ axis and the number of tied observations would be on the $Y$ axis.

The concepts of central tendency and variability have no meaning for ordinal variables. However, the median, though not quantitatively interpretable, can be of interest. For instance, if it is known that, in terms of the continental United States, South Carolina is the state with the median population, then it is known what state it is that ranks in the middle.

If a variable is measured at the ordinal level, it is a common practice to transform the scores by rank ordering them (see Chapters 5 and 18). For reasons discussed in Chapter 18, the mean and the standard deviation of these ranks are of statistical interest. Given a sample of $n$ scores, their mean rank must be

$$\bar{X} \text{ of ranks} = \frac{n + 1}{2}$$

---

[1]This formula is usually presented as $p(1 - p)$. However, because this text uses $n - 1$ in the denominator for variance, the formula in the text is appropriate.

Given no ties, the standard deviation[2] of ranks is:

$$\text{standard deviation of ranks} = \sqrt{\frac{n^2 + n}{12}}$$

So if $n = 10$, then the mean is 5.5 and the standard deviation is 3.03. If there are ties, the usual formulas for standard deviation must be employed.

# Relationship

The remainder of this chapter concerns the measures of association between variables that are either at the nominal or the ordinal level of measurement. Considered first is the relationship between two dichotomous variables. A dichotomous variable is a nominal variable with only two categories. Next is discussed the general problem of measuring the association between two nominal variables. Finally, the issue of how to measure the association between two variables measured at the ordinal level of measurement is discussed.

## Two Dichotomies

Of key interest is whether there is any relationship between two nominal variables. Is political party related to voting behavior? Is a surgical procedure related to survival? Are people more likely to give blood depending on the type of appeal? All of these questions are concerned with the relationship between two nominal variables.

In Table 8.1 is a table of numbers. The data are taken from a study by Korytnyk and Perkins (1983). They placed 29 male, heavy drinkers in a situation in which the subjects could write graffiti on a wall. Of the subjects, 15 were given tonic water and 14 were given the equivalent of two drinks of alcohol. The two variables being associated are beverage consumed (tonic versus alcohol) and whether the subject wrote graffiti or not. Beverage makes up the rows of the table and graffiti behavior, no and yes, makes up the columns. The numbers in the table are called counts. There are four counts. For instance, there are 14 who received tonic and did not write on the walls, and 7 alcohol drinkers who wrote on the wall.

The entries in a table, called frequencies, are the number of persons in that cell. Because each variable is made up of two categories, the table is called a 2 by 2, or $2 \times 2$, table.

Again the rows of the table are beverage (tonic or alcohol) and the columns

---

[2]The numerator of this formula is often presented as $n^2 - 1$. However, because this text always uses $n - 1$ as the denominator for variance, $n^2 + n$ is appropriate.

*TABLE 8.1*    **A 2 × 2 Table**

|  |  | Graffiti |  |  |
|---|---|---|---|---|
|  |  | No | Yes |  |
|  | Tonic | 14 | 1 | 15 |
| Beverage |  |  |  |  |
|  | Alcohol | 7 | 7 | 14 |
|  |  | 21 | 8 | 29 |

are graffiti (no or yes). It is a common practice to add the counts across both rows and columns. There are 15 tonic and 14 alcohol drinkers, and 21 who did not write on the wall and 8 who did. These sums are commonly called the *margins*. The column margins are 21 and 8. The sum of the row margins (15 + 14) should equal the sum of the column margins (21 + 8), and this provides a useful computational check. The total sum is written in the bottom right-hand corner.

Of special interest is whether there is any association between beverage consumed and graffiti behavior. Stated differently, the question is whether those who drink alcohol are more or less likely to vandalize than those who do not. This question concerns whether the two nominal variables are associated. To measure association the researcher can chose among the percentage difference, the phi coefficient, or the logit difference.

***Percentage Difference.***    The simplest and perhaps most natural measure of association is to compute the percentage of tonic drinkers who write on the wall and the percentage of alcohol drinkers who write on the wall. Using the data in Table 8.1, only 6.7% (1/15) of tonic drinkers write on the wall, and 50.0% (7/14) of the alcohol drinkers write on the wall. This difference between 50.0% and 6.7% is 43.3%. This is the percentage difference measure of association for nominal data. The percentage could be computed for each column. That is, of those who do not write on the wall, 66.7% (14/21) are tonic drinkers, and of those who do write on the wall, 12.5% (1/8) are tonic drinkers. This difference is 54.2%. As this example shows, the percentage difference measure is not necessarily a symmetric measure of association. The percentage difference measure may change if the percentages are calculated across rows or across columns.

The percentage difference measure can be viewed as a regression coefficient. One variable is denoted as the predictor variable and is dummy-coded zero and one. The other variable is the criterion and is dummy-coded zero and 100. If a regression coefficient were computed for these two dummy variables, its value would be identical to the percentage difference measure.

***Phi Coefficient.*** The second measure is phi, which is symbolized by $\phi$. This measure of association is not as commonly used as percentage difference. Phi is a correlation coefficient. So if there is no relationship, phi is near zero, and if there is a near-perfect relationship phi is near 1.00 or –1.00. To understand how phi is computed, consider Table 8.2. The two nominal variables have been designated $X$ and $Y$. The four frequencies are designated $a$, $b$, $c$, and $d$. The phi coefficient is found as follows:

$$\phi = \frac{ad - bc}{\sqrt{(a + b)(c + d)(a + c)(b + d)}}$$

For example, in Table 8.1 phi equals .484. As was stated earlier, the phi coefficient is a correlation coefficient. In Table 8.2 there are dummy variables for both variables. The first row is given a 1 and the second a 0. The first column is given a 1 and the second column a 0. Using these dummy variables, one can compute the correlation between the two variables. This correlation equals phi. Ordinarily, it is much simpler to use the formula presented earlier. As will be seen in Chapter 17, phi is a useful number for computing other statistics. Unlike the percent difference measure but like the correlation coefficient, the measure phi is symmetric.

In Chapter 7, it was stated that $r$ equals the square root of the product of the two regression coefficients. This fact can be used to relate phi to the two percentage difference measures: Phi times 100 equals the square root of the product of the two percentage difference measures. So, for the example, 100 times phi (48.4) equals, within rounding error, the square root of the product of the two percentage difference measures (43.3 × 54.2).

**TABLE 8.2** **Symbols for a 2 × 2 Table**

|  | | Variable Y | | |
|---|---|---|---|---|
|  | | 1 | 0 | |
| Variable $X$ | 1 | $a$ | $b$ | $a + b$ |
|  | 0 | $c$ | $d$ | $c + d$ |
|  | | $a + c$ | $b + d$ | $n$ |

*Logit Difference.*   The percentage difference is based on the regression coefficient and phi is based on the correlation coefficient. A third measure of association in a $2 \times 2$ table is the logit difference. In Chapter 5, the logit transformation of proportions is presented. A logit of a proportion $p$ is

$$\ln\left(\frac{p}{1-p}\right)$$

where ln is the logarithm to base $e$. In Appendix A in the back of the book is a table for the conversion of a proportion into a logit. As stated in Chapter 5, the logit of zero corresponds to a proportion of .5, a positive logit to a proportion greater than .5, and a negative logit to a proportion less than .5. Using the symbols in Table 8.2, the logit for the upper row is $\ln(a/b)$ and for the lower row is $\ln(c/d)$. The logit difference is then

$$\ln\left(\frac{a}{b}\right) - \ln\left(\frac{c}{d}\right)$$

Although it is not intuitively obvious, the logit difference is a symmetric measure. That is, it is true that

$$\ln\left(\frac{a}{b}\right) - \ln\left(\frac{c}{d}\right) = \ln\left(\frac{a}{c}\right) - \ln\left(\frac{b}{d}\right)$$

This is true because each equals

$$\ln\left[\frac{a/b}{c/d}\right]$$

The term in the parentheses is a ratio of two odds and is, therefore, called the *odds ratio*. For the example, it is the odds of tonic drinkers writing on the walls divided by the odds of alcohol drinkers doing so. Thus, the logit difference can be interpreted as the natural logarithm of the odds ratio. This odds ratio formula for the logit difference is simpler than the logit difference formula, because the odds ratio formula involves taking a logarithm only once. The logit difference for the example equals 2.64. The odds ratio is 14 and its natural logarithm is 2.64, which is equivalent to the logit difference measure.

   If any of the frequencies equals zero, the logit difference measure is not defined. To remedy this problem, .5 is added to each of the frequencies before the logit difference is computed.

*Interpretation and Comparison.*   Table 8.3 gives formulas for the percentage difference, phi coefficient, and logit difference using the symbols presented in Table 8.2. Although the measures are quite different, when one of them equals zero, the other two also equal zero. They are all zero only when $ad = bc$ (see Table 8.2).

*TABLE 8.3*    **Formulas for Measuring Association in a 2 × 2 Table[a]**

*Percentage Difference*

$$100\left[\frac{a}{a+b} - \frac{c}{c+d}\right]$$

*Phi*

$$\frac{ad - bc}{\sqrt{(a+b)(c+d)(a+c)(b+d)}}$$

*Logit Difference*

$$\ln\left(\frac{a}{b}\right) - \ln\left(\frac{c}{d}\right)$$

[a]See Table 8.2 for the definition of $a$, $b$, $c$, and $d$.

Because the logit difference involves odds and logarithms, it is not as easily interpreted as a percentage difference. One way to interpret the measure is to take the antilog and interpret the odds ratio. Table 8.4 lists the logit differences in terms of percentage differences. The table answers the following question: For a given logit difference, what would the percentage difference measure be if both the row margins as well as the column margins were equal? There are two facts worth remembering from the table. First, a logit difference of .10 corresponds to 2.5 percentage difference, and a 10.0 percentage difference corresponds to a logit difference of .40. Note that the logit difference of 2.64 for the example corresponds about a 55 percentage difference. The actual percentage difference is 43.3%, but the margins are quite skewed because only 27.6% wrote graffiti on the walls.

The student might wonder why bother with all of the difficulties in computing the logit difference. Both the percentage difference and phi are simpler to compute and interpret than the logit difference; however, the logit difference measure has one very important advantage over the percentage difference and phi. These latter two measures have an upper limit, of 100 and 1, and a lower limit, −100 and −1. In certain cases, these limits are further constrained. The logit difference has no such limits regardless of the margins.

The advantage of not having limits is illustrated in Table 8.5. The table contains hypothetical data relating lung cancer and smoking. The three measures of association have been computed. Now imagine that a second group is sampled, and in this group 94.85% are smokers and 5.25% of the group have died from lung cancer. That is, the percentage of smokers and deaths from lung cancer in the second group are different from those in Table 8.5. Given these percentages for the second group, it is impossible for the percentage

*TABLE 8.4*    **Logit Difference in Terms of Percentage Difference Measure in which the Margins Are 50/50**

| Logit Difference | Percentage Difference |
|---|---|
| .00 | 0.0 |
| .05 | 1.2 |
| .10 | 2.5 |
| .15 | 3.7 |
| .20 | 5.0 |
| .25 | 6.2 |
| .30 | 7.5 |
| .35 | 8.7 |
| .40 | 10.0 |
| .45 | 11.2 |
| .50 | 12.4 |
| .60 | 14.9 |
| .70 | 17.3 |
| .80 | 19.7 |
| .90 | 22.1 |
| 1.00 | 24.5 |
| 1.25 | 30.3 |
| 1.50 | 35.8 |
| 1.75 | 41.2 |
| 2.00 | 46.2 |
| 2.50 | 55.4 |
| 3.00 | 63.5 |
| 5.00 | 84.8 |

difference or phi to be as large as it is for the group in Table 8.5. Only for the logit difference can the measure of association possibly remain stable. It is for this reason that many researchers prefer the logit difference measure. The logit difference measure tends to replicate better across time and settings. The logit difference measure tends not to be as affected by changes in the margins as the percentage difference and phi are.

What is the most appropriate way to measure association in a 2 × 2 table? The answer depends on the purposes of the researcher. If a measure is desired that is simple to compute and easy to interpret, then the percentage difference measure is probably best. If the variables cannot be distinguished as a predictor and criterion, then phi is probably best. If the measure is to be computed for different samples with different margins, then the logit difference is probably best.

## Nominal Variables with More than Two Levels

The measures of association between two nominal variables, at least one of which has more than two levels, are analogous to the measures of association

*TABLE 8.5* **Illustration of the Generalizability of the Logit Difference**

| | Cause of Death | | |
| | Lung Cancer | Other | |
| --- | --- | --- | --- |
| Smoker | 681 | 7831 | 8512 |
| Nonsmoker | 123 | 8391 | 8514 |
| | 804 | 16222 | 17026 |

Percentage Difference: 6.56
Phi: .155
Logit Difference: 1.780

between two dichotomous variables. However, a complete presentation of these techniques is beyond the scope of this book. The reader is referred to more advanced texts that present these measures (Fienberg, 1977; Reynolds, 1977).

One can compute percentages across either rows or columns. So to compute the percentage for each row, an entry is divided by its row total. The sum of the percentages across each row should add to 100. As an example, consider the data in Table 8.6. The data are taken from a vote in the United States Congress in 1836. The issue concerned a matter related to slavery, a vote for the law being proslavery vote. There were three possible vote alternatives: yes, abstain, and no. The 225 congressmen are classified according to the section of the country that they represented. So 61 congressmen from the North voted yes. Beneath each number, in parentheses, is the percentage computed across rows. For instance, the percentage of congressmen voting no from the North is 60 divided by 133 (the row margin) times 100 or 45%. The table clearly shows that support for the proslavery position was located in the South and border states.

## Ordinal Variables: Spearman's Rho

Sometimes the researcher might question whether a given variable is measured at the interval level of measurement. The researcher may believe that the variables are measured at only the ordinal level of measurement. That is, the numbers indicate only the relative positions of persons and not any quantitative difference. The association of variables measured at the ordinal level of measurement can be measured by Spearman's rho. Although other measures,

*TABLE 8.6*    **Frequencies for a 3 × 3 Table of Vote by Region and Row Percentages in Parentheses**

| Region Represented | Vote Yes | Vote Abstain | No | Totals |
|---|---|---|---|---|
| North | 61 (46) | 12 (9) | 60 (45) | 133 |
| Border | 17 (71) | 6 (25) | 1 (4) | 24 |
| South | 39 (57) | 22 (32) | 7 (10) | 68 |
| Totals | 117 | 40 | 68 | 225 |

Data were taken from Benson and Oslick (1969).

such as Kendall's tau and Goodman and Kruskal's gamma, are also employed, rho is by far the most common measure of ordinal association.

Variables measured at the ordinal level of measurement can be transformed by a rank-order transformation. This transformation is described in Chapter 5. Each score is given a rank from one to $n$. If two or more scores are tied, they are each given the average rank. Then these rank-order scores can be correlated using the formula for the correlation coefficient presented in the previous chapter. A correlation coefficient of rank orders is called *Spearman's rho* or the *rank-order correlation* and is denoted as $r_S$. Fortunately all the computational work of correlating the two sets of ranks can be avoided. There is a computational shortcut. It involves first computing the difference between each pair of ranks, $D_i$. It happens that

$$\text{Spearman's rho} = 1 - \frac{6\sum D_i^2}{n(n^2 - 1)}$$

where $n$ is sample size and $D_i$ is the difference in ranks for the $i$th pair of scores. This formula presumes no ties in the ranks. If there are ties, one must use the formula for the correlation coefficient given in Chapter 7.

The six in the formula for Spearman's rank-order coefficient strikes some as odd. Its presence is due to the formula presented earlier in the chapter for the standard deviation of ranks. The denominator of that formula has a twelve in it which brings about the six in Spearman's rho or $r_S$.

Another use of Spearman's rho is to control for nonlinearity. The standard measures of association discussed in the previous two chapters presume that the relationship is linear. If the relationship between variables is nonlinear but does not change direction (see Chapter 7), then Spearman's rank-order coefficient is a useful measure because linearity is not assumed.

In 1884, Francis Galton measured the strength of more than 9000 persons who visited various museums in London. Johnson, McClean, Yuen, Nagoshi, Ahern, and Cole (1985) report the average degree of hand strength that Galton obtained for men from 11 through 25 years of age:

| Age | Hand Strength | Age | Hand Strength |
|-----|---------------|-----|---------------|
| 11  | 33.11         | 19  | 80.07         |
| 12  | 37.53         | 20  | 80.19         |
| 13  | 40.12         | 21  | 81.12         |
| 14  | 48.68         | 22  | 80.46         |
| 15  | 57.99         | 23  | 79.86         |
| 16  | 67.41         | 24  | 81.36         |
| 17  | 73.94         | 25  | 82.27         |
| 18  | 78.25         |     |               |

The scores are rank ordered by age in ascending order and the differences (*D*) and the squared differences (*D*²) are computed:

| Age | Hand Strength | D   | D² |
|-----|---------------|-----|----|
| 1   | 1             | 0   | 0  |
| 2   | 2             | 0   | 0  |
| 3   | 3             | 0   | 0  |
| 4   | 4             | 0   | 0  |
| 5   | 5             | 0   | 0  |
| 6   | 6             | 0   | 0  |
| 7   | 7             | 0   | 0  |
| 8   | 8             | 0   | 0  |
| 9   | 10            | −1  | 1  |
| 10  | 11            | −1  | 1  |
| 11  | 13            | −2  | 4  |
| 12  | 12            | 0   | 0  |
| 13  | 9             | 4   | 16 |
| 14  | 14            | 0   | 0  |
| 15  | 15            | 0   | 0  |

The sum of squared differences in ranks is 22 and so Spearman's rho is

$$r_S = 1 - \frac{(6)(22)}{(15)(224)} = .961$$

It is not surprising that for men from 11 to 25, there is a very strong relationship between age and hand strength. Note that the objects for the correlation are groups of persons at a given age. As discussed in Chapter 7, correlations based on aggregates or groups of persons tend to be larger than correlations based on individuals. The relationship between age and hand strength is much weaker for individuals.

# Summary

Nominal and ordinal variables are commonly used in the social and the behavioral sciences. Because the standard statistics that were developed in the previous chapters are for variables that are measured at the interval level of measurement, new procedures are developed. Because the variables are not at the interval level of measurement, the usual measure of central tendency, the mean is not appropriate. However, the mode can be used as a measure of central tendency for nominal variables and the median for ordinal variables. For neither the nominal nor ordinal levels of measurement does the shape of distribution make much sense, except for plotting the frequencies of a nominal variable in a bar graph.

For nominal variables it is sometimes necessary to collapse or eliminate categories. When numbers are assigned to levels of a nominal variable, the resulting variable is called a *dummy variable*.

There are three major measures of association between two nominal variables. They are the *percentage difference,* the *phi coefficient,* and the *logit difference*. The percentage difference is the simplest measure of association. One variable is denoted as a predictor and the other as the criterion. The percentages are computed for those responding in one category of the criterion for each of the categories of the predictor variable. The percentage difference is the difference between these two percentages. The phi coefficient is a correlation coefficient between dummy variables. For the logit difference, the odds of responding are computed for each category. These odds are logged and then differenced. The logit difference, while more difficult to interpret, is more likely to generalize across different samples. Only the percentage difference measure can be easily generalized for nominal variables with more than two levels.

One standard measure of association between two ordinal variables is *Spearman's rank-order correlation,* also called *Spearman's rho*. Spearman's rank-order correlation $r_S$ is a correlation coefficient between ranks. This measure is based on the difference between the ranks of the two variables. The rank-order correlation can be used to measure the association between variables measured at the interval level of measurement when nonlinearity is suspected.

# Problems

1. For the following sets of categories, by collapsing categories, create a new nominal variable with only two categories.

   a. Protestant, agnostic, atheist, Catholic, Jewish
   b. rainy, clear, cloudy, snowy
   c. radio, television, stereo, tape deck
   d. anger, disgust, happiness, fear

2. a. Imagine that 26 persons agree and 49 disagree with a particular statement. If responses are dummy coded (1 = agree, 0 = disagree), what is the mean and standard deviation of the dummy variable?
   b. If 32 agree and 42 disagree, what would be the mean and standard deviation?

3. If 76 Democrats favor capital punishment and 73 disapprove, and 108 Republicans approve and 111 disapprove, set up the 2 × 2 table with margins. Compute phi, the percentage difference (treating political party as the predictor variable), and the logit difference.

4. For the following table compute the percentage difference (treating gender as the predictor variable), phi coefficient, and logit difference. The column variable is whether the person is a smoker or not. Interpret each measure.

|  | Smoking | |
|  | Yes | No |
|---|---|---|
| Women | 71 | 28 |
| Men | 19 | 48 |

5. For the following table compute the percentage difference (treating age as the predictor variable), phi coefficient, and logit difference. Interpret each measure.

|  | Yes | No |
|---|---|---|
| Over 30 | 28 | 83 |
| Under 30 | 16 | 9 |

6. Taylor and Ferguson (1980) asked 200 students where they went when they wanted to be alone (solitude) and when they wanted to talk with a close friend (intimacy). Their answers were then coded as one of three kinds of territory: primary, where the individual can control access to the area; secondary, where the control of access is shared with others; and public, where the individual has no control over access. The table below gives the number of responses in each category.

|  | *Intimacy* | | |
| --- | --- | --- | --- |
| *Solitude* | *Primary* | *Secondary* | *Public* |
| Primary | 24 | 2 | 35 |
| Secondary | 3 | 0 | 4 |
| Public | 59 | 8 | 65 |

a. Compute the percentage of students who chose each kind of territory for intimacy. (Compute the percentages for each column.)

b. What proportion chose each kind of territory for solitude? (Compute the percentages for each row.)

7. In a study of privacy regulation, Vinsel, Brown, Altman, and Foss (1980) had freshman dormitory residents check a list of techniques they might have used to avoid contact with others. A year later, 19 of the students had left the university (dropouts), while 54 were still enrolled (stay-ins). They found that five of the dropouts and nine of the stay-ins used loud music to avoid contact.

a. Create a 2 × 2 table for the use of music (yes or no) and enrollment status (dropout or stay-in).

b. For the table compute the percentage difference (treating enrollment status as the predictor variable), phi coefficient, and logit difference. Interpret each measure.

8. The following table shows the number of hurricanes that occurred in the years 1886–1981.

| *Month* | *Hurricanes* |
| --- | --- |
| January | 0 |
| February | 0 |
| March | 0 |
| April | 1 |
| May | 3 |
| June | 21 |
| July | 32 |
| August | 135 |
| September | 176 |
| October | 85 |
| November | 18 |
| December | 2 |

Treat the twelve months as categories and construct a bar graph of the data.

9. Harrison (1984) asked dormitory residents who had chosen or been assigned to their dorms whether they wanted to change roommates at the end of the semester. The following table gives the results. Compute the percentage difference (treating choice vs. assignment as the predictor variable), the phi coefficient, and the logit difference. Interpret each measure.

| Dorm | Change Roommate No | Yes |
|---|---|---|
| Chose | 22 | 3 |
| Assigned | 34 | 12 |

10. Compute Spearman's rank-order correlation for the following set of scores of seven persons on variables $X$ and $Y$.

| Person | X | Y |
|---|---|---|
| 1 | 12 | 9 |
| 2 | 15 | 7 |
| 3 | 13 | 6 |
| 4 | 9 | 5 |
| 5 | 6 | 8 |
| 6 | 14 | 4 |
| 7 | 19 | 1 |

11. Below are the selected life expectancies for seven countries for males and females. Compute Spearman's rank-order correlation $r_S$ for the countries:

| | Life Expectancy (years) | |
|---|---|---|
| Country | Males | Females |
| India | 51 | 50 |
| Indonesia | 45 | 48 |
| Brazil | 58 | 63 |
| Bangladesh | 50 | 47 |
| Nigeria | 40 | 43 |
| U.S.A. | 70 | 78 |
| Japan | 73 | 79 |

Interpret the value of the rank-order correlation.

12. For the numbers in problem 11 compute the ordinary correlation $r$ between the life expectancy for males and females in seven countries. Why are $r$ and $r_S$ different?

PART 3

# *Inferential Statistics*

# 9 | *Statistical Principles*

In the previous eight chapters many topics have been covered. Formulas for means, variances, regression, and correlation coefficients have been presented. Methods have been presented to *describe* in rather rich detail a sample of numbers. *Descriptive statistics* are used to produce quantitative summaries of numbers.

Yet there is more to data analysis than just describing the data at hand. Besides describing the data, statistics are used to test ideas, theories, and hypotheses about the population. The testing of hypotheses, or more generally the testing of models, is called *inferential statistics*.

In this chapter and the next two, the groundwork is prepared for inferential statistics. There are many important statistical concepts that are essential to the understanding how models are tested. The chapter begins with a discussion of the way in which numbers are chosen to form a sample. Then, the idea that a statistic has a distribution is presented. The next topic concerns criteria that are used to determine which statistic is best. The final topic is the binomial distribution.

All of the topics and concepts in this chapter relate to statistical theory. Many of the quantities referred to in this chapter are computed from hypothetical distributions. Though somewhat abstract, the ideas in this chapter are essential for the intelligent comprehension of the remainder of the book.

## *Sample and Population*

When we step on the scale to check our weight, we care about the number that appears. Our own numbers or scores are important to us. However, we are generally not interested in the numbers or scores of particular other persons. Similarly, we do not care if John's IQ is 123 or whether Paul shocked the person in the Milgram experiment at 450 volts. For a sample of numbers, the specific numbers by themselves are ordinarily of little interest. They become interesting if they are viewed as representative of the numbers that some larger group of persons would provide. The magic of statistics is that the numbers of a few can be used to gauge the numbers of many. If a researcher takes an election survey and finds that a candidate is losing by 5%, the candidate does not really care that he or she lost in the survey. It is the views of all the voters that is important, but a survey of a few persons may reflect the views of many voters.

This process of going from the few to the many is valid only if a set of traditional assumptions are true: One assumption is that there is a set of objects called the *population* and a number is attached to each object. It is also assumed that the set of objects is infinite in size. From the population a small set of observations called the *sample* is gathered. Thus the sample is actually a subset of all the possible observations. The population is infinite in size, whereas the sample is finite.

Although classical statistical theory is based on the assumption that the population is infinite in size, in practice most populations in the social and behavioral sciences are finite in size. For instance, in an election survey the population is the set of potential voters, which is finite in size. It has been found that the theory based on infinite populations can be safely applied to large but finite populations.

The optimal size for a sample depends on several factors. Sometimes a small sample of ten objects is plenty, whereas in other cases hundreds of objects are needed. The choice of sample size depends on the statistic being computed, the resources available, and the degree of confidence required in the conclusions.

Sample data are used to infer properties of the population. For instance, the mean of a sample is computed and used to infer the mean of the population. A quantity computed from the sample is called a *statistic*. So the mean, variance, and correlation coefficient when computed from sample data are called statistics. A quantity computed using all the members of a population is called a *parameter*. The population mean and variance are parameters. Roman letters $\overline{X}$, $s$, and $r$ are used to designate statistics, while Greek letters $\mu$ (mu), $\sigma$ (sigma), and $\rho$ (rho) are used to designate parameters.

The value of a parameter is almost never known and, as a result, it remains a hypothetical value that is estimated by a statistic. It would be nearly impossible, for example, to interview all the voters before an election. So from the population a subset or a sample of voters is selected. A major part of statistical theory concerns how sample data can be used to describe accurately the population from which they are drawn. Ideally the sample should correspond as closely to the population as possible so that the statistics computed from sample data will be as close to the population parameters as possible. However, it is impossible for a finite sample to mirror an infinite population exactly. So whenever one samples from the population, there will be *sampling errors*. Sampling and sampling error go hand in hand.

The term *sampling error* is unfortunate because an error seems to imply a mistake that could have been prevented. Sampling error cannot be prevented. It is necessarily built into sampling. When a few observations are used to represent the many, errors necessarily follow.

Imagine working for a company that produces a certain type of machine that is ordered by more than 1000 factories throughout the world. You want to project how many of your machines will be ordered during the upcoming year. It would be wasteful of time and money to contact managers at the 1000 factories and ask each how many of the product they plan to order for the next year. You cannot afford to interview every member of the population of 1000 factories. Practical considerations force you to choose only a *sample* of factories. Any statistic computed from the sample would not be the same as the population parameter. Thus your estimate of demand will not be what it would be if you surveyed all the 1000 factories. In essence, your survey may be in error. At issue are the procedures that can reduce the amount of sampling error and procedures for determining the likely amount of sampling error. It must be realized that sampling error is inherent in these procedures. So the goal must be to minimize and measure it.

# *Random and Independent Sampling*

In order to reduce the amount of sampling error and to estimate its probable extent, one must sample from the population in certain prespecified ways. Sampling must be random and independent. Random and independent sampling does not prevent sampling errors. It merely reduces their size and permits the determination of the likely amount of sampling error.

There are two strategies for controlling the amount of sampling error. First, the objects are selected randomly. *Random sampling* requires that every object from the population is equally likely to be chosen to become a member of the sample. Second, the objects are chosen independently. *Independent sampling* requires that if a given object is chosen, it in no way increases or decreases the probability that any other object is subsequently chosen. So for a voter survey, persons are randomly chosen. One should not sample haphazardly by interviewing persons who happen to pass by on a busy street and are willing to be interviewed. A "grab sample" of friends and acquaintances is not a random sample.

Random and independent sampling is the cornerstone of classical sampling theory. There are two major ways to obtain a random sample. The first way is to use a method that makes the decision of which objects or persons to include in the sample random. For instance, one can flip a coin or roll a die to make a decision randomly. The second procedure to achieve random sampling from a population of human subjects is to use a random number table. Such a table is presented in Appendix B.

To use a random number table, first a list of the population members is made. Next, these persons are assigned sequential numbers from 1 to $N$, where $N$ is the population size. Assume that the number $N$ has $k$ digits. A number is picked from a random number table, not necessarily the first number in the table. From this random number the first $k$ digits are examined, and it is determined whether anyone in the population has the number. If someone does, that person is included in the sample. The first $k$ digits of the next random number are used to select the next person into the sample. This process is repeated until the desired sample size is achieved.

Independence requires that the probability of a person being sampled not change if anyone else is sampled. For instance, if the population is a city, using all the persons living on one block violates the independent sampling requirement even if the block is chosen at random: If one person is sampled, the person's next-door neighbor must be sampled. However, if one die is repeatedly rolled, a sample of rolls is independent. Rolling a six on one trial does not change the probability of rolling a six on the next trial. The inferential statistical methods discussed in the subsequent chapters are based on the assumption that the data were gathered by means of independent and random sampling methods. The major way that the independent sampling assumption is violated is by measuring a person more than once.

When sampling from a population, one can sample with or without replacement. When sampling without replacement, once an object is chosen to be a member of the sample, it cannot be chosen again. When sampling with replacement, an object can be chosen again. With infinite populations, there is no practical difference between sampling with and without replacement. For small populations, observations can be independent only if one samples with replacement. If two cards are sampled without replacement from a deck of cards, the sampling is not independent. Picking an ace as the first card decreases the probability of picking an ace as the second card if that first ace is not returned to the deck.

There are two major reasons why objects should be randomly and independently sampled. The first is to reduce the amount of sampling error. In the absence of any other information, random and independent sampling provides statistics that are as close to the parameter as possible. The second reason is that random and independent sampling permits the quantification of the amount of sampling error. Thus, it is known how close, in theory, the statistic is to the population parameter.

In reality samples used in social and the behavioral sciences are not randomly or independently formed. For instance, in most psychology experiments students sign up to serve as subjects. Such samples are not random. Moreover, subjects when they are sampled randomly are hardly ever sampled with replacement. (Only in survey research, such as election surveys, are persons randomly and independently sampled.) Although persons are not randomly sampled, it can be argued that the response from a person is a random sample of that subject's behavior. So the responses can be assumed to be a random sample of a population of responses.

# *Sampling Distribution*

A statistic is a number that is computed from a sample. If the same statistic were computed from a different sample taken from the same population, almost certainly a different result would be obtained and neither would exactly equal the population parameter. This variation from sample to sample is called sampling error. For instance, if the average height in a class of 20 students were measured from a random sample of 5 persons, the mean of this sample of 5 will almost certainly not equal the mean of another sample of 5. In any given sample, the people chosen will be a bit taller or shorter than the class average. Sampling error goes hand in hand with statistical estimation. A sampling error is not an intentional mistake; rather, it is an inevitable outcome. In other words, error in the estimation of population parameters is the inevitable price that must be paid for the ease and economy afforded by sampling.

The sampling distribution of a statistic can be conceptualized as follows: If the mean were computed from two different samples with the same *n* drawn from the same population, two different values would be obtained. If an infinite number of samples of size *n* were drawn and for each the mean were computed, a frequency distribution of the sample means of size *n* could be created. The distribution of this infinite set of means is referred to as the *random sampling distribution* of the mean, or more simply as the *sampling distribution* of the mean. In general, any statistic has a sampling distribution. A *sampling distribution of a statistic* is a theoretical distribution based on an imaginary repeated sampling and computation of a statistic.

A sampling distribution has two important properties: its mean and its standard deviation. The mean of the sampling distribution equals what the statistic tends to be on the average. The standard deviation describes how variable the statistic is when repeatedly calculated from different samples of the same size. So, the mean of the sampling distribution states what the statistic is estimating and the standard deviation states how close it comes to that value on the average. The standard deviation of the random sampling distribution of a statistic is called the *standard error of the statistic*. The standard error quantifies the amount of sampling error in the statistic. It measures the degree to which the statistic would be likely to change if another sample were drawn and the statistic were recomputed. In other words, the standard error of a statistic measures how variable a statistic is when it is recomputed using a different sample of the same size.

For most statistics, the standard error decreases as the sample size increases. In fact, if a statistic did not have this attribute, it would be deemed a poor statistic. The relation between sample size and the standard error can be illustrated for the sample mean. If the population variance is 100, the standard error of the sample mean takes on the following values for the given sample sizes:

| $n$ | Standard Error of $\overline{X}$ |
|-----|----------------------------------|
| 10  | 3.16 |
| 25  | 2.00 |
| 100 | 1.00 |
| 150 | .82 |

(The exact formula for the standard error of the mean is presented in Chapter 11.) The standard error of 3.16 for the sample size of 10 implies that the typical difference of the sample mean from the population mean is 3.16 units. When the $n$ is 150, the sample mean differs from the population mean by about .82 unit. Normally the effect of increasing the sample size suffers from the "law of diminishing returns." Doubling the sample size does not cut the standard error in half. To cut the standard error of the mean in half, the sample size must be quadrupled. This can be seen in the above table: The standard error for a sample of 25 is twice that for a sample of 100 subjects.

As the sample size increases, the statistic does not vary as much from sample to sample. However, even with large sample sizes, the sample statistic still does not exactly equal the population parameter. Sampling and sampling error go hand in hand. The standard error quantifies the amount of sampling error. The standard error states how close the statistic is to the parameter, on the average, not in any particular instance. So if an election survey shows that a candidate is leading an election by 8% and the standard error is only 5%, it does not guarantee that the candidate is ahead. The 5% standard error is the average or typical amount of error to be expected given the sample size. The actual error in a particular survey may be zero or even 20%. The error in any particular study is never known. Only known is the average or standard error across many studies of the same sample size.

# Properties of Statistics

In computing measures of central tendencies various issues arose. For instance, as was explained in Chapter 3, there are three measures of central tendency: the mean, the median, and the mode. Is there any reason to prefer one measure over the other? Also the variance is divided by $n - 1$ and not $n$. Why is this? These questions can be answered once two important properties of statistics are defined.

## *Bias*

The mean of the sampling distribution can be used to define an important property of a statistic—namely, bias. A statistic that is supposed to estimate a population parameter is said to be an *unbiased* estimate of that parameter if the mean of the random sampling distribution of the statistic equals the parameter. Unbiased statistics exactly estimate on the average what they purport to be estimating. If a statistic is unbiased, the statistic itself does not necessarily equal the parameter; it only does so on the average. A positively biased estimate statistic is one that tends, on average, to overestimate the parameter value.

The sample mean $\bar{X}$ is an unbiased estimate of the population mean, $\mu$. If the distribution is symmetric, the median is also an unbiased estimate of the population mean. If the distribution is symmetric, the mean is also an unbiased estimate of the population median.

The sample variance $s^2$ is an unbiased estimate of the population variance, $\sigma^2$. The formula for $s^2$ has $n-1$ in the denominator and not $n$. The reason for this is to make $s^2$ an unbiased estimator of $\sigma^2$. If the denominator of $s^2$ were $n$ instead of $n-1$, $s^2$ would be a biased estimator.

A fact not very well known is that the sample correlation coefficient $r$ is a slightly biased estimate of the population correlation when the population correlation $\rho$ is nonzero. For sample sizes of five or more the sample correlation coefficient slightly underestimates positive values of $\rho$ and overestimates negative values of $\rho$. For moderate and large samples, the bias is trivially small and can be safely ignored. Although the correlation coefficient is biased, the regression coefficient is not.

## *Efficiency*

A second important property of a statistic is efficiency. One statistic is said to be relatively more *efficient* than another statistic if its standard error is smaller than that of the other statistic. Thus, if one statistic's standard error is smaller than another's, the former is said to be more efficient than the latter.

To choose between two statistics, say the mean and the median, their efficiency must be considered. Which of the two statistics is more efficient depends on the shape of the distribution. Thus it is necessary to consider what the underlying distribution is before determining which statistic is most efficient.

As will be discussed in the next chapter, in data analysis it is generally assumed that the population distribution is normal. For this reason, the discussion of efficiency will presume that the population distribution is normal.

When the population distribution is normal, $\bar{X}$ and $s^2$ are unbiased estimators of $\mu$ and $\sigma^2$, respectively. Moreover, they are the most efficient unbiased estimators of the parameters, again when the distribution is normal.

Although the sample mean and variance are optimal when the distributions are normal, they can be very inefficient when outliers are present. In particular, the sample standard deviation is a very inefficient estimate of the population standard deviation when there are outliers in the sample. A statistic whose efficiency is not affected much by nonnormality or outliers is said to be *robust*. The median is more robust than the mean, and the interquartile range is more robust than the standard deviation.

# Binomial Distribution

One important population distribution is the binomial distribution, which can be used to describe a series of random events. Before getting into the mathematics of this distribution, consider the following example.

Imagine someone flipping a coin four times and counting the number of heads. That number can vary from zero, no heads at all, to all four flips being heads. One might wonder what the probability is of obtaining exactly three heads in four flips. To determine this and other probabilities, the binomial distribution is used.

First, the probability of flipping a head on a single flip must be determined. That probability is .5 and is denoted as $p$. The probability of not flipping heads must be $1 - p$, which is denoted as $q$. So the probability of a success (flipping heads) is denoted as $p$, and the probability of a failure (flipping tails) is denoted as $q$.

Second, it must be assumed that the trials (flips) are independent. That is, having flipped a heads in trial one does not increase or decrease the probability of flipping a heads in trial two. It seems reasonable to believe that coin flips are in fact independent. However, other events are not. If a trial is picking a card from a deck, then the chances of picking an ace are 1/13 or .077. If the card from trial one is not replaced, then the chances of picking an ace on trial two are affected by what happened on trial one. If an ace were picked on trial one, then the chances of picking an ace in trial two are 1/17 or .059. But if some other card besides an ace were picked in trial one, then the chances of picking an ace in trial two are 4/51 or .078.

If there are a set of independent trials with a known probability, the probability of a given outcome can be determined. Assuming that there are $n$ trials, the probability of $x$ successes, given a binomial distribution, is:

$$\frac{n!}{x!(n-x)!} \, p^x q^{n-x}$$

The term $n!$ is read as $n$ factorial. It equals

$$n(n-1)(n-2) \ . \ . \ . \ (3)(2)(1)$$

So 5! equals $(5)(4)(3)(2)(1) = 120$. By convention $0! = 1$.

Using the binomial formula, the probability of obtaining three heads in four flips can now be computed. The terms for the binomial formula are

$p$(the probability of a success) $= .5$
$q$(the probability of a failure) $= 1 - .5 = .5$
$n$(the number of trials) $= 4$
$x$(the number of successes) $= 3$

Putting the terms in the formula yields:

$$\frac{4!}{3! \; 1!} \; .5^3 .5^1 = \frac{(4)(3)(2)(1)}{(3)(2)(1)(1)} (.125)(.5) = .25$$

So the probability of flipping three heads in four trials is .25.

As a second example, consider the probability of a subject getting eight of ten answers correct on a recognition test with five alternatives if the subject is just guessing. The probability of a correct guess on each trial is 1/5 or .20. So the terms for the formula are

p(the probability of a success) $= .20$
$q$(the probability of a failure) $= 1 - .20 = .80$
$n$(the number of trials) $= 10$
$x$(the number of successes) $= 8$

Putting these terms into the binomial formula yields:

$$\frac{10!}{8! \; 2!} \; .20^8 .80^2 = .0000737$$

If a variable has a binomial distribution, its population mean equals $np$ and its population variance equals $npq$. So for the prior example with $n = 10$, $p = .20$, and $q = .80$, the mean is 2.0 and the variance is 1.6. These are *population parameters*. The mean and variance of sample of subjects would not exactly equal these hypothetical values because of sampling error.

# *Summary*

*Descriptive statistics* are used to summarize the scores in a sample. Examples of descriptive statistics are the mean and the standard deviation. *Inferential statistics* are used to test models.

A *population* is an infinite set of objects and the *sample* is a subset of objects chosen from the population. A *statistic* is a number, like the mean or variance, computed from sample data. A *parameter* is a number computed from all the possible values of the population. The variation of a statistic from sample to sample is called *sampling error*.

A *random* sample is a set of numbers chosen so that each object is equally likely to be chosen to be a member of the sample. Sampling is said to be *independent* if choosing one object in no way changes the probability that some other object will be chosen. Random sampling can be accomplished by rolling a die or using a random number table. Random and independent sampling minimizes the amount of sampling error. Sampling can either be with or without replacement. When sampling *without replacement,* once the object is sampled, it cannot be sampled again. When sampling *with replacement,* an object can be sampled again.

The same statistic could be computed from many different samples drawn from the same population. A statistic has a *random sampling distribution* or, more simply, a *sampling distribution.* The standard deviation of the sampling distribution is called the *standard error* of the statistic.

A statistic is *unbiased* if the mean of its random sampling distribution is equal to the parameter that it is supposed to be estimating. Both $\bar{X}$ and $s^2$ are unbiased statistics. One statistic is more *efficient* than another if the variance of its random sampling distribution is less than the other statistic's variance. A statistic is said to be *robust* if an outlier in the sample does not dramatically change the value of the statistic. The median is a more robust estimate of the population mean than the sample mean. The interquartile range is a more robust estimator of variability than the sample variance.

The binomial distribution is used to describe the probability of an event happening $x$ times in $n$ trials. The probability of a success must be known, and trials must be independent.

# Problems

1. Which of the following schemes are random samples of pages from the phone book?

   a. page 10 through 53
   b. every fifth page
   c. opening the book 50 times and picking a page
   d. using a random number table to pick 60 pages

2. Imagine two statistics $p$ and $q$ that are both estimators of the same parameter. Presume that $p$ and $q$ are estimated from a sample of size 50. Also presume that $p$ is unbiased and its standard error for an $n$ of 50 is .88. The estimate $q$ is biased tending to be .01 unit too high and its standard error is .44. Which of the two statistics would you prefer to use and why?

3. Define in words each of the following for the variable of reaction time.

   a. a sample mean of 555 milliseconds (ms)
   b. a population mean of 531 ms
   c. a sample standard deviation of 112 ms

  d. a population standard deviation of 123 ms
  e. a standard error of the mean (based on ten observations) of 12.3 ms

4. Indicate whether each of the following statements is true or false.

   a. The mean of a random sample equals the mean of the population.
   b. The mean of a random sample will tend to equal the mean of the population.

5. Imagine two statistics $k$ and $q$ that both estimate parameter theta. Assume that the sampling distribution of $k$ for $n = 100$ has a mean of theta and a variance of 20, and the sampling distribution of $q$ for $n = 100$ has a mean of theta and a variance of 25.

   a. Are $k$ and $q$ unbiased estimates of theta?
   b. Which is more efficient, $k$ or $q$?
   c. What are the standard errors of $k$ and $q$?

6. Given that the population correlation $\rho$ equals zero, the quantity $r/\sqrt{(1 - r^2)}$ has a sampling distribution with a mean of zero and a variance of approximately $1/(n - 2)$.

   a. Is $r/\sqrt{(1 - r^2)}$ an unbiased estimate of $\rho/\sqrt{(1 - \rho^2)}$?
   b. What is the standard error of $r/\sqrt{(1 - r^2)}$?

7. For a flat distribution why is the sample median an unbiased estimate of the population mean?

8. If the population is normally distributed, it can be shown that the squared interquartile range multiplied by .55 is essentially an unbiased estimate of the population variance. What estimate of $\sigma^2$ would you prefer: $s^2$ or the adjusted interquartile range statistic? Why?

9. An unbiased estimate of the population mean is any randomly sampled observation in the sample. Why is the sample mean preferable to the single-score estimate of the mean?

10. Explain why each of the following is not an independent sample of college students.

    a. all students in one randomly chosen dormitory
    b. students waiting in line for a movie

11. For the population of 9, 8, 12, and 6, the following are all 16 samples with replacement of sample size 2.

$$(9, 9), (9, 8) (9, 12), (9, 6)$$

$$(8, 9), (8, 8) (8, 12), (8, 6)$$

$$(12, 9), (12, 8) (12, 12), (12, 6)$$

$$(6, 9), (6, 8), (6, 12), (6, 6)$$

Compute from each $\bar{X}$ and make a frequency table of the random sampling distribution of $\bar{X}$. Compute the mean and standard deviation of the sampling distribution.

12. Imagine that a random sample of 50 out of 12,938 students from a university is needed. State how a random sample could be drawn using a random number table.

13. Imagine a sample of three numbers from a population. An estimate, called $U$, of the population mean is

$$.5X_1 + .25X_2 + .25X_3$$

It turns out that $U$ is an unbiased estimate of the population mean with a standard error of $.612\sigma$ where $\sigma$ is the population standard deviation. The sample mean has a standard error of $.577\sigma$. What statistic is more efficient: $\bar{X}$ or $U$? Why?

14. Which of the following can be considered a random and independent sample of students from a classroom?

   a. All students whose names begin with A and K
   b. All students who sit in the first row.
   c. All students who volunteer to be in a study.
   d. The first ten students who come to class one day.

15. What is the probability of rolling a single die six times and obtaining a five three times?

16. If one rolls two dice, the probability of rolling an eight is 5/36. What then is the probability of rolling an eight on four out of five rolls?

17. Jim feels down on his luck. He has bet a number ten times straight on roulette and lost. The chances of hitting the number are 1 out of 38. What is the probability of losing ten times in a row?

18. If the probability of getting divorced is .42, what is the probability that in a sample of seven, five couples will get divorced?

# 10 | *The Normal Distribution*

In Chapter 2 the concept of distribution was introduced. A distribution of scores refers to the relative frequency of the various scores. In this chapter the distribution that is commonly assumed in data analysis—the normal distribution—is discussed. Not only is the normal distribution used in data analysis, it also underlies various data transformations.

## *Properties of a Normal Distribution*

The *normal distribution* is a symmetric, unimodal distribution that looks like a bell. It is a hypothetical distribution dreamed up by mathematicians that approximates the distribution of many naturally occurring variables. The upper and lower limits of the distribution are plus and minus infinity. Although all values are theoretically possible, very large or very small values are practically impossible. Because the normal distribution is a continuous distribution, any values, not just integers, are possible. Figure 10.1 shows two examples.

The normal distribution is not one distribution but actually a family of distributions. They differ only in their mean and variance. For instance, in Figure 10.1 both normal distributions have the same mean but one (the more peaked one) has less variance than the other. Although these distributions differ in their variance, their basic shape is exactly the same. If the mean and the variance of a normal distribution are known, then the exact shape of the distribution is known.

Although the two distributions in Figure 10.1 look quite different, they are both normal distributions. What is meant that they both have the same shape? Two distributions are said to have the same shape if there is a no-stretch transformation of one distribution that makes it possible to superimpose that distribution on the other.

*FIGURE 10.1*    **Examples of normal distributions.**

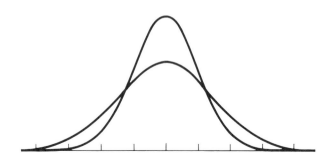

For the normal distribution the interval from the mean to one standard deviation above the mean contains about 34% of the scores, as is shown in Figure 10.2. In the interval from one standard deviation above the mean to two above the mean are about 13.5% of the scores. From two to three standard deviations are about 2% of the scores, and about .1% of the scores are greater than three standard deviations above the mean.

Stated alternatively, the interval from one standard deviation below the mean to one standard deviation above the mean contains about 68% of the scores. In the interval from two standard deviations below the mean to two standard deviations above the mean are about 95% of the scores. In the interval from three standard deviations below the mean to three above are about 99.7% of the objects. These facts hold for any normal distribution, regardless of the value of the mean and the variance.

Again, the normal distribution is a theoretical distribution dreamed up by mathematicians. Real numbers at best only roughly conform to its shape. For instance, the distribution of intelligence test scores is often assumed to be

*FIGURE 10.2*    **Probabilities for the normal distribution.**

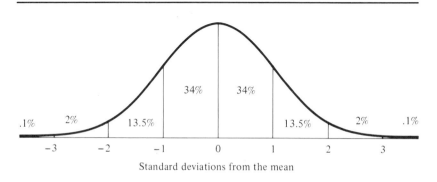

Standard deviations from the mean

normal. However, close inspection of the distribution reveals that there are more persons with very low intelligence than there would be if the distribution of intelligence were perfectly normal. Also, if the distribution were truly normal, an intelligence test score of −20 would be possible. The normal distribution is not "normal" in the sense that it is typical; actually it is quite atypical. Real, rather than theoretical, data are almost never exactly normally distributed. They may be close to normal but they are almost never exactly normal.

Even so, it is still reasonable to assume that the data are exactly normally distributed. There are four reasons for doing so. First, the normal distribution has some mathematically useful properties. For instance, $\bar{X}$ and $s^2$ are unrelated, which is not true of any other distribution. Another important fact is that if the variable $X$ is normally distributed, then the distribution of $\bar{X}$ is also normal. If $X$ has any other distribution, the mean of observations drawn from $X$ has a different type of distribution. Data analysis is sufficiently complicated mathematically even if the normal distribution is assumed. If nonnormality is permitted, much of the algebra becomes quite difficult and, in certain cases, practically impossible.

Second, many variables that social and behavioral scientists study tend to be roughly, if not exactly, normally distributed. There is an important statistical theorem called the *central limit theorem* that explains this fact. The central limit theorem states that if a score is made up of a sum or average of $n$ numbers, then as $n$ gets larger the score tends to have a normal distribution even if the components are not normally distributed. Because many measurements in the social and behavioral sciences are often the sum of a large number of responses, it is all but inevitable that they have a distribution that approaches normality.

Third, even if the numbers are not normally distributed, in many cases they can be transformed to become "more normal" by the transformations discussed in Chapter 5. For instance, skewed distributions can be made more normal by applying one-stretch transformations. Thus, nonnormal data can be made "more normal" through data transformation.

Fourth, even if the data are not normally distributed, the errors resulting from nonnormality are often not that costly. For instance, using the statistical technique called analysis of variance (discussed in Chapters 14 and 15) with nonnormal data results in surprisingly few errors in most cases. The reason for this is the previously mentioned central limit theorem. So, the costs of falsely assuming normality are often only minimal.

A normal distribution has two parameters: its mean symbolized by $\mu$ and its variance symbolized by $\sigma^2$. Regardless of the shape of the distribution, the sample mean ($\bar{X}$) and the sample variance ($s^2$) are unbiased estimators of the population mean $\mu$ and the population variance $\sigma^2$, respectively. If the distribution is normal, $\bar{X}$ is a more efficient estimate of $\mu$ than either the median or the mode even though the median and the mode are unbiased estimates of the population mean. The median and the mode have wider standard errors than the sample mean. In fact, it can be shown that $\bar{X}$ is the most efficient estimator of the population mean given a normal distribution.

Also, given normality, $s^2$ is a more efficient estimator of $\sigma^2$ than any other unbiased estimator of $s^2$. Thus, $\bar{X}$ and $s^2$ are preferred estimators of $\mu$ and $\sigma^2$ when the observations have a normal distribution. Because normal distributions are very commonly presumed in statistical work, both $\bar{X}$ and $s^2$ are the estimators of choice.

# Standard Normal Distribution

In Chapter 5, the $Z$ score transformation was presented. Its formula is

$$Z = \frac{X - \bar{X}}{s}$$

Because the statistics $\bar{X}$ and $s$ are used to compute $Z$ scores, this is a sample-based transformation. Imagine that $X$ is normally distributed with a known population mean of $\mu$ and population variance of $\sigma^2$. The $Z$ score in the population is computed by the formula

$$\frac{X - \mu}{\sigma}$$

Thus observations are adjusted by the parameters $\mu$ and $\sigma$ and not by the sample-based statistics $\bar{X}$ and $s$. The above $Z$ score has a normal distribution like $X$, but unlike $X$, the mean of the $Z$ scores is always zero and the variance is always one. A normal distribution with a mean of zero and variance of one is called the *standard normal distribution* or *Z distribution*. Any normally distributed variable can become a standard normal variable by subtracting the population mean from each score and dividing this difference by the population standard deviation.

The standard normal distribution is a normal distribution in which the scores are expressed in standard deviation units. So if a variable has a $Z$ distribution, a score of 1.5 indicates that the object is one-and-a-half standard deviations above the mean. A score of −1 indicates that the object is one standard deviation below the mean.

It is important to understand the difference between the $Z$ transformation and the $Z$ distribution. The $Z$ transformation can be applied to any set of numbers regardless of their distribution. The $Z$ transformation does not alter the basic shape of a distribution, only its mean and standard deviation. So a $Z$ score transformation does not make a nonnormal distribution normal, as is sometimes mistakenly thought. However, when the $Z$ transformation is applied in the population to a normally distributed variable, the result is the standard normal distribution. Though related, the $Z$ transformation and the $Z$ distribution are not the same.

It bears repeating that any normally distributed variable can be transformed into a variable with a standard normal distribution. Simply subtract the population mean and divide the difference by the population standard deviation. The resulting variable has a normal distribution with a mean of zero and a variance of one.

## Determining Probabilities

The standard normal distribution is used to determine the probability of various types of events. To determine such probabilities Appendix C is used. In this appendix are listed the probabilities that a score is in the interval from zero to the value labeled $Z$ in the table. For example, the probability of obtaining a $Z$ score between 0.00 and 1.00 is .3413, given the mathematical properties of the standard normal distribution. It is then the case that the probability of being in the interval between the mean and one standard deviation above the mean is .3413 for any normally distributed variable. Thus, the table of probabilities for the standard normal distribution can be used to answer questions about any normal distribution for which the mean and variance are known.

To use Appendix C to determine the probability that a score will fall in the interval between zero and .50, first locate .50 in the left-hand column of the table. Then the number to the right, .1915 gives the probability.

The probability that $Z$ exactly equals any particular value is zero. Because a normal distribution is a continuous distribution, any value is possible—not just integers. For continuous distributions that can take on any value, only the probability of an interval is nonzero. The probability of being exactly at zero or at any particular value is zero.

Because the normal distribution is symmetric, it happens that the probability of something being in the interval between 0 and $k$ is the same as it being in the interval between $-k$ and 0. So because the probability of being between 0 and .80 is .2881 (see Appendix C), then the probability of being between $-.80$ and 0 is also .2881.

Another useful fact concerns the probability of an event happening that is greater than some value, say $k$. This question can be reformulated as the difference between two probabilities. What is the probability of a score being greater than zero minus the probability of a score being between zero and $k$? The probability of a score being greater than zero is .5 because the normal distribution is symmetric and so one-half the scores must be above its mean of zero. As an example, what is the probability that a $Z$ score is greater than .80? Because from Appendix C the probability of $Z$ being between 0 and .80 is .2881, the probability of $Z$ being greater than .80 is .5 minus .2881 or .2119. These facts are illustrated graphically in Figure 10.3.

To compute the probability of sampling within some interval of a variable that is normally distributed, the numbers are first transformed into $Z$ scores. The question is reformulated so that it can be answered using the probabilities

*FIGURE 10.3*     **Facts about the Z distribution.**

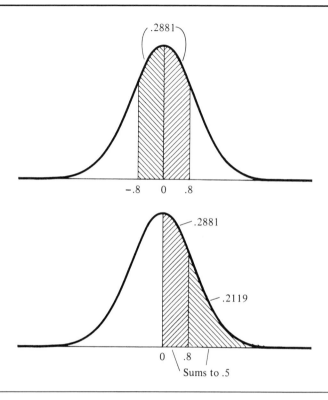

given in Appendix C. Consider some examples using men's height as the variable, which is assumed to have a mean of 70 inches, a standard deviation of 3 inches, and a normal distribution (Stoudt, 1981).

    a. What is the probability that a man is between 70 and 72 inches tall? First, the numbers 70 and 72 are converted into Z scores. These Z scores are

$$\frac{70 - 70}{3} = 0$$

and

$$\frac{72 - 70}{3} = .67$$

The question now becomes what is the probability of obtaining a Z score between 0 and .67. The answer from Appendix C is .2486, as shown here graphically.

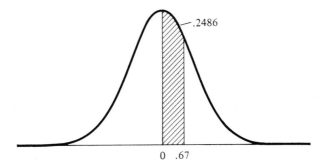

b. What is the probability that a man is between 64 and 70 inches tall? Both 64 and 70 are converted into Z scores:

$$\frac{64 - 70}{3} = -2.0$$

and

$$\frac{70 - 70}{3} = 0.0$$

The question becomes: What is the probability of a Z score being between −2.0 and 0? Because Z is symmetric, the question can be rephrased: What is the probability of sampling someone between 0 and 2.0? The answer from Appendix C is .4772, as shown graphically.

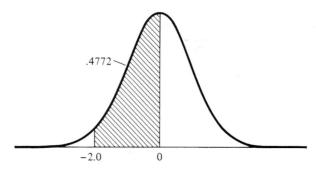

c. How likely is it that a man is between 65 and 72 inches tall? The numbers 65 and 72 are converted into Z scores:

$$\frac{65 - 70}{3} = -1.67$$

and

$$\frac{72 - 70}{3} = .67$$

To answer this question, it is divided into two separate parts. The probability of being between 0 and 1.67 is .4525 and between 0 and .67 is .2486. Both of the probabilities can be ascertained from Appendix C. So the probability of being between the interval −1.67 and .67 is .4525 + .2486, which equals .7009, as shown graphically.

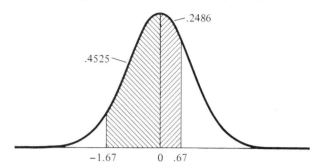

d. How likely is it that a man is between 72 and 75 inches tall? First, 72 and 75 are converted into $Z$ scores:

$$\frac{72 - 70}{3} = .67$$

and

$$\frac{75 - 70}{3} = 1.67$$

The question now is what is the probability of a $Z$ being between .67 and 1.67. From Appendix C, the probability of $Z$ being between 0 and .67 is .2486 and the probability of $Z$ being between 0 and 1.67 is .4525. Thus, the probability of $Z$ being between .67 and 1.67 is .4525 − .2486 = .2039, as shown graphically.

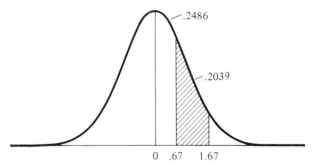

e. How likely is it for a man to be shorter than 67 inches? First, 67 is converted into a $Z$ score:

$$\frac{67 - 70}{3} = -1.0$$

This is equivalent to asking the probability of a $Z$ score greater than 1.0. The probability that $Z$ is greater than zero is .5. From Appendix C, the probability that $Z$ is the interval between zero and 1.0 is .3413. Thus, the probability that $Z$ is greater than 1.0 is .5 − .3413 = .1587, as shown graphically.

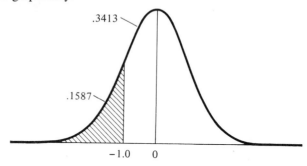

f.  What is the probability that a man will be 77 inches or taller? The score 77 is converted to a $Z$ score:

$$\frac{77 - 70}{3} = 2.33$$

The question is: What is the probability that a $Z$ score will be greater than 2.33? Because .50 is the probability of $Z$ being greater than zero, and because .4901 is the probability of being between 0 and 2.33, the probability of $Z$ being greater than 2.33 is .50 − .4901 = .0099, as shown graphically.

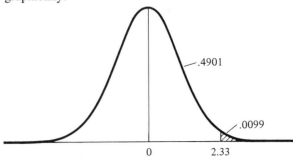

## Determining Percentile Ranks

The standard normal distribution can also be used to determine a score's percentile rank. The percentile rank states the percentage of objects that the given object scores higher than. For instance, Scholastic Aptitude Test (SAT) scores are often expressed in terms of percentile ranks; for example, John has an 87 percentile rank on verbal SAT. He scored higher than 87 percent of the people taking the test.

If one knows a person's score on a test, the test's mean and standard deviation, and one can assume that the variable is normally distributed, the person's percentile rank can be determined. The rule is simple: convert the score to $Z$ and determine the probability value in Appendix C. If $Z$ is positive then add .5 to probability and then multiply by 100 to get the percentile rank. If the $Z$ is negative, the probability is subtracted from .5 and multiplied by 100.

The process can also be reversed. If the percentile rank is known, the $Z$ score can be determined. One could use Appendix C, but it is simpler to use Appendix A. One first finds the percentile rank in the column denoted Proportion and then reads the $Z$ score from the column labeled Probit.

The percentile ranks are used, in part, to determine the recommended daily allowances (RDA) of vitamin intake. Nutritionists survey the population of healthy individuals and calculate the mean and variance of intake for a particular vitamin. Using the mean and variance and assuming normality, the researchers calculate the value of a score with a 97.5 percentile rank. This score is used as the recommended daily allowance (RDA) for many vitamins.

# Data Transformations

In Chapter 5 various data transformations are considered. Two of the transformations, the probit and percentile rank, described in that chapter can be clarified at this time.

## Proportions

The probit transformation is used for proportions and percentages. It is a two-stretch transformation used to remove the lower limit of zero and the upper limit of 1.00 in proportions and 100 in percentages. This transformation is based on the standard normal curve. The probit of a proportion refers to the $X$ axis of the normal curve for that proportion. So the probit transforms a proportion into a $Z$ score.

The probit transformation is illustrated in Figure 10.4. In the top diagram the shaded area contains .50 of the distribution, and the reading on the $X$ axis is 0.00. So the probit of .50 is zero. In the middle diagram the shaded area contains .975 of the distribution and the value on the $X$ axis is 1.96. So the probit of .975 is 1.96. In the bottom diagram the shaded area contains .25 of the distribution and the value on the $X$ axis is −.67. So the probit of .25 is −.67. (Sometimes a value of five is added to the probit to eliminate negative values.)

*FIGURE 10.4*   **Probit examples.**

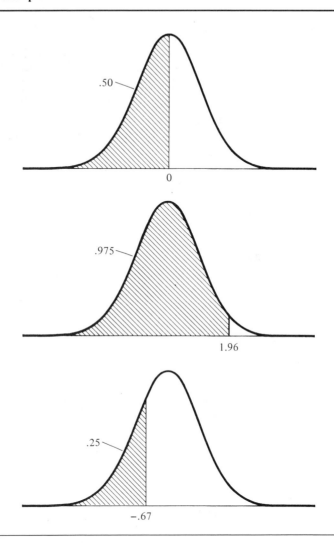

## Ranks

The normal distribution is also used to transform scores that have been rank ordered, but the underlying process that generated the scores is assumed to be normal. The transformation is called the *normalized ranks transformation*. It involves two steps. First, the ranks are transformed into percentile ranks by the formula presented in Chapter 5. The formula is

$$100\left[\frac{R - .5}{n}\right]$$

where $R$ is the rank order of the score and $n$ the sample size. Second, these percentile ranks are converted into Z scores using the probit transformation in Appendix A. The resulting values are the scores' normalized ranks.

The normalized ranks transformation can also be used to normalize the scores of any variable that is not normally distributed. First the scores are rank ordered from the smallest to largest. Then each score's percentile rank is computed. Finally, using the probit transformation in Appendix A, the percentile ranks are converted into Z values. A frequency distribution of the transformed scores is more nearly normally distributed than the untransformed scores.

The use of the normalizing transformation should be applied cautiously. If the obtained distribution appears implausible and there is good reason to believe that the variable is normally distributed, then a normalizing transformation may be helpful. It should not be routinely applied to nonnormal data, however.

As an example, consider the following sample.

$$15, \ 19, \ 21, \ 21, \ 34, \ 50, \ 52$$

These scores rank ordered are

$$1, \ 2, \ 3.5, \ 3.5, \ 5, \ 6, \ 7$$

Converting the scores to percentile ranks yields

$$7, \ 21, \ 43, \ 43, \ 64, \ 79, \ 93$$

Using the probit values in Appendix A, the normalized ranks are

$$-1.476, \ -.806, \ -.176, \ -.176, \ .358, \ .806, \ 1.476$$

# Summary

The *normal distribution* is a symmetric, unimodal, bell-shaped distribution. Because of its relative mathematical simplicity, it is commonly used in statistical work. Also because of the central limit theorem, many statistics have approximately a normal distribution. The *central limit theorem* states that the sum of numbers tends to be normally distributed as more numbers are summed.

The *standard normal distribution* is a normal distribution with a mean of zero and a variance of one. The standard normal is often referred to as the Z *distribution*. This distribution can be used to answer questions about the likelihood of certain types of events as well as to compute percentile ranks.

The normal distribution is used for various data transformations. Proportions can be transformed by the *probit transformation* and ranks by the *normalized ranks transformation*.

# Problems

1. Find the following probabilities from the $Z$ distribution.

   a. between 0.0 and .77
   b. between −.11 and 0.0
   c. between .33 and 1.35
   d. greater than .72
   e. between 0.0 and 2.04
   f. between −1.22 and 0.0
   g. between −1.48 and −.99
   h. less than −.38
   i. between −.46 and 1.13

2. Let IQ be a normally distributed variable with a mean of 100 and a standard deviation of 15. Find the probability that someone's IQ is

   a. between 100.0 and 115.0
   b. between 110.0 and 120.0
   c. less than 95
   d. between 90.0 and 100.0
   e. between 90.0 and 95.0
   f. less than 123

3. If $X$ is a normally distributed variable with a mean of 12 and a variance of 16 what is the probability of the following sets of events?

   a. $X$ between 10.0 and 12.0
   b. $X$ less than 11.0
   c. $X$ greater than 12.5
   d. $X$ between 11.0 and 14.0

4. Find the probit transformations of the following probabilities.

   a. .66     b. .10     c. .55     d. .34

5. Convert the following rank-ordered scores into normalized ranks.

   1, 2, 3, 4, 5, 6, 7, 8, and 9

6. Answer the following questions, assuming that the distribution is normal.

   a. How likely is it for someone to score at least 1.5 standard deviations above the mean or more?

b. How likely is it for someone to score lower than 1.75 standard deviations below the mean?

c. What is the probability of someone scoring between .5 standard deviations below the mean to .5 standard deviations above the mean?

7. Below are the rank-order scores of ten cities in the United States as rated by Rand McNally (Boyer & Savageau, 1985) on three dimensions.

| City | Transportation | Economics |
|---|---|---|
| Atlanta | 2 | 5 |
| Boston | 6 | 4 |
| Chicago | 5 | 10 |
| Cincinnati | 8 | 8 |
| Dallas | 7 | 1 |
| Denver | 4 | 2 |
| New York | 1 | 7 |
| Phoenix | 10 | 3 |
| Pittsburgh | 9 | 9 |
| San Francisco | 3 | 6 |

Convert the ranks to normalized ranks and average the normalized ranks for each city. Compare these averages to the means of the original ranks for each city.

8. Given that $X$ is normally distributed with a mean of $\mu$ and a variance of $\sigma^2$, it is true that $\bar{X}$ has a mean of $\mu$ and a variance of $\sigma^2/n$. Given this fact, if $X$ has a mean of 50 and a variance of 81, what is the probability that $\bar{X}$ will be between 51 and 49 if $n$ is 36?

9. Given that scores on the Scholastic Aptitude verbal test have a mean of 500 and a standard deviation of 100 and a normal distribution, what percentage of the population is outscored if the following scores are obtained?

a. 600    b. 700    c. 500
d. 750    e. 450    f. 350

10. If the probability that of being five units or more above the population mean is .25 and the distribution is normal, what is the standard deviation of the variable?

11. Given that weight for females 18–24 is normally distributed with a mean of 132 and a standard deviation of 27 (Stoudt, 1981), compute percentile ranks for the following weights:

a. 132    b. 150    c. 95
d. 139    e. 180    f. 100

12. For the following percentile ranks, determine the corresponding $Z$ values:

    a. 60     b. 31     c. 29
    d. 48     e. 79     f. 93

13. Explain the following statement: The $Z$ transformation does not make a distribution normal, but the normalized ranks transformation does.

14. Convert the following scores into normalized ranks.

    418, 423, 425, 425, 430, 435, 435, 440, 441

# 11 | *Special Sampling Distributions*

If a random sample of 25 persons was drawn from the population of college students and each person's athletic ability and intelligence were measured, a correlation between the two variables could be computed. Whatever the value of the correlation, a different value would have been obtained if a different sample of 25 persons had been drawn. Hence the correlation coefficient with a sample size of 25 varies from sample to sample. It therefore has a distribution of its own, called a sampling distribution.

When a statistic, such as a correlation coefficient, is computed, it is necessary to know its sampling distribution to be able to interpret it. If all statistics had radically different sampling distributions, data analysis would be an impossible task. Fortunately most statistics used in data analysis have one of four distributions. These four sampling distributions serve as important reference points in data analysis. Moreover, the sampling distributions of other statistics are closely approximated by these four distributions. That is, the statistics are not exactly distributed as one of the four, but one of the four can be used as a reasonable approximation. This chapter considers these four special sampling distributions.

The material presented here is quite abstract. It may be more useful to some students merely to skim the chapter now. Then, when the distributions are presented later, this chapter can be used as a reference.

## *The Standard Normal Distribution*

The first sampling distribution is one that was presented in Chapter 10: the standard normal distribution. The $Z$ or standard normal distribution is defined as follows: If $X$ is a normally distributed variable with mean $\mu$ and variance $\sigma^2$, then

$$\frac{X - \mu}{\sigma}$$

is also normally distributed, with a mean of zero and variance of one. The standard normal distribution is a normal distribution with a mean of zero and a variance of one.

Many statistics, especially those involving means, have a standard normal distribution when the distribution of scores from which the means were computed is normal.

If $X$ is normally distributed with mean $\mu$ and variance $\sigma^2$, the sample mean $\overline{X}$ is also normally distributed with mean $\mu$ and variance of $\sigma^2/n$, where $n$ is sample size. The quantity $\sigma/\sqrt{n}$ is called the standard error of the mean. This is a very important fact:

standard error    standard deviation divided by the
of the mean   =   square root of the sample size

This fact is only guaranteed if the observations are sampled randomly and independently. However, the formula for the standard error of the mean is true of any distribution, not just the normal.

In words, the variability of the mean equals the variability of the observations used to form the mean divided by the square root of the sample size. How far the sample mean is away from the population mean depends on the inherent variability in the population and the sample size. The larger the sample size, the closer on the average is the sample mean to the population mean. Consider separately the two special cases of $n$ equal to one and $n$ equal to infinity. If $n$ is one, the mean is a single observation and its variability is $\sigma^2$ divided by one, which remains $\sigma^2$. If $n$ is very large or infinite, $\sigma^2/n$ equals zero. This implies that the sample mean is identical to the parameter $\mu$ when the sample size is quite large.

The relationship between sample size and the standard error of the mean can be examined graphically. Consider the variable IQ, which is assumed to be normally distributed, with a mean of 100 and a standard deviation of 15. The standard error of the mean is $15/\sqrt{n}$. In Figure 11.1 are two sampling distributions of $\overline{X}$ for sample sizes of 10 and 40. Note how much more variable the sample mean is when $n$ is 10 and than when $n$ is 40.

If the population mean is subtracted from the sample mean and if this difference is divided by its standard error, the following quantity is obtained.

$$\frac{\overline{X} - \mu}{\sigma/\sqrt{n}}$$

If $X$ is normally distributed, the above expression has a standard normal or $Z$ distribution. Even if $X$ does not have a normal distribution, $\overline{X}$ has approximately a normal distribution given the central limit theorem, discussed in the

*FIGURE 11.1*    **Sampling distribution of intelligence mean ($\mu = 100$, $\sigma^2 = 225$) for sample sizes of 10 and 40.**

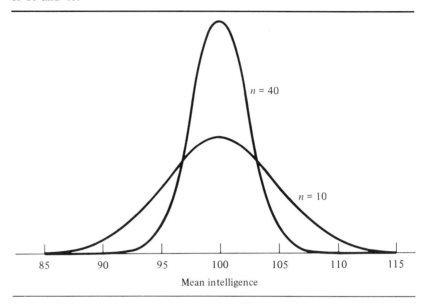

previous chapter. And, the larger $n$ is, the closer to normal is the distribution of $\bar{X}$.

Other quantities have standard normal distributions. Imagine that two samples are drawn from the same normally distributed variable with variance $\sigma^2$ and two sample means are computed. The two sample means are denoted as $\bar{X}_1$ and $\bar{X}_2$, and their sample sizes are $n_1$ and $n_2$, respectively. The quantity $\bar{X}_1 - \bar{X}_2$ is normally distributed with a mean of zero and a variance of

$$\sigma^2 \left( \frac{1}{n_1} + \frac{1}{n_2} \right)$$

given that the observations are normally distributed and independently and randomly sampled. It then follows that

$$\frac{\bar{X}_1 - \bar{X}_2}{\sigma \sqrt{\dfrac{1}{n_1} + \dfrac{1}{n_2}}}$$

has a $Z$ distribution.

Table 11.1 summarizes these facts about normally distributed variables. The quantities in the table have a normal distribution given that $X$ has a normal distribution.

*TABLE 11.1*     **Sampling Distributions That Are Normal**

| Variable | Mean | Standard Deviation |
|:---:|:---:|:---:|
| $X$ | $\mu$ | $\sigma$ |
| $\dfrac{X - \mu}{\sigma}$ | $0$ | $1$ |
| $\bar{X}$ | $\mu$ | $\dfrac{\sigma}{\sqrt{n}}$ |
| $\bar{X} - \mu$ | $0$ | $\dfrac{\sigma}{\sqrt{n}}$ |
| $\dfrac{\bar{X} - \mu}{\dfrac{\sigma}{\sqrt{n}}}$ | $0$ | $1$ |
| $\bar{X}_1 - \bar{X}_2$ | $0$ | $\sigma\sqrt{\dfrac{1}{n_1} + \dfrac{1}{n_2}}$ |
| $\dfrac{\bar{X}_1 - \bar{X}_2}{\sigma\sqrt{\dfrac{1}{n_1} + \dfrac{1}{n_2}}}$ | $0$ | $1$ |

# t *Distribution*

The second sampling distribution that is commonly used in statistical analysis is the $t$ distribution. Consider a normally distributed variable $X$. If a score is sampled from this population, and if the population mean is subtracted from this score, and then if this difference is divided by the population standard deviation, the resulting quantity is

$$\frac{X - \mu}{\sigma}$$

As was previously discussed, this quantity has a standard normal or $Z$ distribution. But what would happen if the population standard deviation or $\sigma$ is replaced with $s$, the sample estimate of the standard deviation? The quantity

$$\frac{X - \mu}{s}$$

does not have a Z distribution but rather has a *t* distribution.

The *t* distribution looks very much like the Z distribution. Like Z it has a mean of zero, is symmetric, and has bounds of plus and minus infinity. However, the *t* distribution is not as peaked as Z at zero, and so its tails are somewhat fatter than Z. These fatter tails make the variance of *t* greater than one. A *t* distribution looks like a bloated Z distribution.

Actually how closely the *t* distribution approaches the Z distribution depends on how closely *s* approaches $\sigma$. Recall that *t* differs from Z in that *s* is substituted for $\sigma$. Because *s* is a statistic and so it has sampling error, the quantity *s* does not equal $\sigma$. How close *s* is to $\sigma$ depends on what are called the *degrees of freedom* used to estimate *s*. The degrees of freedom for *s* are usually *n* − 1. As the degrees of freedom get larger, the *t* distribution approaches Z. For very large degrees of freedom, *t* and Z are virtually indistinguishable. Although there is only one Z distribution, there are many *t* distributions. These different *t* distributions are denoted by *t*(*df*), where *df* stands for degrees of freedom.

Many quantities have a *t* distribution. In fact, all of the statistics in Table 11.1 that have a standard normal or Z distribution have a *t* distribution when the sample standard deviation is substituted for the population standard deviation.

So the quantities

$$\frac{X - \mu}{\frac{s}{\sqrt{n}}}$$

and

$$\frac{\bar{X}_1 - \bar{X}_2}{s\sqrt{\frac{1}{n_1} + \frac{1}{n_2}}}$$

have a *t* distribution. In each case the *t* distribution involves a statistic that has a normal distribution minus its population mean divided by its estimated standard error. The facts concerning the *t* distribution are summarized in Table 11.2.

The *t* distribution is useful for testing hypotheses about means, and these facts will be used in Chapters 12 and 13. Also, it will be seen in Chapter 16 that the *t* distribution is used to test hypotheses concerning correlation and regression coefficients.

*TABLE 11.2*    **Various Statistics That Have $t$, $\chi^2$, and $F$ Distributions**[a]

$t$ Distribution

$$\frac{X - \mu}{s} \qquad \frac{X - \mu}{\frac{s}{\sqrt{n}}} \qquad \frac{\bar{X}_1 - \bar{X}_2}{s\sqrt{\frac{1}{n_1} + \frac{1}{n_2}}}$$

$\chi^2$ Distribution

$$Z^2 \qquad \frac{(n-1)s^2}{\sigma^2}$$

$F$ Distribution

$$\frac{s_1^2}{s_2^2} \qquad \frac{\chi_1^2/df_1}{\chi_2^2/df_2}$$

[a]The variable $X$ has a normal distribution with a mean of $\mu$ and a variance of $\sigma^2$. The samples from which $\bar{X}$ and $s^2$ are computed are randomly and independently drawn.

# Chi Square Distribution

Like $t$, the chi square distribution is closely related to $Z$. Consider the variable $X$, which has a normal distribution. To measure relative position a $Z$ score can be computed. To measure how deviant or unusual a $Z$ score is, it could be squared: $Z_i^2$. So if $\mu = 20$, $\sigma^2 = 100$, and $X_i = 22$, then $Z_i = (22 - 20)/10 = .2$ and $Z_i^2 = .04$. The variable $Z^2$ has a *chi square distribution*. The chi square distribution is symbolized by $\chi^2$.

All chi square distributions have a positive skew and a lower limit of zero. The lower limit must be zero since $Z^2$ must be positive.

It is possible to take repeated random and independent samples from $X$. Many $Z$ scores and $Z^2$ could be computed. The sum of $k$ independent $Z^2$ values has a chi square distribution with $k$ degrees of freedom. Thus, $\chi^2$ is not one distribution, but rather a family of distributions that differ by their degrees of freedom which equal the number of $Z$'s that are squared. The term $k$ is called the degrees of freedom of the $\chi^2$ distribution. The different chi square distributions are symbolized by $\chi^2(k)$.

A $\chi^2$ distribution with $k$ degrees of freedom has a mean of $k$ and a variance of $2k$. So a $\chi^2$ with 10 degrees of freedom has a mean of 10 and a variance of 20. The shape of the distribution of $\chi^2$ is positively skewed with a lower bound of zero. As the degrees of freedom get larger, the skew becomes less pronounced and $\chi^2$ with $k$ degrees of freedom approaches a normal distribution with a mean of $k$ and a variance of $2k$. So $(\chi^2 - k)/\sqrt{2k}$ has approximately a $Z$ distribution, if $k$ is appreciable, say greater than 20.

It can also be shown that

$$\frac{(n-1)s^2}{\sigma^2}$$

has a $\chi^2$ distribution with $n-1$ degrees of freedom. In words, the sample variance times $n-1$ divided by the population variance has a chi square distribution. Facts about $\chi^2$ are presented in Table 11.2.

The main use of the $\chi^2$ distribution is testing models concerning frequency data. The chi square distribution is used for this purpose in Chapter 17. Chi square is also used to test hypotheses of differences between medians which is described in Chapter 18, and differences between correlations which is described in Chapter 16. Often when $\chi^2$ is used, it is used to approximate a sampling distribution.

# F Distribution

Again let $X$ be a normally distributed variable. Two random, independent samples of $X$'s of sizes $n_1$ and $n_2$ are chosen from the population. The variances, $s_1^2$ and $s_2^2$, are computed from each sample. If the ratio of $s_1^2/s_2^2$ is computed, the quantity would have an $F$ distribution. Like $\chi^2$, $F$ is positively skewed with a lower bound of zero. (Because variances are always nonnegative, their ratio must be nonnegative.) Its peak comes near the value of one.

Like $\chi^2$, $F$ is actually a family of distributions. To determine which $F$ distribution is being referred to, one needs to know the degrees of freedom of the numerator, $s_1^2$, and the degrees of freedom of the denominator, $s_2^2$. The number of degrees of freedom equals the denominator of the formula for the variance for each sample. So for an $F$, the degrees of freedom on the numerator and the degrees of freedom on the denominator must be determined. A given $F$ distribution is denoted as $F(df_n, df_d)$ where $df_n$ are the degrees of freedom on the numerator and $df_d$ the degrees of freedom on the denominator.

The $F$ distribution is closely related to $\chi^2$. It can be shown that $F$ is a ratio of two independent $\chi^2$ variables, each divided by its degrees of freedom.

The main use of the $F$ distribution is to test hypotheses about means. A procedure for doing so, called analysis of variance, is described in Chapters 14 and 15.

# Relation Between Sampling Distributions

The four major sampling distributions, though distinct, are closely related. One aspect that ties the four together is the normal distribution. The $Z$ distribution is itself normal. The $\chi^2$ distribution is based on the sum of squared scores that are normally distributed. The $t$ distribution also presumes that the numerator is normally distributed. Finally, the $F$ distribution can be viewed as the ratio of two independent variances whose scores are normally distributed. Hence, the normal distribution is the starting point for all four of these distributions.

But the four distributions are more closely linked. In some cases their distributions are identical. There are four major equivalences between pairs of the major sampling distributions. First a $\chi^2$ with one degree of freedom is identical to a $Z^2$ value. So the probabilities in Appendix C can be used to determine the probability of various $\chi^2$ events. For instance, what is the probability of obtaining a $\chi^2$ value with one degree of freedom larger than 1.0? The answer to this question lies in finding the probability of obtaining a value of $Z$ greater than 1.0 or less than −1.0. The answer, using Appendix C, is $1 - (2)(.3413) = .3174$.

The second fact linking the distributions is that a $t$ with an infinite number of degrees of freedom equals $Z$. This holds because if $t$ has an infinite number of degrees of freedom its denominator becomes $\sigma$, and so $t$ becomes $Z$.

The third fact is that a $t$ with $q$ degrees of freedom when squared equals an $F$ with one degree of freedom on the numerator and $q$ on the denominator. This fact is not so obvious. If a $t^2$

$$\frac{(\bar{X} - \mu)^2}{s^2/n}$$

is examined, both the numerator and the denominator estimate $\sigma^2/n$, and so both are estimates of the same population variance.

The final fact is that an $F$ with $q$ and infinite degrees of freedom is identical to a $\chi^2$ distribution with $q$ degrees of freedom which is divided by $q$. This is due the fact that $F$ equals the ratio of two $\chi^2$'s divided by their degrees of freedom, and so

$$F(df_n, df_d) = \frac{\chi^2(df_n)/df_n}{\chi^2(df_d)/df_d}$$

A $\chi^2/df$ with an infinite degrees of freedom equals one. Substituting this fact into the denominator of the above equation, the result is

$$F(df_n, \infty) = \frac{\chi^2(df_n)}{df_n}$$

These facts are summarized in Table 11.3.

*TABLE 11.3*    **Equivalences Between the Four Major Sampling Distributions**

$$\chi^2(1) = Z^2$$
$$Z = t(\infty)$$
$$t(q)^2 = F(1, q)$$
$$F(q, \infty) = \chi^2(q)/q$$

# *Summary*

Every statistic has a sampling distribution. There are four major sampling distributions. Many statistics' sampling distributions exactly or nearly exactly correspond to one of the four distributions. They are $Z$, $t$, $\chi^2$, and $F$.

The $Z$ or standard normal distribution is a normal distribution with a mean of zero and a variance of one. If the observations have a normal distribution, the sampling distribution of the sample mean also has a normal distribution. The standard deviation of the sampling distribution of the mean is the standard deviation of the observations divided by the square root of the sample size.

The $t$ distribution is identical to $Z$, but the denominator is the sample standard deviation and not the population standard deviation. The $t$ distribution looks like $Z$ but it is less peaked and has fatter tails. Like $Z$, $t$ has a mean of zero and is symmetrically distributed.

The $\chi^2$ distribution with $k$ degrees of freedom is a positively skewed distribution with a mean of $k$ and a variance of $2k$. A $\chi^2$ statistic can be viewed as the sum of $k$ independent $Z^2$ values. The value $k$ is called the degrees of freedom.

The $F$ distribution is the ratio of two independently computed variances drawn from the same normally distributed population. Like $\chi^2$, $F$ is positively skewed with a lower limit of zero. The peak in the $F$ distribution is near one.

These sampling distributions of $Z$, $t$, $\chi^2$, and $F$ are routinely used in testing statistical models. It is this topic of testing models that is presented in the next chapter.

# *Problems*

1. Let $X$ be a normally distributed variable with a mean of 40 and a standard deviation of 9. What is the distribution of the following statistics?

   a. $\bar{X}$          b. $s^2$
   c. $(\bar{X} - \mu)/s$     d. $(X - \mu)^2/\sigma^2$

2. If $X$ is a variable with a mean of 20 and a variance of 49, determine the standard error of the mean for sample sizes of

   a. 100     b. 10     c. 1000     d. 50

3. If $Y$ has a mean of 80 and a variance of 64, what would $n$ have to be for the standard error of the mean to be 1.00?

4. Describe the distribution of the sample mean with a sample size of 49 if the numbers are drawn from a normal distribution with a mean of 10.0 and a variance of 64.

5. If $X$ is normally distributed with a mean of 20 and a variance of 100, determine the probability that $\overline{X}$ is greater than 22 if

   a. $n = 25$  b. $n = 50$  c. $n = 100$  d. $n = 200$

6. Using the $Z$ distribution determine the following probabilities.

   a. $\chi^2(1) > 1.44$  b. $\chi^2(1) > 4.00$  c. $t(\infty) > 1.00$

7. Compute the standard error of the mean for the following cases.

   a. $\sigma^2 = 100$, $\mu = 50$, $n = 25$
   b. $\sigma^2 = 9$, $\mu = 0$, $n = 16$
   c. $\sigma^2 = 25$, $\mu = -5$, $n = 64$
   d. $\sigma^2 = 81$, $\mu = 4$, $n = 100$

8. Using the facts in Table 11.3, show that $t(\infty)^2 = \chi^2(1)$.

9. Describe how $\chi^2$ and $F$ are similar and different in their shape. Consider the following.

   a. skew                    b. central tendency
   c. upper and lower limit   d. variability

# 12 | *Testing a Model*

How many times have you wished for one more hour to study for a midterm exam to increase your chances of getting an A? All you needed was that one extra hour of study. But does one more hour of study really make that much of a difference?

To address this question, a teacher could instruct students to come to a two-hour midterm examination with their notes and study materials. Then half the students would be given an opportunity to study and the other half would engage in some irrelevant activity such as watching soap operas. After an hour they would all take the midterm. The teacher would then see whether those who had the extra hour of study did better on the midterm than those who did not have the extra time. The teacher would know if one hour of study makes a difference, or more accurately, how much of a difference.

What was just described is a research study. Research can be used to help answer important questions such as the following.

1. Does divorce affect children's social development?
2. Does psychotherapy improve one's mental health?
3. Does television violence make children more aggressive?
4. Does bilingual education retard or accelerate the performance of children in schools?

Research is more than "men in tweed suits, cutting up frogs, paid for by huge government grants" (Woody Allen, in the movie *Sleeper*). Research helps us in understanding the world around us. Research in the behavioral and social sciences often involves testing statistical models.

## What Is a Model?

A statistical model is a formal representation of a set of relationships between variables. Statistical models contain an outcome variable that is the focus of study. In studies of weight change, the outcome is weight change; in studies of psychotherapy, it is adjustment; in studies of education, one often-studied outcome is reading skill. In research, the outcome of interest is called the *dependent variable*. A dependent variable is what is supposed to change in response to changing events. In statistical models, it is written on the left-hand side of the equal sign.

The variable that brings about changes in the dependent variable is called the *independent variable*. Examples of independent variables are type of psychotherapy, drug dosage, and age. The dependent variable is assumed to be some function of the independent variable. How the independent variable affects the dependent variable is represented on the right-hand side of the equation.

Sometimes the designation between independent and dependent variable depends on the variables under study and the researcher's theoretical orientation. For instance, researchers study the relationship between self-esteem and academic performance. Some designate self-esteem as the independent variable and academic performance as the dependent variable. Others reverse the designations.

Other variables that cause the dependent variable to vary besides the independent variable are represented by the *residual variable*. The residual variable represents the degree to which the researcher is ignorant about what causes the dependent variable. The residual variable is sometimes referred to as error or noise.

In simple equation form the model is

$$\begin{array}{c}\text{dependent}\\\text{variable}\end{array} = \begin{array}{c}\text{effect of the}\\\text{independent}\\\text{variable}\end{array} + \begin{array}{c}\text{residual}\\\text{variable}\end{array}$$

By far the vast majority of models in the social and behavioral sciences take on this general form. The only major difference is that most models have more than one independent variable on the right-hand side, but the basic specification of the model remains the same.

In this model the independent variable and the residual variable are *added* together to cause the dependent variable. This is not the only way that the independent and the residual variable could combine. For instance, they could multiply. However, an additive formulation is by far the simplest and most common formulation. Most of the standard statistical models assume that the effect of the independent variable and the residual variable add together.

Instead of expressing the model as an equation, the model could be just as easily specified by a diagram; arrows could be drawn from cause to effect, as follows:

$$\begin{array}{c}\text{independent}\\\text{variable}\end{array} \longrightarrow \begin{array}{c}\text{dependent}\\\text{variable}\end{array} \longleftarrow \begin{array}{c}\text{residual}\\\text{variable}\end{array}$$

A representation of a model that uses arrows is called a *path diagram*.

To better understand a statistical model, consider the following example. A researcher, investigating the effect of owning a personal computer on grade-point average, made arrangements to give a personal computer to each of 30 students. Another 30 students served as a comparison group and they did not receive computers. One year later, the researcher measured the grade-point averages of the two groups. The independent variable is owning or not owning a personal computer, and the dependent variable is grade-point average. The residual variable represents any other causes of grade-point average besides owning a personal computer. The residual variable is the way of accounting for the fact that all students with computers (or without computers) do not have the same grade-point average.

Statistical models are a bit more complicated than the independent variable and the residual variable causing the dependent variable. In most models a constant is added to every person's score. In equation form,

$$\frac{\text{dependent}}{\text{variable}} = \text{constant} + \frac{\text{effect of the}}{\text{independent}} + \frac{\text{residual}}{\text{variable}}$$

In many models the constant term corresponds to the population mean of the dependent variable.

The residual term is a necessary part of a statistical model. It is also called the disturbance, error, or noise. The mean of the residual variable is set to zero. This is not a mathematical necessity but is merely a convention. Also, it is very often assumed that the residual variable has a normal distribution with a given variance. It should be noted that it is the residual and not the dependent variable that is assumed to have a normal distribution. Also, it is commonly assumed that the variance of the residual does not vary as a function of the independent variable. Many of the assumptions of the model refer to the residual variable. In sum, the residual is a normally distributed variable with a zero mean.

# Model Comparison

In this chapter the logic of model testing is presented. It is first illustrated for one type of model and then the general procedure is discussed. A very simple model is one in which the dependent variable equals a constant plus the residual variable.

$$\frac{\text{dependent}}{\text{variable}} = \text{constant} + \frac{\text{residual}}{\text{variable}}$$

It is this model that will be considered in this chapter. There is no independent variable effect in the model. This model will be called the *complete model* because later an even simpler version of the model is considered. The model has two parameters: the constant and the standard deviation of the residual variable. The constant in this model is the population mean of the dependent variable, and the standard deviation of the residual variable is the standard deviation of the dependent variable.

In models, a parameter can be fixed or free. If the parameter is free, it must be estimated from the data. If it is fixed, then the researcher sets the parameter to some a priori value. In this chapter, the constant is set to some a priori value, as in the following examples.

1. Eighty-seven persons are asked to learn pairs of words like "cat-package." They are then presented the word "cat" and are asked to recall whether the other word was "package" or "glass." Because there were two alternatives for each word pair, the probability of being correct is .5. There are ten such trials, and if subjects were only guessing, they would be expected to be correct on five of the ten trials. The dependent variable is the number correct out of ten and the a priori constant is 5.0.
2. Twenty persons aged 50 were asked at what age they would ideally prefer to retire. The researcher sought to compare the preferred age of retirement of persons to the standard retirement age of 65. The dependent variable is preferred retirement age and the a priori constant is 65.0.
3. Robinson and Hastie (1985) had 40 undergraduates read a mystery story "The Poisoned Philanthropist," in which there are five suspects. The subject had to estimate the probability that any given suspect was guilty. For each subject the five probabilities are summed. The dependent variable is total probability and the a priori constant is 1.00. (The mean probability of the subjects was over 2.00.)
4. In an extrasensory perception study, twelve proclaimed psychics were asked to guess whether a head or a tail results when a coin is flipped. The coin was flipped 30 times. By chance, each psychic would be correct 15 times. The dependent variable is the number of correct judgments and the a priori constant is 15.0.

When the constant is fixed and not free, the researcher is specifying a simple and restricted version of the model:

$$\frac{\text{dependent}}{\text{variable}} = \text{constant} + \frac{\text{residual}}{\text{variable}}$$

This model is restricted in that the constant is not free to take on any value but instead is fixed or set to some a priori value. A model in which a parameter of the complete model is fixed is called the *restricted model*. In the restricted model under consideration, the constant parameter is fixed or restricted to some a priori value. The hypothesis of interest is whether the parameter equals the value to which it is restricted. This hypothesis is referred to as the null hypothesis. The *null hypothesis* is the constraint on the complete model that is present in the restricted model. (It is common to symbolize the null hypothesis as $H_0$.) For instance, for the model that the psychics are guessing, the null hypothesis is that the constant equals 15.0. Although in testing the restricted model the interest is primarily the null hypothesis, it is not uniquely tested. Rather, the plausibility of a model, of which the null hypothesis is a part, is evaluated. The *alternative hypothesis* is the hypothesis that is true if the null hypothesis is false. (It is common to symbolize the alternative hypothesis as $H_A$.) It states that the constant is free to take on any value. So, for the psychic example, the alternative hypothesis is that the constant does not equal 15.0.

Model testing is always model comparison. The restricted model is compared to the complete model. The restricted model is a simpler model which is identical to the complete model except that one of the parameters in the restricted model is fixed to some value. If the restricted model is not contradicted by the data, then the restricted model is retained for reasons of simplicity, and the more complicated complete model is not considered. However, if the restricted model is contradicted by the data, the restricted model is rejected, and the more complicated complete model must be adopted.

To illustrate the difference between the complete and restricted models, consider the three presented in Table 12.1. Model I is the simplest of the three. In it, the dependent variable is not caused by any independent variable. In Model II the dependent variable is caused by variable $A$, and in Model III it is caused by both variables $A$ and $B$. Considering III as the complete model, II would be a restricted model for III. The restriction present in Model II is that

*TABLE 12.1*   **Illustration of Complete and Restricted Models**

---

Model I

dependent variable = constant + residual

---

Model II

dependent variable = constant + effect of independent variable $A$ + residual

---

Model III

dependent variable = constant + effect of independent variable $A$ + effect of independent variable $B$ + residual

---

independent variable *B* does not cause the dependent variable. Considering II as the complete model and I as the restricted model, the restriction present in Model I is that variable *A* has no effect. So, Model II can be considered either as a complete or restricted model: If II is compared to III, it is a restricted model; if compared to I, it is a complete model.

Consider the hypothetical data in Table 12.2 from the experiment with twelve psychics. A coin was flipped 30 times and each time each "psychic" guessed whether it came up heads or tails. Because there are two sides to a coin, pure guessing would lead to accuracy on 15 trials (or 1/2 times 30). So, if there were a large number of supposed psychics who were only guessing, they would on the average be correct on 15 out of 30 trials. But there is not a large number—only twelve. The question is whether the numbers in Table 12.2 are compatible with the view that the psychics are fakers who are just guessing. The mean of the twelve numbers is 16.0 and the standard deviation is 2.04.

Although the psychics did not do a stunning job at the task, their results seem to be better than chance. Only one had an exactly chance performance of 15. Of the remaining eleven, there were eight who did better than chance and only three who did worse than chance. The mean of the twelve is 16.0, a full one "guess" better than chance. The conclusion might be drawn that the psychics did better than chance.

However, if they were merely guessing, then about half the time they would appear to do better than chance and about half the time they would appear to do worse than chance. Even if it is believed that the twelve were just guessing, it is totally unrealistic to expect each psychic to be correct exactly 15 out of 30 trials or even for the sample mean of the twelve psychics to be exactly 15. Just because the sample mean is greater than the chance value of 15.0 does not necessarily refute the view that the supposed psychics were just guessing. Sampling error is to be expected, and so it would be expected that they would score better than chance about half the time. At issue is whether the value of 16.0 obtained by the psychics is within the limits of reasonable sampling error.

In the restricted model, the constant is set at 15.0. The restricted model presumes that the psychics are guessing. In the complete model the constant may be any value, and so it is compatible with the view that the psychics are not guessing.

If the restricted model were true (that psychics are guessing), then the

*TABLE 12.2*  **Guesses of Twelve Psychics (Hypothetical Data)**

| | | | |
|---|---|---|---|
| 15 | 17 | 19 | 13 |
| 16 | 16 | 18 | 16 |
| 19 | 14 | 13 | 16 |

sample mean of the number of correct guesses should be near 15.0. It happens that this particular sample mean is 16.0, one unit greater than the a priori value of 15.0. At issue is whether 16.0 is near enough to 15.0 to be explained by sampling error. The standard deviation of the guesses can be used to gauge how near 15.0 the sample mean should be. Assuming that the psychics were guessing, the smaller the standard deviation, the nearer the sample mean should be to 15.0. Also as the sample size gets larger, the sample mean should be nearer to 15.0. So, as the standard deviation gets smaller and the sample size larger, the sample mean should approach its a priori value.

Both the standard deviation and the sample size are in the formula for the standard error of the sample mean minus an a priori constant. As is presented in the previous chapter, the standard error of the mean minus a constant equals the standard deviation of the observations divided by the square root of the sample size. The difference between the sample mean and the a priori mean can be divided by its standard error to obtain

$$\frac{\bar{X} - M}{s/\sqrt{n}}$$

where $\bar{X}$ is the sample mean, $M$ the a priori mean, $n$ the sample size, and $s$ the standard deviation of the observations. This value normalizes the difference between the sample mean and the presumed population mean to take into account sample size and variability.

For the psychic example, $\bar{X}$ is 16.0, $M$ is 15.0, $n$ is 12, and $s$ is 2.04. The sample mean minus its a priori value divided by its standard error is as follows:

$$\frac{16.0 - 15.0}{2.04/\sqrt{12}} = 1.698$$

Thus, the sample mean is 1.698 standard errors above the mean. The question now is just how unlikely is this type of outcome. If $\bar{X}$ was ten standard errors above or below the a priori constant, it would be known almost for certain that the psychics were not guessing because it is virtually impossible to obtain a value ten standard errors above the mean. Alternatively, if it were only one standard error or less above the mean, it is still plausible to believe that they are guessing. But the value of 1.698 standard error above the mean for the psychic example is ambiguous. The "psychics" did better than chance, but it is not clear whether their success might have been due to sampling error.

# The Test Statistic and Its Sampling Distribution

The quantity $(\bar{X} - M)/(s/\sqrt{n})$ is called the *test statistic*. Of prime concern is how unusual is a test statistic of 1.698. To determine exactly how unlikely a value like 1.698 is, the distribution of the quantity $(\bar{X} - M)/(s/\sqrt{n})$ must be known. The quantity $(\bar{X} - M)/(s/\sqrt{n})$ is computed from sample data and so it is a statistic. As described in Chapter 9, the distribution of a statistic is called a *sampling distribution*. Given the restricted model, if the residual variable is normally and independently distributed, then $(\bar{X} - M)/(s/\sqrt{n})$ has a $t$ distribution with $n - 1$ degrees of freedom. Figure 12.1 shows the theoretical $t$ distribution for eleven degrees of freedom.

As can be seen in Figure 12.1, the $t$ distribution is a symmetric unimodal distribution whose mean is zero. Its variance depends on its degrees of freedom and is always greater than one for finite degrees of freedom. As the degrees of freedom increase, the variance of $t$ approaches one. Consequently, the tails of the $t$ distribution are a bit fatter than the standard normal or $Z$ distribution. The number of degrees of freedom for $t$ in this case is $n - 1$. A $t$ value may be denoted by $t(df)$ where $df$ stands for degrees of freedom. So for the psychic example, the $df$ are twelve minus one, or eleven.

Because $t$ is a continuous distribution, the probability that $t(11)$ exactly equals any particular value, such as 1.698, is zero. What is needed is not the probability that $t$ is 1.698 but rather the probability of obtaining a value of 1.698 or greater. At issue is the probability of obtaining a value at least as large as the test statistic.

There are two ways the restricted model could be wrong. The population mean could be larger than the a priori value or it could be smaller. For the psychic example, they could do better than chance (better than 15), which they did, or they could have performed worse than chance (worse than 15). So if the null hypothesis is wrong, there are two directions or sides that it could be wrong.

If the null hypothesis is false, either the psychics could do better than chance or worse than chance. Only one of the two may be plausible. For the particular example, it does not seem very reasonable that the psychics could

**FIGURE 12.1**    **The $t$ distribution with 11 degrees of freedom.**

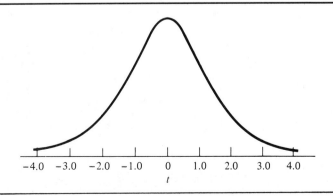

be operating at a level worse than chance. However, if such a result did occur, it should be considered as unusual. So even though there is little reason to expect the psychics to do worse than chance, it remains a possibility, and both alternative hypotheses need to be considered: that $(\bar{X} - M)/(s/\sqrt{n})$ is very positive or very negative. For the psychic example the probability of obtaining a value greater than 1.698 or less than −1.698 must be determined.

If the restricted model were clearly false, the value of $(\bar{X} - M)/(s/\sqrt{n})$ would tend to be either very positive or very negative. From Figure 12.1 it can be seen that if the restricted model were false, the value of *t* would fall in either *tail* of the *t* distribution. It is for this reason the test is called *two-tailed*. Although it is not recommended, a researcher might wish to consider only one direction or tail. For instance, only the probability that $(\bar{X} - M)/(s/\sqrt{n})$ is greater than 1.698. Such a test is called a *one-tailed test*. A one-tailed test is not recommended because if the value of the test statistic is quite unusual but in the wrong direction, most researchers would still consider it significant. Also, almost all computer programs output two-tailed *p* values.

As has been stated, under the restricted model, the test statistic $(\bar{X} - M)/(s/\sqrt{n})$ has a *t* distribution with *n* − 1 or eleven degrees of freedom. Using the *t* distribution it can be determined how likely a value greater than 1.698 or less than −1.698 actually is. Such a value would occur by chance about 12% of the time or about one out of eight times. It must now be decided whether 12% of the time is sufficiently unusual to reject the restricted model.

# Significance Level and p Value

Researchers who test statistical models have established a fairly standard, though arbitrary, criterion for judging how unusual a result must be to reject the restricted model. They have, by informal convention, required that the result and even more extreme results must occur no more than 5% of the time before the restricted model is rejected. The question now becomes: Given the restricted model, how often will the absolute value of $(\bar{X} - M)(s/\sqrt{n})$ be 1.698 or more? If it would occur 50% of the time, the result would not be considered unusual. But if it only occurs once in 20 times, the result would be unusual.

This 5% criterion is said to be the *significance level*. It is the standard of proof that is required for the restricted model to be deemed implausible. Other standards are also used. A common alternative standard is the .01 (or 1%) significance level. A result is judged to be improbable if it would occur by chance only once in 100 times. More stringent levels of once in 1000 are sometimes used, and less stringent rules of once in ten and even once in five are infrequently used. The choice of the significance level depends on the type of error that the researcher is more willing to accept. (See later section on errors in model comparison.)

The significance level is symbolized by the Greek letter alpha ($\alpha$). Conventionally, alpha is fixed to .05 or once in 20. *The usual significance level used in research is .05.* There is nothing magical about .05, just as there is nothing magical about setting the legal definition of being drunk at 0.1% blood alcohol. Some cutoff must be set and for various reasons the .05 is the value taken for alpha. The .05 significance level often means that the test statistic must be more than twice as large as its standard error to be significant.

A *p value* is the probability of obtaining a value equal to or more extreme than the test statistic. The *p* value for the psychic experiment is .12. If the *p* value is less than or equal to the significance level, then the null hypothesis is rejected and the test statistic is said to be *statistically significant*. If the *p* value is greater than the significance level, the null hypothesis is retained and the test statistic is said to be *statistically insignificant*. For the psychic example, because .12 is greater than .05, the null hypothesis is retained.

The significance level is usually set at .05. How is the *p* value determined? Computer packages routinely calculate the *p* value of the test statistic. Without a computer, one can use Appendix D to determine the approximate *p* value of a test statistic distributed as *t*.

To use Appendix D, one first determines the degrees of freedom of the *t* statistic. For this model the degrees of freedom equal the sample size less one or $n - 1$. One locates in the first column in Appendix D the degrees of freedom, n − 1. If the exact value for degrees of freedom is not in the first column, one uses the closest value that is smaller than the actual degrees of freedom. One "rounds down" to the nearest value. For instance, 105 is not in the table and so 100 would be used. One then reads across the row and finds the value that is the closest to the test statistic without being larger than the test statistic. One then reads up the column to determine the approximate *p* value. The exact *p* value is always less than or equal to the approximate *p* value. Thus, this method results in a conservative estimate of the *p* value. The numbers in the table are called *critical values* because they are the values that the test statistic must exceed to be statistically significant.

The 1.698 value for the psychic example does not exceed the .05 level of significance. The null hypothesis is retained and the 1.698 value is judged to be not significant. So on the basis of this data set, there is no reason to believe that the "psychics" have any special powers beyond mere guessing.

If one wishes to consider only one tail of the *t* distribution (a one-tailed test), the sign of the test statistic must match the prediction of the researcher before it is tested. If it does match, then the *p* value should be divided in half. (Note, the *p* value and not the significance level is divided by two.) For instance, if it is considered only that the psychics would do better than chance and not worse, then the test statistic must be positive. Because it is, the *p* value would be .06. If $\bar{X}$ had been less than 15.0 (even a lot less), the restricted model would be retained. As was stated earlier, one-tailed tests are not recommended.

For the quantity $(\bar{X} - M)/(s/\sqrt{n})$ to be distributed as $t$ with $n - 1$ degrees of freedom, it must be assumed that the residual variable has a normal distribution and that observations are randomly and independently sampled.

Even if the normality assumption is violated moderately, there is only a slight effect on the $p$ values. Thus, unless the numbers are highly skewed or bimodal, one need not worry about the normality assumption. The reason for this is the central limit theorem described in Chapter 10. As sample size increases, the distribution of $\bar{X}$ approaches a normal distribution regardless of the shape of the distribution of the scores used to compute $\bar{X}$.

The random and independent sampling assumptions are more important for testing hypotheses concerning the constant. Random sampling ensures that persons are representative of the population. The independence assumption requires that persons do not interact with one another, be observed only once, and that the person be the sampling unit.

# The Summary of the Logic of Model Comparison

The logic of model testing involves the following steps. First, a model is specified from theory that contains the parameter of interest. This is the *complete model*. A restricted version of the same model is constructed with some reasonable constraint on the parameter of interest. The constraint is called the *null hypothesis*, and the model with the constraint is called *the restricted model*.

The model describes the behavior in the population. Sample data are gathered from the population. The researcher computes a statistic from the sample data. The statistic, called the *test statistic*, has a distribution such as $Z$, $t$, $\chi^2$, or $F$ if the restricted model is true. Given the distribution, the probability of obtaining a value as or more extreme than the test statistic can be determined. This probability is called the *p value*. The $p$ value can be exactly computed by using a computer program or it can be approximated by using tables. If the $p$ value is less than the *significance level*, which is usually set at .05, the null hypothesis is rejected and the test statistic is said to be *statistically significant*. If the $p$ value is greater than the significance level, the restricted model and null hypothesis are retained and the test statistic is said to be *not significant*. This information as applied to the test for psychic powers is summarized in Table 12.3.

Model testing can be viewed as a series of "let's assume" and, "given all this" statements. First, *let's assume* that the null hypothesis is true. Second, *let's assume* a restricted model which contains the null hypothesis is true. Third, from the data a number called the test statistic is computed. Fourth, *given all this*, the test statistic has a sampling distribution. Fifth, *given all*

***TABLE 12.3***    **Steps in Model Testing Illustrated for the Test of Psychic Powers in Table 12.2**

Complete Model

$$\frac{\text{number}}{\text{correct}} = \text{constant} + \frac{\text{residual}}{\text{variable}}$$

Restricted Model

$$\frac{\text{number}}{\text{correct}} = 15.0 + \frac{\text{residual}}{\text{variable}}$$

| Null Hypothesis | Alternative Hypothesis |
| :---: | :---: |
| constant $= 15.0$ | constant $\neq 15.0$ |

Test Statistic

$$t(n - 1) = \frac{\overline{X} - M}{s/\sqrt{n}}$$

$$t(11) = \frac{16.0 - 15.0}{2.04/\sqrt{12}} = 1.698$$

*p* Value
exact .12
approximate .20

*this,* if the test statistic's *p* value is less than or equal to the significance level (usually set at .05), the null hypothesis is rejected. The complete model is never directly tested, rather it is the restricted model that is tested. If the restricted model is judged to be implausible, it is rejected and the complete model is adopted.

A second example is used to apply these ideas. A researcher in a school district wants to determine whether the children in the district score above national norms on a test. The norm on the test is 200. The scores of nine children randomly and independently sampled are 240, 230, 220, 190, 220, 200, 250, 230, and 190. The complete model is

test score = constant + residual

and the restricted model is

test score = 200 + residual

The mean of the nine scores is 218.89 and the standard deviation is 21.47. The test statistic is

$$t(8) = \frac{218.89 - 200}{21.47/\sqrt{9}}$$

which equals 2.639. Using Appendix D, the test statistic 2.639 yields a *p* value of .05. (The exact *p* value is .030.) Because the *p* value is less than or equal to the significance level of .05, the test statistic of 2.639 is statistically significant at the .05 level. The null hypothesis that the constant equals 200 is rejected. Because 218.89 is above 200, it is thus concluded that the children in the district score above the national norm of 200.

For this example, the truth of the assumption of random sampling is essential in the test of the restricted model. If the students were not randomly sampled but only the district's brightest students were studied, the conclusion that students in the district score above the norm would be unjustified. Also, if the students shared answers on the test, then the sample data would not be independent, and thus the conclusion would be unjustified.

# Errors in Model Comparison

There are two major types of errors that can be made in the comparison of statistical models. To understand these errors, four hypothetical yet possible results of testing a restricted model must be considered. The restricted model can be actually true or it can be false. For instance, the psychics could be just guessing or they could be true psychics. Of course, one never knows with perfect certainty whether any model is valid or not and so some idealized knowledge is being considered. The results of the statistical analysis can lead to rejection of the restricted model or its retention. For instance, the psychics could be operating significantly above chance or they could be operating at chance levels. Table 12.4 lists these four outcomes.

*TABLE 12.4*    **Four Possible Results of Model Testing**

| Reality | Statistical Analysis | |
| --- | --- | --- |
| | Retain Restricted Model | Reject Restricted Model |
| Restricted Model True | Retain a True Model | Type I Error (alpha) |
| Restricted Model False | Type II Error (beta) | Reject a False Model |

For two of the outcomes in Table 12.4, the correct conclusion is drawn. In the top left-hand cell, the restricted model is correctly retained. For instance, the psychics are fakers and it is also concluded that their mean is not significantly above chance. In the bottom right-hand cell, a false restricted model is correctly rejected. For instance, the psychics are true psychics and it is also concluded that their performance is above chance. In both cases the sample data and the statistical test mirror reality. In the best of all worlds, one would hope to make the correct decision every time. However, statistical tests of models do not allow for inferences about the nature of reality with total certainty. Statistical logic never brings with it certainty; rather, statistical logic results in only a probability.

There are two types of errors. The first is the error of falsely rejecting a restricted model that is actually true. This is called a Type I error. For instance, if the psychics were fakers, it might be falsely concluded that their performance is above chance. The second error is to retain the restricted model that is actually false. This is a Type II error. For instance, if the psychics were true psychics, it might be mistakenly concluded that they are not performing significantly above chance levels.

The probability of making a Type I error is called alpha and is identical to the significance level. Alpha is usually set at .05 or one out of 20. The probability of making a Type II error is symbolized by *beta, β*. Its value is not set by the researcher like alpha, but rather it is largely determined by the number of persons in the study. Beta is then smaller if more persons are studied. The probability of correctly concluding that the restricted model is false is called *power*. Power then equals one minus the probability of making a Type II error. In Chapters 13 and 16, methods for determining power are presented.

There are two other important errors in model testing that need to be considered. One is to draw the incorrect conclusion when the restricted model is rejected. If the *p* value of the test statistic is less than or equal to the significance level, then the null hypothesis is rejected. But what is rejected is the restricted model and not necessarily the null hypothesis. It can be that some other aspect of the restricted model is false. For instance, it might be that the assumption that the residual variable has a normal distribution is false. Just because the restricted model is false does not imply that the null hypothesis is false. This error will be referred to as an *assumption violation*.

Second, when the restricted model is rejected, the null hypothesis is rejected. Just because the null hypothesis is rejected, it does not mean that the result necessarily supports the researcher's theoretical position. For instance, it may be that the psychics do not perform at chance levels, not because they do better than chance but because they do worse than chance. If it is concluded that the psychics did better than chance, an error would be made. This is an error about the direction in which the null hypothesis is false. This error will be referred to as *choosing the wrong direction*, and can be avoided by careful examination of the data.

# *Remainder of the Book*

So far only the simplest model in which the dependent variable equals the constant plus the residual variable has been considered. The restricted version of this model constrains the constant to equal some fixed value determined a priori. The remainder of the book considers more complex models. These more complex models are outlined in Table 12.5.

In Chapter 13 a model is presented in which the dependent variable is caused by an independent variable, and the independent variable is a nominal variable with only two levels. In Chapter 14 the independent variable is still a nominal variable, but it may have more than two levels. In Chapter 15 there are two nominal independent variables that both cause the dependent variable. In Chapter 16 both the independent variable and the dependent variable are measured at the interval level. In Chapter 17, both the independent and dependent variables are not at the interval level of measurement, but rather are at the nominal level of measurement. Finally, in Chapter 18, the dependent variable is at the ordinal level of measurement.

When the independent variable is at the nominal level of measurement, it is possible to have the same persons in all conditions or have different persons.

*TABLE 12.5*    **Taxonomy of Models**

---

*Dependent Variable at the Interval Level of Measurement*

No independent variable
   Test of the constant, Chapter 12

Nominally measured independent variable
   One independent variable
      Dichotomous: two-sample *t* test, Chapter 13
      Multilevel: one-way analysis of variance, Chapter 14
   Two independent variables: two-way analysis of variance, Chapter 15

Intervally measured independent variable
   Regression, Chapter 16

*Dependent Variable at the Nominal Level of Measurement*

No independent variable: chi-squared goodness-of-fit test, Chapter 17

Nominally measured independent variable: chi-squared test of independence, Chapter 17

*Dependent Variable at the Ordinal Level of Measurement*

Nominally measured independent variable
   Dichotomous: Mann-Whitney *U* test, Chapter 18
   Multilevel: Kruskal-Wallis analysis of variance, Chapter 18

Ordinally measured independent variable: rank-order coefficient, Chapter 18

---

When different persons are in each group, the design is said to have *independent groups*. When the same persons are in each group, the design is said to have *nonindependent groups;* nonindependent groups with two groups are commonly referred to as *paired groups;* and multiple-group designs in which groups are nonindependent are called *repeated measures* designs.

A nonindependent group design can come about even when different persons are at each level of the independent variable. Whenever there is some factor that links together observations across the different conditions, the design can be considered a nonindependent groups design. So, if persons are from the same family, litter, or class, the design can be considered a nonindependent groups design.

The statistical procedures presented in Table 12.5 presume that the groups are independent. In Table 12.6 are the statistical procedures for nonindependent groups. For each procedure, the independent variable is a nominal variable. Different statistical tests are used for nominal, ordinal, and interval-dependent variables.

# Summary

A *model* is a formal set of relationships between variables. The *dependent variable* in the model is the outcome and the *independent variable* is presumed to bring about the change in the dependent variable. Most models have a *constant* that is added to every score and a *residual variable* that is added to the constant. The residual variable represents all other causes of the dependent variable besides the independent variable.

The model under consideration is called the *complete model*. In model testing a *restricted* version of the model is proposed that is identical to the

---

*TABLE 12.6*   **Statistical Procedures for Nonindependent Groups**

*Intervally Measured Dependent Variable*

   Dichotomous independent variable: paired t-test, Chapter 13
   Multilevel independent variable: repeated-measures analysis of variance,
      Chapter 15

*Nominally Measured Dependent Variable*

   McNemar test, Chapter 17

*Ordinally Measured Dependent Variable*

   Dichotomous independent variable: sign test, Chapter 18
   Multilevel independent variable: Friedman two-way analysis of variance,
      Chapter 18

---

complete model, except that there is one constraint on one parameter of the complete model. This constraint is referred to as the *null hypothesis*.

In this chapter the complete model assumes that the dependent variable is equal to a constant plus the residual variable. In the restricted model the constant is fixed or set to some a priori value. If the restricted model were true, then the sample mean should differ from the a priori value within the limits of sampling error. Given the restricted model and the assumptions of random sampling, independence, and normality, the following quantity is distributed as $t$ with $n - 1$ degrees of freedom.

$$\frac{\bar{X} - M}{s/\sqrt{n}}$$

where $\bar{X}$ is the sample mean, $M$ the a priori constant, $s$ the sample standard deviation, and $n$ the sample size. The quantity $(\bar{X} - M)/(s/\sqrt{n})$ is called the *test statistic*.

The probability of obtaining a value as or more extreme than the test statistic is called the *p value*. If the *p* value is less than or equal to the significance level, then it is concluded that the null hypothesis is false and the test statistic is said to be *statistically significant*. If the *p* value is greater than the significance level, the null hypothesis is retained and the test statistic is said to be *not statistically significant*. The standard significance level is .05.

There are two major errors in model testing. A *Type I error* is rejecting the restricted model when, in fact, it is true. A *Type II error* is a failure to reject the restricted model when it is not true. The probability of making a Type I error is denoted as *alpha* and is set by the significance level. The probability of making a Type II error is denoted as *beta* and is determined by the sample size and other factors. Two other errors are (a) rejecting the restricted model not because the null hypothesis is false but because the assumptions are false, and (b) interpreting that the null hypothesis is false in the wrong direction.

# *Problems*

1. For the following degrees of freedom, find the critical value for the following significance levels for the *t* distribution.

|  | *df* | *Alpha* |
|---|---|---|
| a. | 12 | .01 |
| b. | 23 | .05 |
| c. | 76 | .001 |
| d. | 209 | .02 |
| e. | 17 | .10 |
| f. | 48 | .05 |

2. For the following $t$ values and degrees of freedom determine the $p$ value.

   a. $t(24) = -1.583$     b. $t(78) = 1.990$     c. $t(24) = 3.145$
   d. $t(19) = -3.117$     e. $t(28) = 2.963$     f. $t(77) = 1.942$

3. A prison official wishes to determine whether the inmates in a prison score above a national norm on a personality test. The scores of nine randomly chosen inmates are

   $$15, \ 18, \ 23, \ 41, \ 19, \ 25, \ 31, \ 43, \ 51$$

   The norm is 25. Are prisoners above the norm?

4. In a memory experiment, guessing would lead to a score of 10. The scores of six subjects are

   $$9, \ 15, \ 12, \ 17, \ 13, \ 10$$

   Is it reasonable to assume that subjects are guessing at this task?

5. What value would $(\overline{X} - M)/(s/\sqrt{n})$ have to equal or exceed (ignoring sign) to be significant at the .05 significance level for the following degrees of freedom?

   a. 15     b. 59     c. 25     d. 190

6. Explain the difference between a Type I and a Type II error.

7. Test the null hypothesis that the population means equals 50.

   $$63, \ 51, \ 43, \ 55, \ 60, \ 36, \ 40, \ 57, \ 54$$

8. Eight married couples were asked what proportion of the housework each did. The proportions were summed for both members. Test a restricted model that the constant is 100.

   $$109, \ 121, \ 98, \ 95, \ 105, \ 112, \ 123, \ 134$$

9. For the following two models, which is the restricted and which is the complete model?

   $$\frac{\text{dependent}}{\text{variable}} = \text{constant} + \frac{\text{effect of the}}{\text{independent}} + \frac{\text{residual}}{\text{variable}}$$

   $$\frac{\text{dependent}}{\text{variable}} = \text{constant} + \frac{\text{residual}}{\text{variable}}$$

   What is the restriction in the restricted model?

10. Imagine a psychologist who is interested in subliminal perception. Stimuli, either an A or B, are flashed on a tachistoscope. The subject responds

by saying whether an A or B was flashed. Each subject is presented with 15 trials. The number correct for ten subjects are

$$10, 15, 12, 6, 11, 9, 8, 12, 11, 8$$

a. Determine the constant if subjects were guessing.
b. Test the restricted model that subjects are only guessing in this task.

11. Fifteen different groups of subjects were asked to estimate the population of Phoenix, Arizona, in units of 100 thousands. The estimates are as follows:

| 9 | 8 | 12 | 10 | 8 |
|---|---|----|----|---|
| 6 | 9 | 6 | 9 | 6 |
| 11 | 11 | 7 | 7 | 5 |

The correct answer is 8 (hundred thousand). Do groups on this task tend to significantly over- or underestimate the population of Phoenix?

12. A company advertises that its cars get 30 miles per gallon gas mileage. An inquiring car dealer measures the miles per gallon of 20 cars. She obtains the following:

$$30, 25, 28, 30, 27, 34, 41, 25, 28, 30,$$
$$28, 35, 31, 34, 32, 31, 26, 31, 24, 32$$

What should she conclude about the manufacturer's claim?

# 13 | *The Two-Group Design*

The prototypical research study is the two-group design. Persons are in one of two groups. For example, one group receives an experimental treatment and another group serves as a control group. Examples of treatments are a new drug to cure cancer, an instructional program for disadvantaged children, a procedure to change attitudes, a pain relief strategy for childbirth, and an exercise program. The *control group* is a group of persons who are assumed to be identical to those in the treatment group except that individuals in the control group do not receive the treatment.

In this chapter a model for the analysis of the two-group design is presented. Also discussed are the design considerations concerning the assignment of persons to treatment groups and the formation of the two groups. Measures of the differences between the two groups are presented and statistical power considerations are discussed.

## Model

The model for the two-group design is fairly simple. The model is

$$\begin{matrix} \text{dependent} \\ \text{variable} \end{matrix} = \text{constant} + \begin{matrix} \text{effect of the} \\ \text{independent} \\ \text{variable} \end{matrix} + \begin{matrix} \text{residual} \\ \text{variable} \end{matrix}$$

The restricted model is identical to the above model except that the independent variable has no effect on the dependent variable. Hence

$$\begin{matrix} \text{dependent} \\ \text{variable} \end{matrix} = \text{constant} + \begin{matrix} \text{residual} \\ \text{variable} \end{matrix}$$

The restricted model in this chapter is identical to the complete model discussed in the previous chapter.

Consider the terms in the complete model. The dependent variable is what changes or varies. It is the outcome that the treatment variable is designed to alter. The constant is the average score of persons in both groups. (The test of the constant was extensively discussed in Chapter 12.) The independent variable is a nominal variable with two levels—that is, a dichotomy. For instance, one level is the treatment group and the other is the control group. The independent variable is the variable that causes the dependent variable to change or vary. The residual variable represents variation in the dependent variable that is not explained by the independent variable. The residual variable is forced to have a mean of zero.

As an example, consider an experiment in which a researcher randomly assigns ten infants to one of two groups. All infants spend 20 minutes with a stranger. Then the infants are put into a situation with a number of fear-arousing stimuli. For five of the ten infants the stranger is present (present condition), and for the other five the stranger is absent (absent condition). The researcher measures the number of fear responses of the ten infants.

| Present | Absent |
|---------|--------|
| 6 | 12 |
| 4 | 6 |
| 3 | 8 |
| 7 | 10 |
| 4 | 7 |

The means are 4.8 for present and 8.6 for absent.

The central question with the two-group design is whether the independent variable affects the dependent variable. If the independent variable did not affect the dependent variable, as in the restricted model, then the *population* means for the two groups are equal. Because of sampling error, the two *sample* means are not exactly equal even if the restricted model is true. For the example it is not known whether the difference between the means of 4.8 and 8.6 can be explained by sampling error.

The two groups will be designated 1 and 2. The sample means will be designated $\bar{X}_1$ and $\bar{X}_2$, with sample sizes of $n_1$ and $n_2$, respectively. At issue is the amount of sampling error in the quantity $\bar{X}_1 - \bar{X}_2$ given that the population means are equal.

In Chapter 9 the idea that statistics vary was presented. The standard deviation of a statistic is called the standard error. As shown in Chapter 11, the standard error of the difference between two means randomly and independently sampled from the same population is

$$\sigma\sqrt{\frac{1}{n_1} + \frac{1}{n_2}}$$

where $\sigma$ is the population standard deviation of the observations and $n_1$ and $n_2$ are the sample sizes of the two sample means.

To estimate the standard error of the difference between two means an estimate of $\sigma$ is needed. In terms of the model, $\sigma$ is the population standard deviation of the residual variable in the restricted model. The variance of the residual variable can be estimated by computing the variance of scores within each of the groups. Thus the variance is computed for each of the two groups. These variances are denoted as $s_1^2$ and $s_2^2$. Both of these are unbiased estimates of $\sigma^2$, the variance of the residual variable. Some way is needed to average or pool these variances to produce the most efficient estimate of the variance. When averaging variances the most *efficient* way to do so is to weight each variance by its denominator, $n - 1$. That is, weighting by $n - 1$ results in an estimate with the smallest standard error. The most efficient estimate of $\sigma^2$ is called $s_p^2$, given as follows:

$$s_p{}^2 = \frac{(n_1 - 1)s_1{}^2 + (n_2 - 1)s_2{}^2}{n_1 + n_2 - 2}$$

For the example, the variance for the present group is 2.7 and the variance for the absent group is 5.8. The pooled variance or $s_p^2$ is

$$\frac{(4)(2.7) + (4)(5.8)}{5 + 5 - 2} = 4.25$$

Now that there is an estimate of $\sigma^2$, the standard error of the difference between two means sampled from the same population can be estimated. That estimate is

$$\sqrt{\frac{(n_1 - 1)s_1{}^2 + (n_2 - 1)s_2{}^2}{n_1 + n_2 - 2}\left[\frac{1}{n_1} + \frac{1}{n_2}\right]}$$

This formula states how variable the difference between means would be if the two sets of observations were drawn from the same population. Such an assumption is made in the restricted model. So to evaluate whether the independent variable causes the dependent variable (the complete model), a model in which the independent variable has no effect is tested. Given this restricted model, the population means of the two groups are equal. To test the restricted model and the null hypothesis of equal population means, the difference between sample means is compared to its standard error. For the example the standard error of the difference between two means is

$$\sqrt{4.25\left[\frac{1}{5} + \frac{1}{5}\right]} = 1.304$$

The difference between the means is $4.8 - 8.6 = -3.8$, and its standard error is 1.304. Their ratio is $-3.8/1.304 = -2.914$.

If it were known how the quantity

$$\frac{\bar{X}_1 - \bar{X}_2}{s_p\sqrt{\dfrac{1}{n_1} + \dfrac{1}{n_2}}}$$

was distributed, it could be more precisely estimated how unusual the difference between the means is relative to its standard error. For the example, the question is how unusual is $-2.914$, the mean difference divided by its standard error. It happens that

$$\frac{\bar{X}_1 - \bar{X}_2}{s_p\sqrt{\dfrac{1}{n_1} + \dfrac{1}{n_2}}}$$

has a $t$ distribution with $n_1 + n_2 - 2$ degrees of freedom, given a series of assumptions that are discussed in the following section.

As discussed in the previous two chapters, the $t$ distribution closely resembles the $Z$ or standard normal distribution except that it is less peaked and has fatter tails. The tails are fatter because the denominator of $t$ is the statistic $s$, whereas the denominator of $Z$ is the parameter $\sigma$. How fat the tails of $t$ are depends on how precise the estimate of the variance is, and that precision depends on the degrees of freedom. There is then a family of $t$ distributions, which vary by their degrees of freedom.

For the two-group study, the degrees of freedom are $n_1 + n_2 - 2$. Because $n_1 + n_2$ equals the number of persons in the study, the degrees of freedom are the total number of persons in the study less two. It is less two because the means for the two groups are estimated.

In the two-group study, to test the restricted model that the independent variable has no effect on the dependent variable, the test statistic is computed

$$\frac{\bar{X}_1 - \bar{X}_2}{s_p\sqrt{\dfrac{1}{n_1} + \dfrac{1}{n_2}}}$$

The test statistic is then compared to the critical values, ignoring sign, in Appendix D for the appropriate degrees of freedom. As is explained in the previous chapter, if the exact degrees of freedom are not in the table, one rounds *down* to the nearest value and then determines whether the test statistic, ignoring sign, is larger than any critical value for the degrees of freedom. If it is, the null hypothesis of equal population means is rejected, and the test statistic is said to be statistically significant. The $p$ value is determined by noting the largest value in the table that the test statistic exceeds. The $p$ level is given by the column heading. If the test statistic, ignoring sign, is smaller than all values in the table, then the difference between means is not statistically significant and the null hypothesis of equal population means is retained.

For the example, the *df* are eight, and a $-2.914$ value is statistically significant at the .02 level of significance. If a computer is used to compute the test statistic, the exact $p$ value is .0195. The null hypothesis of equal means is rejected.

What has just been described is a two-tailed test. The null hypothesis is rejected if $t$ is either very positive or very negative. Instead, the researcher may wish to perform a one-tailed test, which requires that he or she specify a priori which mean should be larger. For the example, theory might say that fear responses should be lower when a familiar adult is present. If the sample means confirm the prediction, one proceeds as in a two-tailed test, but the $p$ value is cut in half. As was explained in Chapter 12, one-tailed tests are not recommended because the researcher would probably still believe the result was statistically significant even if the result were not in the predicted direction. For instance, it could have happened that fear responses increased when the stranger was present.

# *Assumptions*

There are three major assumptions for the two-group $t$ test, all of which refer to the residual variable:

1. normal distribution,
2. homogeneous variance, and
3. independence of observations.

The score on the residual variable for a given person is estimated by taking each person's score and subtracting the group mean.

## *Normality*

The residual term must have a normal distribution for

$$\frac{\bar{X}_1 - \bar{X}_2}{s_p \sqrt{\dfrac{1}{n_1} + \dfrac{1}{n_2}}}$$

to have a $t$ distribution under the restricted model. To test this assumption a histogram is constructed for the set of observations minus the group mean and determine whether their shape is normal. (The normality assumption refers to residual variable and not to the dependent variable itself.) If the distribution is skewed, then the one-stretch transformations discussed in Chapter 5 should be considered; or if it is bounded on both sides, a two-stretch transformations may be needed. When any transformation of the dependent variable is contemplated, it must be determined whether transformation will render the dependent variable uninterpretable.

In practice, the normality assumption is not usually examined. With small sample sizes, it is difficult to detect that the distribution is nonnormal. With large samples, the effect of nonnormality does not disturb the $t$ test very much. The reason for this is the central limit theorem: As the sample size increases, the distribution of $\bar{X}$ becomes more normal even though the distribution of the scores may not be normal. Given the central limit theorem, it is also true that the distribution of $\bar{X}_1 - \bar{X}_2$ approaches normality as $n_1$ and $n_2$ increase.

## Equal Variances

The two-sample $t$ test requires a pooling or an averaging of the two-sample variances, $s_1{}^2$ and $s_2{}^2$. The equal variance assumption requires that the population variances of both groups are equal to the same value. Although the means may differ, the variances are assumed not to. A procedure is needed for determining whether the sample variances are significantly different from one another. That is, a way is needed to determine whether the sample variances differ by more than the amount expected given sampling error. It turns out that the ratio of the two sample variances is distributed as $F$ given the null hypothesis that their population variances are equal. The $F$ test is presented in the next chapter.

If the variances differ significantly, there are a number of strategies available. First, one might consider transformations to promote equal variances. For instance, if the data are skewed, the one-stretch transformations described in Chapter 5 may make the variances in the groups more nearly equal. Second, if $n_1$ is nearly equal to $n_2$, the problem can be safely ignored, because the $t$ test is only slightly affected by unequal variances. However, if the variances and sample sizes are unequal, caution must be exercised in interpreting $p$ values. It must be determined which group has the larger variance. The $t$ test results in too many Type I errors if the group with the larger variance also has the smaller $n$. The $t$ test results in too few Type I errors if the group with the larger variance has the larger $n$.

## Independence

The scores of persons on the residual variable are assumed to be uncorrelated. *Independence* requires that if one residual score is positive, the residual score of any other observation is no more likely to be positive or negative. There are a number of factors that aid in determining whether the observations are likely to be independent from each other. They concern (a) whether repeated observations are taken from the same person, (b) what the sampling unit is, and (c) whether there is social contact between the persons that generate the observations. Below is a consideration of each of these conditions.

First, whatever it is that generates the data is referred to as a *unit*. The unit may be a person, animal, or group of persons. For the two-group $t$ test each observation must be from a different unit. So each unit, be it a person or nerve cell, is measured only once. There must not be repeated measures from the same unit. For instance, assume that a person is measured before undergoing therapy and after and so each person is measured twice. These two observations are not likely to be independent. Also, if a behavior modification study is conducted using the same person, then the same person provides all the data and the scores are not likely to be independent. There are analysis procedures for these kinds of data structures, but they are different from the two-group *t* test.

Second, independence can be enhanced through the design of the study. The sampling unit of the study should be the unit that provides the observation. That is, each unit should enter the study singly. For instance, if married couples were in the study and both members provide data, then the independence assumption is likely to be violated because a husband is likely to be more similar to his wife than to someone else's wife. The observations must not come in pairs as in couples, friends, littermates, or twins. If they do, other statistical methods must be used.

Third, to achieve independent observations persons in the study must not influence others' responses. Once subjects enter the study, they should, if possible, be kept isolated so that they do not influence each other. They should not communicate with each other or know any other subject's response on the dependent variable. If they do communicate or observe each other, their observations are likely to be correlated because they may imitate or influence each other.

The effect of nonindependent observations is to bias the estimate of residual variance and, therefore, the standard error of the difference between means. Usually, though not always, the direction of bias in the two-group design is to make the estimate of the standard error too small, which makes researchers falsely confident that the means are significantly different. Unlike the normality and equal variance assumptions, even moderate violation of the independence assumption has very serious consequences. The failure to meet the independence assumption invalidates the *p* values.

One solution to the problem is to design the research so that observations are independent. If this is not possible, it may be possible to find a different way of analyzing the data to meet the assumption. There is one case in which observations are nonindependent, but data can be reanalyzed to meet the independence assumption. It is the case of paired observations, which is now discussed.

# *Paired* t *Test*

Some two-group studies contain observations in which pairs of observations are linked. Each observation in one group is paired or linked to one other observation in the other group. Consider some examples:

1. Twenty-five persons enter a stop-smoking program. The number of cigarettes smoked before entering the program and six months after completion of the program are measured. There are two groups of observations: those before the treatment and at a six-month follow-up. The observations are paired, that is, each person provides two scores, one in the pretreatment group and another in the posttreatment group.
2. A researcher is interested in the different ways in which fathers and mothers treat their infant children. A total of 40 infants are observed, each with its father and mother. Again, the observations are paired. Each infant provides two data points, one in the mother group and one in the father group.
3. Pairs of rats from the same litter are used in an experiment on learning. One rat from the litter has an operation that is supposed to facilitate learning. The other rat does not have the operation. A total of 20 pairs are studied. Each litter provides two observations, one of which is in the operation group and the other in the nonoperation group.

These three examples illustrate the key element of the paired design. Each observation is linked to one and only one other observation in the other group. Thus, each of $n$ observations in one treatment group is linked to one of $n$ observations in a second group. The degree to which the observations are linked can be measured by a correlation coefficient.

When observations are linked in this way, the independence assumption is violated because the linked observations are likely to be correlated. This lack of independence makes the two-group analysis that has been described in this chapter no longer valid because normally the $t$ test will yield more Type I errors than it should. It happens that the one-group $t$ test described in the previous chapter can be applied to the paired two-group design.

The key idea is to compute a difference score, always subtracting the scores in the same way. For example, the pretreatment score is always subtracted from the posttreatment score. The test that the mean of the difference score equals zero is equivalent to the hypothesis that the two groups have the same mean. The use of the one-sample $t$ test with difference scores is called a *paired t test*.

In a paired $t$ test, each of the $n$ pairs of scores is differenced. The mean of the differences, $\bar{X}_D$, and the standard deviation of the differences, $s_D$, are computed. Then, the quantity

$$\frac{\bar{X}_D}{s_D/\sqrt{n}}$$

has a $t$ distribution with $n - 1$ degrees of freedom, given the restricted model. Recall that $n$ is the number of pairs and not the number of scores. If the $t$ is statistically significant, the restricted model that the independent variable has no effect on the dependent variable is rejected.

# Computational Formulas

Earlier $s_p^2$ was defined as the pooled or average variance across the two groups. Its formula is

$$\frac{(n_1 - 1)s_1^2 + (n_2 - 1)s_2^2}{n_1 + n_2 - 2}$$

Given the definition of $s_p^2$ the above formula can be rewritten as

$$\frac{\sum X_1^2 - (\sum X_1)^2/n_1 + \sum X_2^2 - (\sum X_2)^2/n_2}{n_1 + n_2 - 2}$$

This is the formula generally used to compute $s_p^2$. So for the example, the formula for $s_p^2$ is

$$\frac{126 - (24)^2/5 + 393 - (43)^2/5}{5 + 5 - 2} = 4.25$$

The formula for $1/n_1 + 1/n_2$ can be more simply computed by

$$\frac{n_1 + n_2}{n_1 n_2}$$

These computational formulas can be entered into the formula for $t$ resulting in the following formula.

$$\frac{\bar{X}_1 - \bar{X}_2}{\sqrt{\left(\dfrac{n_1 + n_2}{n_1 n_2}\right)\dfrac{\sum X_1^2 - (\sum X_1)^2/n_1 + \sum X_2^2 - (\sum X_2)^2/n_2}{n_1 + n_2 - 2}}}$$

Ordinarily $t$ is computed to three decimal places.

The computational formula for the paired $t$ test is

$$\frac{\bar{X}_D}{\sqrt{\dfrac{\sum D^2 - (\sum D)^2/n}{n(n - 1)}}}$$

where $D$ is the difference between linked scores and $n$ the number of linked scores.

# Effect Size and Power

Even if the restricted model is rejected, it is not known how large the treatment effect is. Statistical significance cannot be equated with scientific significance because statistical significance depends on theoretically unimportant factors such as sample size. For instance, consider two studies that attempt to reduce cigarette smoking. It is possible for the $t$ statistic for one study to be 8.433, yet the treatment reduces cigarette smoking by two cigarettes. Whereas in a second study the *t* statistic could be only 2.108, yet the program reduces the level of the smoking by 20 cigarettes. This could happen if the first study has 16,000 subjects, the second only 10 subjects, and the pooled standard deviation is 15.

## Effect Size

The most commonly used measure of how much the treatment affects the dependent variable is a measure called *effect size* or *Cohen's d*. The quantity *d* is defined as

$$\frac{\mu_1 - \mu_2}{\sigma}$$

The numerator is the difference between the population means. The denominator is the standard deviation of the residual variable. The size of *d* can range from negative to positive infinity, but values larger than two are quite rare. Most values of *d* vary from zero to one.

Cohen's *d* is like a *Z* score in that its denominator is a standard deviation. It measures how different the means of the two groups are relative to the standard deviation within groups. Cohen (1977) describes three different effect sizes. They are

small *d* = .2
medium *d* = .5
large *d* = .8

A small effect is so small that to detect it one needs a statistical analysis. An example of an effect size of this magnitude is the difference in height between 15- and 16-year-old girls (Cohen, 1977). A medium effect is one that is large enough to see without doing statistical analysis. It is reflected by the difference in height between 14- and 18-year-old girls. A large effect is so large that statistics are hardly even necessary. It is reflected by the size of the difference in height between 13- and 18-year-old girls.

To better understand the meaning of the *d* measure of effect size, imagine that you are considering which of two movies to see one night. Assume that you have access to a survey that was done that measured the extent to which college students enjoyed each of the two movies. If there was sufficient information in the survey you could measure the *d* for the two movies. The value of *d* would indicate the degree to which one movie was enjoyed by more college students than the other. If *d* was small, say .2, that would indicate that if you saw both movies, the probability that you would prefer the one others found to be enjoyable would be .56. If *d* was .5, the probability that you would prefer the more popular movie would be .64 and if *d* was .8, the probability would be .71. (The probabilities of .56 for small, .64 for moderate, and .71 for large are determined from the standard normal distribution.)

In research areas where empirical data are lacking, one must make an intelligent guess of the value of $d$ in order to estimate power. If previous studies have been conducted, $d$ can be estimated by

$$\frac{\overline{X}_1 - \overline{X}_2}{s_p}$$

or the mathematically equivalent formula of

$$t\sqrt{\frac{1}{n_1} + \frac{1}{n_2}}$$

When the sample sizes are equal ($n_1 = n_2 = n$), $d$ equals

$$t\sqrt{\frac{2}{n}}$$

So for the example, the effect size equals $-2.914\sqrt{2/5}$, or $-1.84$. If the paired $t$ test is used, the estimate of $d$ is

$$t\sqrt{\frac{2(1 - r)}{n}}$$

where $t$ is the paired $t$, $n$ the number of pairs, and $r$ the correlation between the paired scores.

## Power

One reason for determining the value of $d$ is that $d$ must be known to ascertain the power of the two-sample $t$ test. In the previous chapter, power is defined as the probability of rejecting the restricted model when it is false. It also equals one minus the probability of making a Type II error. The power of the two-group or two-sample $t$ test depends on three factors: the difference between means, the residual variance, and the sample sizes. The difference between means can be increased by choosing more extreme treatments. Instead of comparing one week of psychotherapy versus none, one year could be compared to none. Although power can be enhanced in this way, generalizibility may suffer because extreme groups may be atypical of everyday treatments.

The residual variance can be reduced by choosing to study persons who are relatively similar. Animal researchers minimize variability by choosing organisms from the same strain. Variability can also be reduced by carefully measuring the dependent variable. A third way to reduce the residual variance is to use a paired design. The residual variance of the paired design is reduced to the degree that there is a correlation between paired observations. A paired design tends to have more power than an unpaired design.

Increasing sample size enhances power in two ways. First, it increases degrees of freedom of the *t* test, so the difference between means need not be as large to be significant. Second, it reduces the standard error of the mean because $1/n_1 + 1/n_2$ is part of the formula. If the total sample size is fixed, the way to minimize the standard error of the mean difference is by having $n_1$ equal to $n_2$.

For a given value of Cohen's *d*, a given *n*, and a given alpha, power can be determined. In Table 13.1 is the power for the two-sample *t* test for small, medium, and large effect sizes. They are given for the .05 level of significance. The *n* in the table is the sample size in each of the two groups. So, the total sample size of the study is 2*n*. The entry in the table is the power multiplied by 100. So if a researcher contemplates doing a study with 20 persons in each group and the effect size is moderate, from Table 13.1 the chances of rejecting the null hypothesis is .33. This means that for every three times that the experiment is done, the null hypothesis is rejected once.

For a given *d*, alpha, and level of power desired, the *n* that is needed for that power can be determined. These sample sizes are given in Table 13.2. For instance, if *d* is .5 with an alpha of .05 and power of .80, a researcher would need 64 subjects in each of the two conditions.

Adjustments need to be made to *d* if a paired *t* test is planned and Tables 13.1 or 13.2 are employed. In this case the new *d'* value is equal to $d/\sqrt{(1 - r)}$, where *r* is the degree of correlation between the paired observations. Also, if the sample sizes are unequal, the *n* in the tables must be adjusted. The new *n*, denoted *n'*, equals $2n_1n_2/(n_1 + n_2)$.

# Design Considerations

Before the results of a two-group experiment can be interpreted, various design issues must be considered. Two important questions are, first, the rule

**TABLE 13.1**   **Power Tables for the Two-Sample *t* Test,[a] with Alpha = .05 and $n = n_1 = n_2$**

|  | Effect Size (Cohen's *d*) | | |
| --- | --- | --- | --- |
| *n* | .2 | .5 | .8 |
| 10 | 7 | 18 | 39 |
| 20 | 9 | 33 | 69 |
| 40 | 14 | 60 | 94 |
| 80 | 24 | 88 | 99 |
| 100 | 29 | 94 | 99 |
| 200 | 51 | 99 | 99 |

[a]Taken from Cohen (1977).
NOTE: Entry in the table is the probability of rejecting the null hypothesis times 100 for a given effect size and sample size.

*TABLE 13.2*    **Sample Size Required for a Two-Sample *t* Test[a] to Achieve a Given Level of Power for a Given Effect Size and Alpha of .05**

|          | Effect Size (Cohen's *d*) | | |
| -------- | ---- | ---- | ---- |
| Power    | .2   | .5   | .8   |
| .25      | 84   | 14   | 6    |
| .50      | 193  | 32   | 13   |
| .60      | 246  | 40   | 16   |
| .70      | 310  | 50   | 20   |
| .80      | 393  | 64   | 26   |
| .90      | 526  | 85   | 34   |
| .95      | 651  | 105  | 42   |
| .99      | 920  | 148  | 58   |

[a]Taken from Cohen (1977).
NOTE: Entry in the table is the sample size for each of the two groups.

by which persons are assigned to groups and, second, the manner in which the two groups are formed.

There are two basic ways in which persons are assigned to groups. They can be assigned randomly or on the basis of some variable.

*Random assignment* requires that each person has the same probability of being assigned to a given group. Random assignment can be accomplished by coin flip, dice roll, or a random number table. With random assignment, each person has an equal probability of being assigned to a given group. In the absence of treatment effects, the difference between the means is totally explained by sampling error. However, if the means differ by a statistically significant amount, that difference can be attributed to the independent variable. The advantage of a random assignment rule is that it is known that the treatment means differ either due to sampling error or due the independent variable.

A nonrandom rule is one in which persons are assigned to groups on the basis of some variable. For instance, persons are assigned to a surgical procedure on the basis of some clinical test. To analyze the design correctly with a nonrandom assignment rule, that variable must be controlled in analysis. One way in which this can be accomplished is through multiple regression, which is described in advanced statistical texts. Most of the time when assignment is nonrandom, however, it is not known exactly which variable made the groups different and so it is not known which variable to control in the analysis. If the variable that determines assignment to levels of the independent variable cannot be controlled, then when the means differ it is not known whether the treatment made them different or whether the variable that assigned persons to groups made the groups different. A random assignment rule is preferable to a nonrandom rule in order to establish the causal connection between the independent variable and the dependent variable.

It is important to distinguish random *assignment* from random *selection*. Random selection refers to the entry into the study, whereas random assignment refers to the entry into levels of the independent variable. Random selection of persons means that the sample is representative of the population from which it is sampled. Random assignment yields strong causal inference.

The second major design consideration is the formation of the two groups. There is more than one way to study the effects of the independent variable. For instance, consider a study of the effects of jogging: Two groups of persons would be formed, a jogging group and the other a control (that is, no jogging) group. There are many ways to form the two groups:

1. Marathon runners are compared to persons who engage in no physical exercise.
2. Persons who jog ten miles a week are compared to those who swim four times a week.
3. Rats who run mazes for two hours a day are compared to rats who are confined to a cage all day.

The advantage of plan 1 is that the maximum effect of jogging could be estimated, but the disadvantage is that it does not estimate the potential benefit of jogging to most persons. Plan 2 would test the effect of jogging over an alternative exercise plan, but it probably would have very low power. Plan 3 would allow for randomization and exactly measure the effect of exercise, but it would have dubious generality to humans. No one plan is best for all purposes, and each has serious drawbacks. So, when a two-group experiment is undertaken, its interpretation depends on how subjects are assigned to levels of the independent variable and how the two groups are formed.

# *Illustrations*

In this section four different examples are considered. These examples illustrate the computation required for the two-group design.

## *Example 1*

One group consists of ten persons in a smoking cessation program, and the other group contains ten persons who were put on a waiting list. The two groups were formed randomly. The dependent variable is the number of cigarettes smoked per day two weeks after the program is completed. The scores of the treatment group are

$$0, 15, 12, 9, 10, 0, 0, 25, 5, 3$$

and the control group

$$18, 23, 15, 10, 8, 16, 13, 10, 20, 16$$

The mean for the treatment group is 7.9, and for the control group the mean is 14.9. The pooled variance is

$$\frac{1209 - 79^2/10 + 2423 - 149^2/10}{10 + 10 - 2} = 43.767$$

The standard error of the difference between means is

$$\sqrt{43.767\left[\frac{1}{10} + \frac{1}{10}\right]} = 2.959$$

The test of no effect of the treatment is

$$t(18) = \frac{7.9 - 14.9}{2.959} = -2.366$$

which with 18 degrees of freedom is statistically significant at the .05 level of significance. Thus, the program lowered the level of cigarette smoking to an extent that cannot be explained by sampling error. Because groups were formed randomly, the difference can be attributed to the program and not to any other variable. The value of Cohen's $d$, using the formula $t\sqrt{2/n}$, is $-2.366\sqrt{2/10} = -1.06$.

## Example 2

Five persons undergo a drug treatment to reduce blood pressure and five others receive an inert drug. There are two groups: a drug and a placebo group. Their changes in blood pressure are

Drug:   $-15, -17, -14, -6, 4$

Placebo:   $0, -6, 8, 9, -7$

The means are $-9.6$ and $.8$ for the drug and placebo groups, respectively. The sums of squared scores are 762 and 230 in the drug and treatment groups, respectively. The pooled variance is

$$\frac{762 - (-48)^2/5 + 230 - 4^2/5}{5 + 5 - 2} = 66.000$$

The $t$ test value is

$$t(8) = \frac{-9.6 - .8}{\sqrt{66.000\left[\frac{1}{5} + \frac{1}{5}\right]}} = -2.024$$

The $t$ value of $-2.024$ is not significant at the .05 level. It is, however, significant at the .10 level and some researchers refer to this level of significance as *marginal significance*. There is, then, not very compelling evidence from this study that the drug reduces blood pressure. The effect size equals $-2.024\sqrt{2/5}$, which is $-1.28$. Even though the effect size is $-1.28$, the sample size makes the power so low that the result is not statistically significant.

## Example 3

Of 28 people involved in a study on attitude change, 13 received a message from a high-status source and 15 from a low-status source. The resulting attitude changes for the two groups are

High-status source: 5, 6, 9, 3, 0, 4, 10, 6, 9, 5, 6, 5, 7

Low-status source: $-1$, 0, 3, $-4$, $-6$, $-2$, $-1$, 0, 3, 6, $-3$, $-2$, $-1$, $-2$, 1

A positive change indicates change in the direction consistent with the message, whereas a negative change indicates the reverse. The means for the high- and low-status groups are 5.77 and $-.60$. The sums of the squared scores for the two groups are 519 and 131. The pooled variance is

$$\frac{519 - 75^2/13 + 131 - (-9)^2/15}{13 + 15 - 2} = 8.150$$

The $t$ test value is

$$t(26) = \frac{5.77 - (-.60)}{\sqrt{8.150\left[\frac{1}{13} + \frac{1}{15}\right]}} = 5.888$$

This result is statistically significant at the .001 level. There was more attitude change that was consistent with the message in the high-status than in the low-status group. The value of Cohen's $d$ is

$$5.888\sqrt{\frac{1}{13} + \frac{1}{15}} = 2.23$$

## Example 4

Each of ten six-year-old children interact with a different four-year-old child. Measured is the degree of social responsiveness by each person in the conversation. The hypothesis is that six-year-olds are more socially responsive than four-year-olds. The data are paired because two persons of different ages interact. The scores are

| Pair | Six-Year-Old | Four-Year-Old |
|------|--------------|---------------|
| 1 | 6 | 5 |
| 2 | 5 | 4 |
| 3 | 4 | 5 |
| 4 | 7 | 6 |
| 5 | 6 | 3 |
| 6 | 7 | 5 |
| 7 | 3 | 4 |
| 8 | 6 | 3 |
| 9 | 8 | 6 |
| 10 | 5 | 4 |

The correlation between scores is .45. The differences between each four-year-old and each six-year-old are 1, 1, –1, 1, 3, 2, –1, 3, 2, and 1, and the mean is 1.2. The variance of the different scores is

$$\frac{32 - 12^2/10}{9} = 1.956$$

The $t$ test value is

$$t(9) = \frac{1.2}{\sqrt{1.956/10}} = 2.713$$

This value of $t$ is statistically significant at the .05 level. Thus six-year-olds are more socially responsive than four-year-olds. The value of $d$ is

$$2.713\sqrt{\frac{2(1 - .45)}{10}} = .90$$

(The correlation between the two scores is equal to .45.)

# Summary

The complete model for the two-group design involves a dichotomous independent variable that causes the dependent variable. In the restricted model the independent variable has no effect on the dependent variable. This model is evaluated by computing the difference between the means divided by the standard error of the difference between means. This standard error equals the pooled standard deviation of the two groups times the square root of $1/n_1 + 1/n_2$. When the restricted model is true, the difference between means divided by its standard error has a $t$ distribution, with $n_1 + n_2 - 2$ degrees of freedom. The test of the restricted model presumes that the residual variable has a normal distribution, that the variances in the two groups are equal, and that the observations are independent.

When observations are paired, differences are computed and the mean of the differences evaluates the equality of the group means. The size of the treatment effect is measured by *Cohen's d,* which is called a measure of *effect size.* With the sample size, alpha, and the effect size, the power of the *t* test can be determined. The interpretation of a significant *t* test depends upon design considerations. If the units are randomly assigned to levels of the independent variable, then significant differences on the dependent variable can be attributed to the independent variable.

In the next chapter the independent variable may take on more than two levels.

# Problems

1. Determine the minimum value of *t* needed to achieve the given significance levels with the corresponding degrees of freedom.

   | Alpha | df |
   |---|---|
   | a.  .05 | 26 |
   | b.  .01 | 6 |
   | c.  .10 | 44 |
   | d.  .02 | 62 |
   | e.  .001 | 132 |
   | f  .05 | 77 |

2. The following scores are taken from a study that compared two different methods of increasing vocabulary. The scores of ten persons, five under each method, are:

   A:   16, 19, 20, 18, 24

   B:   12, 15, 16, 15, 14

   Is there any evidence that one method is superior to the other?

3. Compute a paired *t* test to evaluate the effectiveness of a weight loss program.

   | Person | Before | After |
   |---|---|---|
   | 1 | 163 | 150 |
   | 2 | 149 | 143 |
   | 3 | 236 | 240 |
   | 4 | 189 | 180 |
   | 5 | 176 | 160 |
   | 6 | 216 | 205 |

4. For the following *t* values compute *d.*

   a.  $t(20) = 1.380$, $n_1 = 11$, $n_2 = 11$
   b.  $t(98) = 2.110$, $n_1 = 50$, $n_2 = 50$

   c. $t(10) = 1.530$, $n_1 = 8$, $n_2 = 4$

   d. $t(54) = -.470$, $n_1 = 30$, $n_2 = 26$

5. Determine the power of the following tests.

   a. $n_1 = n_2 = 20$ and $d = .5$

   b. $n_1 = n_2 = 100$ and $d = .2$

   c. $n_1 = n_2 = 80$ and $d = .8$

   d. $n_1 = n_2 = 10$ and $d = .8$

   e. paired design; $d = .22$, $r = .8$, and $n = 20$

   f. $n_1 = 11$, $n_2 = 100$, and $d = .8$

6. A program is developed to improve the intelligence scores (IQ) of preschool children. Two groups of children are randomly formed. Test whether the program affects IQ:

   Treated group:   109, 123, 141, 119, 133, 117, 118, 120

   Control group:   106, 103, 114, 120, 116, 107, 98

7. Twenty persons are randomly assigned to one of two treatments. In the treatment group, ten persons are taught a series of strategies to improve their memory. The control group learned none of the strategies. The scores on a memory test are

   Memory group:   88, 76, 83, 75, 64, 80, 76, 73, 84, 78

   Control group:   84, 73, 84, 78, 68, 78, 71, 70, 80, 79

   Are the two groups significantly different?

8. Describe the advantages and disadvantages of using the control groups in a study to evaluate the effect of group therapy to reduce cigarette smoking.

   a. individual therapy

   b. hypnosis condition

   c. a film that encourages quitting

9. A psychologist studies the degree of happiness of people at various stages in life. His measure of general happiness varies from 0 to 60. In one study he compared the happiness of married and single men aged 25. Is there a significant difference between the two groups?

| Married | Single |
|---------|--------|
| 58 | 57 |
| 45 | 44 |
| 50 | 59 |
| 54 | 44 |
| 49 | 39 |
| 39 | 60 |
| 50 | 44 |
| 51 | |

10. Nine persons were asked to rate the taste of cola A and cola B on a scale from one to ten. Is one drink significantly preferred to the other?

| Person | Cola A | Cola B |
|--------|--------|--------|
| 1 | 7 | 7 |
| 2 | 8 | 9 |
| 3 | 8 | 7 |
| 4 | 9 | 5 |
| 5 | 10 | 9 |
| 6 | 9 | 7 |
| 7 | 8 | 6 |
| 8 | 8 | 10 |
| 9 | 7 | 8 |

11. For the following studies estimate Cohen's $d$.

    a. $t = 2.910$, $n_1 = 10$, $n_2 = 12$
    b. $t = -.410$, $n_1 = 5$, $n_2 = 5$
    c. a paired design in which $t = 5.910$, there are 8 pairs, and $r = .8$
    d. $t = -.970$, $n_1 = n_2 = 80$

12. A researcher wishes to test whether eight-grade girls outscore eighth-grade boys in vocabulary. She tested 30 boys and 42 girls and found means of 64.53 for boys and 66.42 for girls. The standard deviations are 12.34 for boys and 12.59 for girls. Compute Cohen's $d$ for this study and interpret it. Evaluate whether the sex difference is statistically significant.

13. The data for Example 1 in the chapter are repeated here: The scores of the treatment group are

$$0, \ 15, \ 12, \ 9, \ 10, \ 0, \ 0, \ 25, \ 5, \ 3$$

and for the control group are

$$18, \ 23, \ 15, \ 10, \ 8, \ 16, \ 13, \ 10, \ 20, \ 16$$

Compute the standard deviations for group and evaluate in words the assumption of equal variances and its effect on the $p$ value.

14. Diehl, Kluender, and Parker (1985) tested for the recognition of auditory stimuli on two tasks. Each of 13 subjects received a score on each task, the maximum being 40. Does performance on the tasks significantly vary?

| Subject | Task A | Task B | Subject | Task A | Task B |
|---------|--------|--------|---------|--------|--------|
| DS | 20 | 19 | LD | 30 | 24 |
| MM | 21 | 18 | RL | 23 | 18 |
| JH | 28 | 24 | TW | 29 | 19 |
| JS | 17 | 6 | TA | 30 | 16 |
| MC | 15 | 13 | VS | 34 | 29 |
| CM | 20 | 13 | CJ | 21 | 20 |
| LG | 28 | 22 | | | |

15. For the following effect sizes and designated power, state the necessary sample size needed in each group.

|     | Effect Size | Power |
| --- | --- | --- |
| a. | .5 | .50 |
| b. | .8 | .80 |
| c. | .5 | .95 |
| d. | .2 | .25 |

# 14 | *One-Way Analysis of Variance*

For the model in Chapter 12 the dependent variable is caused by only the residual variable; that is,

$$\frac{\text{dependent}}{\text{variable}} = \text{constant} + \frac{\text{residual}}{\text{variable}}$$

In Chapter 13 added to the model is an independent variable, as follows:

$$\frac{\text{dependent}}{\text{variable}} = \text{constant} + \frac{\text{effect of the}}{\text{independent}} + \frac{\text{residual}}{\text{variable}}$$

In Chapter 13 the independent variable is limited to a dichotomy, so only two groups can be compared. In this chapter the independent variable remains a nominal variable, but it may have more than two levels.[1]

The statistical technique used to analyze the model in which nominal independent variables affect a dependent variable measured at the interval level is called *analysis of variance,* commonly referred to as *ANOVA*. If there is a single nominal independent variable, the technique is called *one-way analysis of variance*. Analysis of variance is the most commonly used data analysis technique in psychology and is also commonly used in education, biology, and engineering. It represents not only a statistical test but also a way of thinking about research. In fact, one social psychologist, Harold Kelley, has suggested that persons in everyday life use something like analysis of variance to understand social reality.

The term "analysis of variance" is potentially confusing because the analysis of *variance* tests hypotheses about *means*. It must be remembered that an analysis of variability provides information about the means.

---

[1]The independent variable in ANOVA can be at the ordinal or interval level of measurement. The key requirement is that the variable have discrete groups. Because a nominal variable is always discrete, it is convenient to say that the independent variable is nominal.

In Table 14.1 is a data set that will be used in this chapter. A psychologist is interested in the effects of various strategies to enhance memory. Subjects were asked to memorize a list of 15 words and were tested one week later. The psychologist seeks to compare four different types of instruction: imagery, story, person, and none. In the imagery condition, subjects were told to picture each word. In the story condition, they were told to make up a story using the words. In the person condition, they were asked to associate each word with a person that they know. The none condition was a control condition. The means of the four groups, each with a sample size of ten, are as follows:

| Imagery | Story | Person | None |
|---------|-------|--------|------|
| 12.3 | 10.6 | 9.4 | 7.3 |

Can the differences between the means be explained by chance? By chance is meant that the four groups are random samples drawn from the same population. Even if the groups were drawn from the same population, the means would still differ because of sampling error. At issue is whether the differences between means can be explained by chance. For example, 12.3 words recalled is greater than 10.6, but this 1.7 difference might have happened by chance. Just how likely would a 1.7 difference arise by chance? This is the type of question that the analysis of variance answers.

# The ANOVA Model

The independent variable, called a *factor* in ANOVA, is denoted as Instructions or more simply as factor I. Ordinarily the factor is given an appropriate descriptive name, such as Drug or Reinforcement Schedule. The

**TABLE 14.1**    **Number of Words Recalled out of a Maximum of 15 Under Four Conditions**

| | Instruction Level (I) | | | |
|---|---|---|---|---|
| | Imagery | Story | Person | None |
| | 12 | 10 | 12 | 6 |
| | 14 | 9 | 8 | 4 |
| | 15 | 10 | 7 | 12 |
| | 10 | 11 | 5 | 8 |
| | 12 | 10 | 11 | 9 |
| | 14 | 13 | 13 | 11 |
| | 15 | 10 | 12 | 4 |
| | 12 | 11 | 10 | 6 |
| | 10 | 13 | 7 | 7 |
| | 9 | 9 | 9 | 6 |
| $\Sigma X$ | 123 | 106 | 94 | 73 |
| $\bar{X}$ | 12.3 | 10.6 | 9.4 | 7.3 |

factor can be abbreviated using a single uppercase letter such as D for Drug or R for Reinforcement Schedule. Categories of the independent variable are referred to as levels or groups. For instance, the factor in Table 14.1 has four levels or groups.

The variance of the means can be computed. This variance is

$$\frac{12.3^2 + 10.6^2 + 9.4^2 + 7.3^2 - (12.3 + 10.6 + 9.4 + 7.3)^2/4}{3}$$

which equals 4.42. The variance of the means provides a quantitative index of how different the four means are. Recall that the variance of the means can be interpreted as one-half the average squared difference among all pairs of means.

The variance within each of the groups can also be computed. The respective variances for the four groups are 4.68, 2.04, 6.93, and 7.34. The average of these variances is 5.25. This is the average or pooled variance within groups and is designated by $s_p^2$ because it is analogous to the pooled variance in Chapter 13.

There are now two measures of variability: the variance of the means, which equals 4.42, and the variance pooled within groups, which equals 5.25. As discussed in Chapter 11, the variance of the mean is a function of sample size. The variance of the mean is $\sigma^2/n$, where $\sigma^2$ is the variance of the observations that are used to make up the mean and $n$ is the sample size. Thus, as sample size increases, the means should be more tightly bunched. So because the variance of the means is an inverse function of sample size, the variance of the means should be corrected by multiplying by the sample size; that is, $ns_{\overline{X}}^2$, where $s_{\overline{X}}^2$ is the variance of the group means and $n$ is the group size. This quantity provides a measure of the variability of group means corrected for sample size.

The complete model for one-way ANOVA is

$$\frac{\text{dependent}}{\text{variable}} = \text{constant} + \frac{\text{effect of the}}{\text{independent}} + \frac{\text{residual}}{\text{variable}}$$
$$\qquad\qquad\qquad\qquad\qquad \text{variable}$$

In the restricted model the independent variable has no effect; thus

$$\frac{\text{dependent}}{\text{variable}} = \text{constant} + \frac{\text{residual}}{\text{variable}}$$

If the restricted model were true, then both $ns_{\overline{X}}^2$ and $s_p^2$ estimate the variance of the residual variable. Thus, if the independent variable has no effect, the value of $ns_{\overline{X}}^2$ and $s_p^2$ should be close together because they both estimate the variance of the residual variable. In the memory example, $ns_{\overline{X}}^2$ equals 44.2, which is substantially larger than $s_p^2$, which equals 5.25. The two statistics do not appear to be estimating the same variance.

If the restricted model, which has no effect of the independent variable, is false, then only $s_p^2$ estimates the variance of the residual variable. The variance of sample means adjusted for sample size estimates the variance of the residual variable plus the variance of the population means times the sample size. So, if the independent variable has an effect, the variance of the means tends to be greater than the pooled variance, and this is the case in the example. So, by computing two variances—one the variance within groups and the other the variance of group means—hypotheses about means can be tested. This is the fundamental logic of analysis of variance.

If an independent variable called factor A affects the dependent variable, then the means for the various levels of factor A should differ from the grand mean in the population. The null hypothesis of one-way analysis of variance is that the population means of the $k$ groups are all equal to each other.

The restricted model in one-way analysis of variance is that the independent variable has no effect on the dependent variable. Under the restricted model, the variance of the group means times the number of persons in each group and the pooled variance within groups both estimate the residual variance. Given assumptions that underlie the restricted model (to be discussed later in this chapter), the ratio of these two estimates is distributed as $F$. If the restricted model is false, the value of $F$ should be large.

The $F$ distribution has a lower bound of zero, an upper bound of positive infinity, and a mode near one. An example of $F$ is presented in Figure 14.1. There is not one $F$ distribution but a family of distributions. The $F$ distribution has two parameters. They are the degrees of freedom on the numerator and degrees of freedom on the denominator. They are symbolized by $df_n$ and $df_d$, respectively. An $F$ test statistic is written as $F(df_n, df_d)$. How to calculate the degrees of freedom for one-way analysis of variance will be explained later.

To find the $p$ value of a test statistic distributed as $F$, the appropriate degrees of freedom for the numerator are located in the top row of Appendix E of the appropriate page, and the degrees of freedom in the denominator are

*FIGURE 14.1*     **Example of the $F$ distribution.**

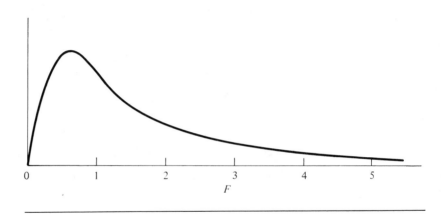

located in the first column. For a value to be significant, it must equal or exceed the tabled value, which is always some value greater than one. The significance level is obtained by choosing the largest value equaled or exceeded. In Appendix E are four significance levels: .10, .05, .01, and .001. (The $F$ distribution was also presented in Chapter 11.)

# Estimation and Testing

The term $X_{ij}$ symbolizes the score for person $i$ in group $j$. The first subscript refers to person and the second to group or level of the independent variable. There are $n_j$ persons in group $j$, and there is a total of $k$ groups. If for each group the group sizes are the same, the group size is denoted as $n$. The total number of scores is symbolized by $N$. So for the example in Table 14.1, $n$ is 10, $k$ is 4, and $N$ is 40.

The sum of all the scores is symbolized by $\Sigma\Sigma X_{ij}$. The double summation signs indicate that scores are first summed in a group and then these group sums are added together. The sum of all the scores in a given group, say $j$, is symbolized by $\Sigma X_{ij}$. So the double summation indicates the sum of all the scores and the single sum indicates the sum of scores in a group.

The mean for group $j$, symbolized by $\overline{X}_{.j}$, is

$$\overline{X}_{.j} = \frac{\Sigma X_{ij}}{n_j}$$

The sum of the groups sizes or $\Sigma n_j$ is denoted as $N$. If group sizes are equal, $N$ equals $nk$ or group size times the number of groups. The mean of the means or the grand mean $\overline{X}_{..}$ equals

$$\overline{X}_{..} = \frac{\Sigma\Sigma X_{ij}}{N}$$

The dot notation indicates that the mean is computed across that subscript.

Alternatively, the grand mean can be computed as a weighted mean of the group means. The grand mean is computed by

$$\overline{X}_{..} = \frac{\Sigma n_j \overline{X}_{.j}}{N}$$

If the sample sizes are equal, the set of $k$ means can be averaged to compute the grand mean, as follows:

$$\overline{X}_{..} = \frac{\Sigma \overline{X}_{.j}}{k}$$

It is useful to compute the grand mean both ways as a check on the computations.

To measure the constant in the complete model, the grand mean is used. To measure the effect of the independent variable, use the following:

$$\bar{X}_{\cdot j} - \bar{X}_{\cdot \cdot}$$

that is, the mean for level $j$ minus the grand mean. To compute the residual score for person $i$ in group $j$, use:

$$X_{ij} - \bar{X}_{\cdot j}$$

that is, the score for that person minus the group mean. The estimated residual score is used to test assumptions of ANOVA that are presented later in this chapter.

## The Analysis of Variance Table

Analysis of variance has a whole set of special terms, which are summarized in a table, such as Table 14.2. There are three rows in a one-way ANOVA table. The label for the top row is groups or in this case factor A. This line of the analysis of variance table represents variance attributable to groups. The term is sometimes referred to as the between-groups term. The second line represents the variation of subjects within groups. It is usually abbreviated as S/A, where the slash indicates within. This term is sometimes referred to as the *within-groups term* or *error term*. Subjects are said to be nested within levels of factor A. That is, each person is a member of one and only one group. The last line represents the total variation. Total is commonly abbreviated as TOT.

There are four columns to an analysis of variance table. The first column is for sum of squares. It can be abbreviated as SS. The sum of squares for groups or $SS_A$ equals

**TABLE 14.2** **Analysis of Variance Table**

| Source of Variation | Sum of Squares | Degrees of Freedom | Mean Square | F |
|---|---|---|---|---|
| Factor A | $SS_A$ | $k - 1$ | $\dfrac{SS_A}{df_A}$ | $\dfrac{MS_A}{MS_{S/A}}$ |
| Subjects within groups (S/A) | $SS_{S/A}$ | $N - k$ | $\dfrac{MS_{S/A}}{df_{S/A}}$ | |
| Total (TOT) | $SS_{TOT}$ | $N - 1$ | | |

$$SS_A = \sum n_j (\overline{X}_{.j} - \overline{X}_{..})^2$$

For this sum of squares, the term that is squared is the estimate of the treatment effect.

The sum of squares for subjects within levels of factor A, or $SS_{S/A}$, is as follows:

$$SS_{S/A} = \sum\sum (X_{ij} - \overline{X}_{.j})^2$$

Each score has its group mean subtracted from it, the difference is squared, and then all the squared differences are summed across the entire set of scores. As stated earlier, the quantity $X_{ij} - \overline{X}_{.j}$ is the estimate of the residual score for person $i$ in group $j$.

The sum of squares total or $SS_{TOT}$ is as follows,

$$SS_{TOT} = \sum\sum (X_{ij} - \overline{X}_{..})^2$$

again where $\overline{X}_{..} = \Sigma\Sigma X_{ij}/N$. It can be shown that

$$SS_{TOT} = SS_A + SS_{S/A}$$

and so the $SS_{S/A}$ can be obtained by subtraction as follows:

$$SS_{S/A} = SS_{TOT} - SS_A$$

Ordinarily the $SS_{S/A}$ is computed indirectly by subtracting the $SS_A$ from the $SS_{TOT}$.

The second column contains the degrees of freedom or *df*. The total degrees of freedom are

$$df_{TOT} = N - 1$$

or the total sample size less one. The degrees of freedom for factor A are:

$$df_A = k - 1$$

or the number of groups less one. The degrees of freedom for *S/A* are

$$df_{S/A} = N - k$$

or the total sample size minus the number of groups.

It is helpful to remember that degrees of freedom partition in the same way as sums of squares. So just as $SS_{S/A} = SS_{TOT} - SS_A$, it is also true that

$$df_{S/A} = df_{TOT} - df_A$$

Any partitioning of the sums of squares can also be done to the degrees of freedom.

The *mean square* equals the sum of squares for the line divided by the degrees of freedom for the line. So to compute $MS_A$, the $SS_A$ is divided by

$df_A$; and to compute $MS_{S/A}$, the $SS_{S/A}$ is divided by $df_{S/A}$. Usually the $MS_{TOT}$ is not computed, but it would equal the variability of the observations ignoring the independent variable.

The $MS_A$ is the variability of the groups means corrected for sample size. The $MS_{S/A}$ is the pooled variance within levels of factor A. Both mean squares estimate the residual variance given that the restricted model is true. The final column in the analysis of variance table is the $F$ test statistic, which equals

$$F(k{-}1,\ N{-}k) = \frac{MS_A}{MS_{S/A}}$$

The degrees of freedom for this $F$ test on the numerator or $df_n$ are $k-1$ and the degrees of freedom on the denominator $df_d$ are $N-k$. The $F$ test evaluates the restricted model. It essentially compares the variability in the group means to the variability of scores within groups. If the restricted model is false, then $F$ tends to be large. If the $F$ is statistically significant, the restricted model is rejected, and the complete model in which the independent variable, factor A, affects the dependent variable is preferred.

In Table 14.3 is the analysis of variance table for the data in Table 14.1. The $F$ ratio of 8.42 is statistically significant at the .001 level. It indicates that "Instruction" affects the dependent variable. Thus, subjects recalled different amounts of material in the four conditions.

Most of the effort in analysis of variance involves the computation of sums of squares. For every analysis at least one sum of squares is computed from others. For instance, in one-way ANOVA the sum of squares within groups or $SS_{S/A}$ equals the sum of squares total, $SS_{TOT}$, minus the sum of squares for the independent variable, $SS_A$. It is helpful to draw a circle diagram that illustrates the partitioning of the sum of squares, as in Figure 14.2. The complete circle represents the sum of squares total. The circle inside the circle represents the sum of squares for factor A, and the remainder represents the sum of squares for persons within levels of factor A. For more complex designs, these diagrams can be especially helpful.

*TABLE 14.3*    **Analysis of Variance Table for the Data in Table 14.1**

| Source of Variation | Sum of Squares | Degrees of Freedom | Mean Square | $F$ |
|---|---|---|---|---|
| Instruction (I) | 132.6 | 3 | 44.20 | 8.42 |
| Subjects within I (S/I) | 189.0 | 36 | 5.25 | |
| Total | 321.6 | 39 | | |

*FIGURE 14.2*     **Circle diagram for one-way analysis of variance.**

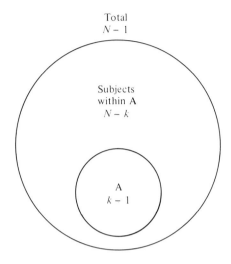

## Computational Formulas for the Sum of Squares

The computation of various ANOVA terms is discussed, first for the case when groups are equal in size and second for the case when group sizes are unequal.

For the computation of the sum of squares, it is necessary to define various totals. They are the grand total

$$T_{..} = \sum\sum X_{ij}$$

and the group total for group $j$

$$T_{.j} = \sum X_{ij}$$

For the example in Table 14.1, the group totals are 123, 106, 94, and 73. The grand total, which is the sum of the group totals, equals 396.

To compute the $F$ test to evaluate the relative plausibility of the complete and the restricted models, both $SS_A$ and $SS_{S/A}$ must be computed. First the *correction term for the mean, C,* is computed as follows:

$$C = \frac{(\sum\sum X_{ij})^2}{N}$$

or using totals, $C = T_{..}^2/N$. In words, the correction term for the mean is the square of the sum of all the observations, divided by the total number of observations in the study. It is called a correction term for the mean because it tends to be large if the mean is large. The sum of all the observations for the memory example is 396, and so the correction term for the mean is $396^2/40 = 3920.4$.

The sum of squares for factor A, or $SS_A$, is

$$SS_A = \frac{\sum T_{.j}^2}{n} - C$$

where again $C$ equals $T_{..}^2/N$.

The $SS_{S/A}$ is computed indirectly, as follows:

$$SS_{S/A} = SS_{TOT} - SS_A$$

The sum of squares total equals

$$SS_{TOT} = \sum\sum X_{ij}^2 - C$$

That is, it equals the sum of the square of every observation, minus $C$.

If group sizes are unequal, the $SS_A$ equals

$$SS_A = \sum \frac{T_{.j}^2}{n_j} - C$$

So the group total is squared and divided by the group's $n$. These are then summed across the $k$ groups, and the mean's correction term $C$ is subtracted.

### Test of the Constant

As shown in Chapter 12, a researcher may have an a priori hypothesis concerning the constant in the model. For instance, if a recognition memory test is used, the constant should equal some a priori value if subjects were guessing. It is possible to test a restricted model that the mean equals some value, within a one-way ANOVA.

The a priori constant is denoted as $M$. The sums of squares for the constant are $N(\overline{X}.. - M)^2$. (Note that if $M$ is zero, this quantity equals $C$, the correction term for the mean.) The sum of squares for the constant divided by the $MS_{S/A}$ is an $F$ test with 1 and $N - k$ degrees of freedom. If the $F$ is statistically significant, the null hypothesis that the population mean equals $M$ is rejected.

As an example, suppose it is desired to test the null hypothesis that the constant in Table 14.1 is 10.0. The grand mean equals 9.9. The sum of squares for the constant is $40(10.0 - 9.9)^2$, which equals .40. From Table 14.3 the $MS_{S/A}$ is 5.25 and so the $F$ is .40/5.25 or .08. Because the $F$ is not significant, the constant is not significantly different from 10.0.

## Assumptions and Interpretation

The assumptions of analysis of variance are basically the same as those of the $t$ test discussed in the previous chapter. These assumptions all refer to the residual variable and they are (a) normal distribution, (b) equal variability, and (c) independence of observations.

Because these assumptions are extensively discussed in the previous chapter, they will not be reviewed here except to repeat that the independence assumption is the one assumption that must be carefully scrutinized. The reader should consult Chapter 13 for a discussion of these assumptions.

If the group means differ significantly, the meaning of those differences depends on design considerations. If persons are randomly assigned to groups, then the significant effect can be attributed to the independent variable. Without random assignment it is not clear what causes what. For instance, consider a study on the effects of jogging. Three groups are formed: nonjoggers, joggers, and dedicated joggers. The dependent variable is weight. Suppose it is found that persons who jog more often weigh less. We cannot conclude that jogging causes a loss in weight; the truth might be that overweight persons simply do not wish to jog. That is, it is the dependent variable that causes the independent variable and not vice versa. Without random assignment of persons to groups, it cannot be unequivocally concluded that the independent variable causes the dependent variable.

# Power and Measure of Effect Size

Recall that the power of a test is the probability of rejecting the restricted model given that the restricted model is false. To evaluate the power the following factors must be considered: First, how different are the population means from one another? The more different they are, the greater the power. So the more truly different the groups are, the more likely the restricted model will be rejected. Second, how many persons are in each group? The more persons in each group, the greater the power. Third, how great is the variability within treatment groups? The more similar persons within the same group are, the greater the power. All three of these factors were considered in more detail in the previous chapter.

The most common measure of effect size for one-way analysis of variance is called *omega squared* or, as it is commonly symbolized, $\omega^2$. Its computational formula is

$$\omega^2 = \frac{SS_A - (k-1)MS_{S/A}}{SS_{TOT} + MS_{S/A}}$$

Omega squared can be interpreted as the proportion of variance in the dependent variable that is explained by the independent variable. So, like a correlation coefficient, the upper limit of omega squared is 1.00. If the estimated value of omega squared is less than zero, then omega squared is set to zero. From the values in Table 14.3,

$$\omega^2 = \frac{132.6 - 3(5.25)}{321.6 + 5.25} = .36$$

It should be noted that omega squared is in squared units. If the standards for large, medium, and small that were set for the $d$ in the previous chapter are used for $\omega^2$, a large $\omega^2$ is .40, a medium value is .20, and a small value is .04.

# *Contrasts*

The complete model postulates some effect of the independent variable on the dependent variable. It does not, however, explicitly predict exactly how the means differ, but only that at least two groups have different population means. The researcher may have a clear idea about exactly how the means differ. Consider some specific examples.

1. A researcher is interested in examining the effect of the day of the week on absenteeism. She believes that absenteeism is higher on Monday than any other day of the week. She conducts a one-way analysis of variance using day of the week as the independent variable.
2. A researcher believes that more study time will improve performance on an examination. She creates four groups who study zero, one, two, and three hours. She then measures performance on an examination. She analyzes her data by a one-way analysis of variance.
3. A researcher creates four groups to investigate the effects of smoking and stress on blood pressure. The four groups are

   > I: No smoking, no stress
   > II: Smoking, no stress
   > III: No smoking, stress
   > IV: Smoking, stress

   He analyzes his data by a one-way analysis of variance, although he wished to compare those who smoked with those who did not.

In all of these examples the researchers had a specific hypothesis about the patterning of the means. However, they did not explicitly test for this pattern. This failure to make such a test can lead to mistaken conclusions. If the $F$ is significant, it does not imply that the researcher's hypothesis is true. Conversely, if the $F$ is not significant, it does not imply that the researcher's hypothesis is false. The overall $F$ test does not directly test the researcher's specific hypothesis. It only evaluates the very general null hypothesis that all the population means are equal to each other.

When there are three or more groups and an explicit hypothesis about how the means differ, a contrast can be used to test that hypothesis. A *contrast* is a set of weights assigned to each level of the independent variable to evaluate an explicit hypothesis. These weights that tap the hypothesis of interest must sum to zero.

A *contrast* is a set of numbers, each of which is paired with one level of the independent variable. To determine the particular contrast weights, the null hypothesis must be explicitly stated. For instance, for the researcher who is interested in Monday absenteeism, the null hypothesis is that Monday does not differ from the other four days of the week. Algebraically, this null hypothesis can be expressed as

$$\mu_{MO} = \frac{\mu_{TU} + \mu_{WE} + \mu_{TH} + \mu_{FR}}{4.0}$$

(The $\mu$ terms are the population means.) The terms can be rearranged and all put on the right-hand side:

$$0 = \mu_{MO} - .25\mu_{TU} - .25\mu_{WE} - .25\mu_{TH} - .25\mu_{FR}$$

The contrast weights are the numbers that multiply the means; that is,

| Monday | Tuesday | Wednesday | Thursday | Friday |
|--------|---------|-----------|----------|--------|
| 1.0 | −.25 | −.25 | −.25 | −.25 |

Note that the five contrast weights sum to zero, as they must.

Sometimes the researcher has quantitative values attached to the levels of the independent variable. To derive the contrast weights, the quantitative values are averaged. The contrast weight for a given level equals the quantitative value minus this average. So, for the second example, persons study zero, one, two, and three hours. The average of the four numbers is 1.5. The contrast weights are

| *Number of Hours Studied* | | | |
|------|------|------|------|
| 0 | 1 | 2 | 3 |
| −1.5 | −.5 | .5 | 1.5 |

Again, the sum of the contrast weights is zero.

For the final example, the researcher seeks to compare smokers (groups II and IV) with nonsmokers (groups I and III). The null hypothesis is $\mu_I + \mu_{III} = \mu_{II} + \mu_{IV}$. Putting all the terms on the right-hand side yields

$$0 = \mu_{II} + \mu_{IV} - \mu_I - \mu_{III}$$

The contrast weights are then

| *Group* | | | |
|-----|-----|-----|-----|
| I | II | III | IV |
| −1 | +1 | −1 | +1 |

To test a contrast, its sum of squares must be computed. The contrast value for the $j$th mean is denoted as $p_j$. Given an equal number of observations in each group (that is, $n_j$ is constant), the sum of squares for a contrast equals

$$\frac{(\sum_{j} p_j T_{\cdot j})^2}{n \sum p_j^{\ 2}}$$

The degrees of freedom for a contrast are equal to one. So, the sum of squares and the mean square are the same. To test the contrast, the mean square for the contrast is divided by the $MS_{S/A}$. Given the null hypothesis, the ratio is distributed as $F$ with 1 and $N - k$ degrees of freedom. Formulas for the sum of squares for a contrast with unequal group sizes are given in more advanced texts (Myers, 1979; Winer, 1971).

A contrast is a dummy variable. To each level of a nominal variable (the independent variable) numerical values (contrast weights) are attached. This dummy variable can be viewed as the predictor variable in a regression equation in which the dependent variable serves as the criterion. If such a dummy variable is created and the regression coefficient is computed and tested using the method described in Chapter 16, the $p$ value is the same as the one-way ANOVA test of the contrast.

# Post Hoc Tests

The use of contrasts to test an explicit hypothesis is called an *a priori test*. If the researcher has no explicit hypothesis about the patterning of the means, he or she can perform what is called a *post hoc test* to determine how it is that the means differ. There are many, many different ways of performing post hoc tests and there is no clear consensus about which is the best technique. The Tukey least significant difference (lsd) test is presented because it is relatively easy to compute. The reader is referred to more advanced texts for descriptions of other post hoc test procedures (Myers, 1979; Winer, 1971).

The Tukey lsd test is called a *protected* test. It can only be done if the $F$ test is significant. Assuming that it is, all possible pairs of means are compared. Each pair of means is tested using the formula

$$\text{Tukey lsd} = \frac{\bar{X}_1 - \bar{X}_2}{\sqrt{MS_{S/A} \left[ \dfrac{1}{n_1} + \dfrac{1}{n_2} \right]}}$$

where $\bar{X}_1$ and $\bar{X}_2$ are the sample means and $n_1$ and $n_2$ are the respective sample sizes. The formula for the Tukey lsd test is a *t* test of the difference between means (see the previous chapter) with $MS_{S/A}$ substituted for $s_p^{\ 2}$. To determine whether the difference is statistically significant, the *t* distribution with $N - k$ degrees of freedom is used, where $k$ is the number of groups.

As an example consider the data in Table 4.1. Because the $F$ is significant, the Tukey lsd test can be used. Because there are four groups, there are six possible comparisons. The results are

| | |
|---|---|
| Imagery with Story: | $t(36) = 1.659$ |
| Imagery with Person: | $t(36) = 2.830$ |
| Imagery with None: | $t(36) = 4.880$ |
| Story with Person: | $t(36) = 1.711$ |
| Story with None: | $t(36) = 3.220$ |
| Person with None: | $t(36) = 2.049$ |

Using Appendix D, for a $t$ with 36 degrees of freedom to be statistically significant at the .05 level of significance, it must be at least 2.030. So, four of the six comparisons are judged as significant.

When group sizes are equal, the smallest difference between means that would be significant—that is, the least (l) significant (s) difference (d)—equals

$$t(.05, \ df) \ \frac{\sqrt{2MS_{S/A}}}{\sqrt{n}}$$

where $t(.05, \ df)$ is the critical value for .05 significance with degrees of freedom of $df$. For the example, the lsd is

$$2.030 \ \frac{\sqrt{2(5.25)}}{\sqrt{10}} = 2.080$$

Hence a difference between a pair of means must be at least 2.080 to be significantly different at the .05 level of significance.

# Illustrations

## Example 1

An experimenter seeks to compare the degree of comfort of two automobiles. Twenty different persons are asked to sit in one of the two automobiles. Comfort is measured on a ten-point scale, with higher numbers indicating greater comfort. The cars are designated as car 1 and car 2. The numbers are

Car 1:   8, 9, 7, 8, 7, 6, 9, 6, 7, 5

Car 2:   10, 9, 8, 10, 9, 8, 7, 9, 9, 8

The means for the two cars are 7.2 and 8.7, and the grand mean is 7.95. Car 2 is rated as more comfortable, but it must be determined whether the difference is statistically significant.

The grand total or $T_{..}$ equals 159, the group totals are 72 and 87, $N$ is 20, $n$ is 10, and $k$ is 2. The correction term for the mean is $159^2/20$ or 1264.05. The sum of squares for cars is

$$\frac{72^2 + 87^2}{10} - 1264.05 = 11.25$$

The sum of each squared score is 1299. The sum of squares total is

$$1299 - 1264.05 = 34.95$$

By subtraction, the sum of squares for subjects within cars is

$$34.95 - 11.25 = 23.70$$

The analysis of variance table is presented in Table 14.4.

The $F$ value of 8.54 is statistically significant at the .01 level. Car 2 is significantly preferred over car 1. The value of omega squared for this example is .27, which is between moderate and large.

If a $t$ test had been done comparing the two groups, it would have been found that the $t$ value is 2.92, which is the square root of 8.54. When there are only two groups in a one-way analysis of variance, the square root of $F$ exactly equals the value that would be obtained in a two-group $t$ test. So, an ANOVA with two levels is equivalent to the two-group $t$ test described in the previous chapter. [In Chapter 11 it was pointed out that $t(df)^2 = F(1, df)$.]

## Example 2

A sociologist studies the degree of satisfaction of workers. She wishes to compare the satisfaction with job of secretaries, janitors, and managers. She uses a 20-point scale to measure job satisfaction where higher numbers indicate greater satisfaction. Her results are

Secretaries:   12, 16, 15, 19, 14

Janitors:   17, 19, 14, 18

Managers:   20, 19, 18

This is an unequal $n$ design because there are five secretaries, four janitors, and three managers. The grand mean is 16.75. The mean for secretaries is 15.2, for janitors 17.0, and for managers 19.0.

**TABLE 14.4**   **Analysis of Variance Table for Car Study**

| Source of Variation | Sum of Squares | Degrees of Freedom | Mean Square | $F$ |
|---|---|---|---|---|
| Car (C) | 11.25 | 1 | 11.250 | 8.54 |
| Subjects within C (S/C) | 23.70 | 18 | 1.317 | |
| Total | 34.95 | 19 | | |

The correction term for the mean is $201^2/12 = 3366.75$. The sum of squares for job is

$$\frac{76^2}{5} + \frac{68^2}{4} + \frac{57^2}{3} - 3366.75 = 27.45$$

The sum of all the squared scores is 3437 and so the sum of squares total is

$$3437 - 3366.75 = 70.25$$

The sum of squares for persons within jobs is $70.25 - 27.45 = 42.80$. The analysis of variance table is presented in Table 14.5. The $F$ is not statistically significant at the .05 level even though omega squared has a moderately large value of .24. So there is no evidence that the workers differ in their satisfaction. The sample sizes are so small that the level of power is quite low even for a moderate effect size. A Tukey lsd post hoc test should not be done because the $F$ is not significant.

## Example 3

Three different types of psychotherapy are to be evaluated. A fourth group which is a control group is also set up. There are five persons in each group. The scores on an adjustment scale (the larger number, the greater the adjustment) are

Therapy Group I:   6, 7, 5, 7, 4

Therapy Group II:   4, 5, 6, 7, 4

Therapy Group III:   3, 5, 4, 3, 6

Control Group:   2, 4, 3, 2, 3,

The means for the three therapy groups are 5.8, 5.2, and 4.2, respectively. The mean for the control group is 2.8, and so the persons receiving psychotherapy are relatively more adjusted. The grand mean is 4.5.

*TABLE 14.5*   **Analysis of Variance Table for Job Satisfaction Study**

| Source of Variation | Sum of Squares | Degrees of Freedom | Mean Square | F |
|---|---|---|---|---|
| Job (J) | 27.45 | 2 | 13.725 | 2.89 |
| Subjects within J (S/J) | 42.80 | 9 | 4.756 | |
| Total | 70.25 | 11 | | |

The correction term for the mean is $90^2/20 = 405$. The sum of squares for groups is

$$\frac{(29^2 + 26^2 + 21^2 + 14^2)}{5} - 405 = 25.8$$

The sum of squared observations is 454, and so the sum of squares total is

$$454 - 405 = 49$$

By subtraction $(49.0 - 25.8)$ the sum of squares for subjects within groups is 23.2. The degrees of freedom for groups are $4 - 1 = 3$, and the degrees of freedom for subjects within groups are $20 - 4 = 16$. The analysis of variance table is presented in Table 14.6. The $F$ for groups is statistically significant at the .01 level of significance, and omega squared is .43.

Although the $F$ is highly significant, it is not known whether the difference is due to the therapy groups being higher than the control group. A contrast must be created to test this hypothesis. The contrast compares the average of the three therapy groups with that of the control group. This results in contrast weights of .33 for the three therapy groups and $-1$ for the control group. The sum of squares for the contrast is

$$\frac{[(.33)(29) + (.33)(26) + (.33)(21) + (-1.0)(14)]^2}{5(.33^2 + .33^2 + .33^2 + 1^2)} = 19.27$$

The $F$ test of the contrast is

$$F(1,16) = \frac{19.27}{1.45} = 13.29$$

which is statistically significant at the .001 level.

# Summary

One-way analysis of variance is a statistical procedure used to test the effect of a nominal variable on an interval variable. The independent variable is a

*TABLE 14.6*  **Analysis of Variance Table for the Therapy Study**

| Source of Variation | Sum of Squares | Degrees of Freedom | Mean Square | $F$ |
|---|---|---|---|---|
| Group (G) | 25.8 | 3 | 8.60 | 5.93 |
| Subjects within G (S/G) | 23.2 | 16 | 1.45 | |
| Total | 49.0 | 19 | | |

multilevel nominal variable. In the restricted model the independent variable has no effect on the dependent variable. Variation of scores is partitioned into two sources: between groups and within groups. For each source of variation its sum of squares and degrees of freedom are computed.

The degrees of freedom for the independent variable are $k - 1$, where $k$ is the number of groups. The degrees of freedom for subjects within groups are $N - k$, where $N$ is the total number of subjects in the study. The degrees of freedom for the total are $N - 1$. The sum of squares for subjects within groups equals the sum of squares total minus the sum of squares groups. A *mean square* equals a sum of squares divided by its degrees of freedom.

The sum of squares, degrees of freedom, and mean squares are summarized in an *analysis of variance table*. The fit of the restricted model is evaluated by an $F$ test. The numerator of the test is the mean square for the independent variable and the denominator is the mean square subjects within levels of the independent variable.

The power of the $F$ test in one-way ANOVA depends on the difference between the populations means, the group size, and the degree of similarity within groups. *Omega squared* is used to measure the proportion of variance in the dependent variable that is explained by the independent variable. *Contrasts* are used to test a priori hypotheses about the exact patterning of the means. The *Tukey lsd test* can be used to test differences between all possible pairs of means. This test is a *post hoc* test, which means that the researcher need not have any hypotheses about how the means differ.

In the next chapter the topic is two-way analysis of variance, involving two independent variables. Two-way ANOVA can be viewed as an extension of one-way ANOVA.

# Problems

1. For the following significance levels and degrees of freedom determine the appropriate $F$ value needed:

|    | Alpha | $df_n$ | $df_d$ |
|----|-------|--------|--------|
| a. | .05   | 1      | 16     |
| b. | .05   | 4      | 56     |
| c. | .05   | 2      | 44     |
| d. | .001  | 1      | 63     |
| e. | .01   | 4      | 123    |
| f. | .10   | 6      | 23     |
| g. | .001  | 8      | 48     |
| h. | .01   | 2      | 72     |

2. If there are five levels of factor A with sample sizes of 10, 9, 8, 6, and 10

in the five groups, complete the following analysis of variance table and give the significance level for $F$.

| Source of Variation | Sum of Squares | Degrees of Freedom | Mean Square | F |
|---|---|---|---|---|
| Factor A | 140.0 | | | |
| Subjects within A (S/A) | | | | |
| Total | 520.0 | | | |

3. A psychologist is interested in the relationship between handedness and athletic ability. He measures the athletic ability of three groups of persons: left-handed, right-handed, and ambidextrous persons. His results are:

   Left-handed:   11, 13, 14, 13, 15

   Right-handed:   10, 8, 7, 10, 14

   Ambidextrous:   12, 8, 6, 11, 15

   Do a one-way analysis of variance to determine whether the groups differ significantly. Compute and interpret omega squared.

4. a. For the data in problem 3, do a Tukey lsd test of the difference between means.
   b. For the data in problem 3, test whether the constant is significantly different from 10.

5. A researcher seeks to measure the degree of allergic reaction to a drug. Is there a significant difference between the groups?

   Group I:   21, 19, 18, 13, 15, 20, 22, 25, 17, 17

   Group II:   12, 10, 20, 14, 18, 8, 12

   Use both analysis of variance and a $t$ test to determine whether the groups significantly differ from one another.

6. Consider a one-way analysis of variance with five levels and twelve subjects in each level. Given that the means for the five groups are

   $$2.0, \ 3.2, \ 4.1, \ 5.2, \ 5.1$$

   create a contrast that compares the first two means with the second two. Compute the mean square for that contrast.

7. For the following analysis of variance table, compute and interpret the value of omega squared.

| Source of Variation | Sum of Squares | Degrees of Freedom | Mean Square | F |
|---|---|---|---|---|
| Factor T | 55.33 | 3 | 18.44 | 7.41 |
| Subjects within T (S/T) | 139.44 | 56 | 2.49 | |
| Total | 194.77 | 59 | | |

8. An investigator wishes to determine the effectiveness of three different treatments in relieving headache pain. The drugs to be studied are aspirin, acetaminophen, and a placebo. Ten different persons take one drug and rate their pain on a ten-point scale after three hours. The scores are as follows:

   Aspirin:  7, 6, 9, 5, 3, 5, 3, 2, 4, 2

   Acetaminophen:  5, 8, 6, 4, 7, 4, 6, 2, 3, 7

   Placebo:  9, 7, 8, 7, 5, 4, 6, 8, 3, 7

   Use analysis of variance to evaluate whether the groups differ. Compute and interpret omega squared.

9. For problem 8 create a contrast that compares the two medicated groups to the placebo group. Create a contrast that compares the aspirin group to the acetaminophen group. Test each contrast.

10. For the following ANOVA table compute and interpret omega squared.

| Source of Variation | Sum of Squares | Degrees of Freedom | Mean Square | F |
|---|---|---|---|---|
| Factor A | 200 | 4 | 50 | 5 |
| Subjects within A (S/A) | 100 | 10 | 10 | |
| Total | 300 | 14 | | |

11. For problem 10 answer the following questions.

   a. How many subjects are there in the study?
   b. How many levels of the independent variable are there?
   c. What are the degrees of freedom for the $F$ test?
   d. May a Tukey lsd post hoc test be done on the means?

12. A researcher seeks to compare the marital satisfaction of women who have married for varying number of years. She finds the following (higher numbers, greater satisfaction).

One year:   56, 48, 57, 41

Two years:   63, 51, 65, 54

Ten years:   70, 61, 55, 58

  a. Use one-way ANOVA to test the effect of length of marriage on satisfaction.

  b. Test whether the average satisfaction score is significantly different from 55.

13. For problem 12 create a contrast to test the hypothesis that satisfaction increases (or decreases) for every year of marriage. Test the contrast by an $F$ test.

14. A researcher seeks to determine if the maturity of a five-year-old's speech depends on the age of his or her partner. He pairs 15 individual five-year-old subjects with one of four types of partners: infant, five-year-old, twelve-year-old, or adult. The scores on a maturity scale are

Infant:   3, 7, 5

Five-year-old:   8, 11, 14

Twelve-year-old:   11, 15, 18

Adult:   14, 12, 17, 19, 15, 13

Use one-way ANOVA to test whether the child adjusts his or her speech for different types of partners. Compute and interpret omega squared for the study.

15. A researcher seeks to determine the effect of a drug on the number of hours of sleep. Four different dosages are compared: none, 10 ml, 20 ml, and 30 ml. The results are

None:   4, 6, 5, 8, 3, 2

10 ml:   6, 8, 9, 6, 8, 4

20 ml:   7, 9, 6, 5, 4

30 ml:   9, 8, 7, 6

Use one-way ANOVA to test the effect of the drug on the number of hours slept.

16. Conduct a Tukey lsd test for the data in problem 15.

# 15 | *Two-Way Analysis of Variance*

The preceding chapter discussed ways of evaluating a model in which a nominal variable affects an interval dependent variable. The method, called one-way analysis of variance, consists of computing the variability of the group means weighted by sample size and comparing it to the variability within levels of the independent variable. In this chapter the topic is the study of the simultaneous effects of two independent variables, both measured at the nominal level. It will be shown that two-way analysis of variance is a relatively straightforward extension of one-way analysis of variance. Two-way analysis of variance is sometimes referred to as two-way ANOVA.

## *Factorial Design*

Consider the study by Ball and Bogatz (1970) on the effect of the first year of "Sesame Street" on preschool children. In their study, they divided children into four different viewing groups: (a) nonviewers, (b) occasional viewers, (c) moderate viewers, and (d) heavy viewers. They also classified children as either disadvantaged or advantaged on the basis of neighborhood. There are two independent variables: four levels of viewing and two levels of socioeconomic background. One dependent variable that they studied was the number of letters of the alphabet learned during the six months after "Sesame Street" went on the air. This variable will be called *letters learned*.

In Table 15.1 the various combinations of the Ball and Bogatz evaluation of the television program "Sesame Street" are laid out. The rows in the table are the two levels of socioeconomic status: advantaged and disadvantaged. The columns are the four levels of viewing: none, seldom, moderate, and heavy. Because there are four levels of the viewing variable and two levels of the socioeconomic variable, there are eight possible combinations, also shown in Table 15.1. These combinations are called *cells*. For instance, the

*TABLE 15.1* **Factorial Design**

| Socioeconomic Status | Viewing | | | | |
|---|---|---|---|---|---|
| | Never | Seldom | Moderate | Heavy | |
| Advantaged | 10 | 10 | 10 | 10 | 40 |
| Disadvantaged | 10 | 10 | 10 | 10 | 40 |
| | 20 | 20 | 20 | 20 | 80 |

cell in the upper left-hand corner contains those children who come from advantaged backgrounds and do not watch "Sesame Street." The cell in the bottom right-hand corner contains those children whose parents are disadvantaged and are heavy viewers of "Sesame Street." The creating of all possible combinations is called *factorial design*.

In this chapter, the two nominal independent variables are called factor A and factor B. Factor A has *a* levels, and factor B has *b* levels. There are a total of *a* times *b* cells in the study. It is usual practice to have an equal number of persons in each of the cells. So, for the "Sesame Street" example in which there are eight cells, if there were 10 children in each cell, there would be a total of 80 children in the study, as is shown in Table 15.1. A table of the *n*'s for the cells is helpful in the computation of two-way analysis of variance.

There are two important reasons for having an equal number of subjects. First, other things being equal, the estimates of the effects of the independent variable are more efficient when the cell sizes are equal. So, to measure more accurately the effect of "Sesame Street," sampling error can be minimized by having equal cell sizes. Second, the computation of the sums of squares becomes much more complicated when the cell sizes are unequal. In fact, there are a number of alternative procedures for estimating the sums of squares. Thus, for reasons of both efficiency and computational ease, equal cell sizes are preferred. All of the discussion in this chapter presumes that cell sizes are equal.

# Definitions

The score $X_{ijk}$ refers to score of person $i$ at level $j$ on factor A and at level $k$ on factor B. There are $a$ levels of factor A and $b$ levels of factor B. There are $n$ persons in each cell and a total of $ab$ cells in the design. The total number of scores is $abn$ or $N$.

To distinguish various summation terms, the following convention will be used in this chapter. The subscript under the summation sign indicates what is summed across. So

$$\sum_{i} X_{ijk}$$

indicates the sum of all the observations in the *jk* cell,

$$\sum_{i} \sum_{j} X_{ijk}$$

indicates the sum of scores for level *k* of factor B, and

$$\sum_{i} \sum_{j} \sum_{k} X_{ijk}$$

indicates the sum of all the scores.

Means can be computed for each cell of the design. They are each based on *n* observations. The cell mean for level *j* of factor A and level *k* of factor B is equal to

$$\overline{X}_{.jk} = \frac{\sum\limits_{i} X_{ijk}}{n}$$

The means for factor A are averaged across levels of factor B. Because there are *b* levels of factor B, there are a total of *bn* observations that are averaged to compute the *a* means for factor A. In terms of a formula, the mean for level *j* of factor A is

$$\overline{X}_{.j.} = \frac{\sum\limits_{i} \sum\limits_{k} X_{ijk}}{bn}$$

The means for factor B are averaged across levels of factor A. Because there are *a* levels of factor A, there are a total of *an* observations that are averaged to compute the *b* means for factor B. In terms of a formula, the mean for level *k* of factor B is

$$\overline{X}_{..k} = \frac{\sum\limits_{i} \sum\limits_{j} X_{ijk}}{an}$$

The grand mean is denoted as $\overline{X}_{...}$ and it equals the sum of observations divided by the total number of observations; that is,

$$\overline{X}_{...} = \frac{\sum\limits_{i} \sum\limits_{j} \sum\limits_{k} X_{ijk}}{abn}$$

The means for the levels of factor A, the means for levels of factor B, and the grand mean can be expressed in terms of the cell means. The grand mean can be shown to equal

1. the sum of the cell means divided by $ab$,
2. the sum of the means for factor A divided by $a$, or
3. the sum of the means for factor B divided by $b$.

All of these formulas should yield the same value for the grand mean, and so they can be used as a computational check.

The mean for the level $j$ for factor A can be computed by averaging all cell means at level $j$. There are $b$ such means. The mean for the level $k$ for factor B can be computed by averaging all the cell means at level $k$. There are $a$ such means.

It is helpful at times to present the cell means, the means for factors A and B, and the grand mean all in one table. Such a table is illustrated in Table 15.2 for the "Sesame Street" example. The numbers in the table are only hypothetical data.

In the last column are the mean for the advantaged group, 8.9 new letters learned, and the mean for the disadvantaged group, 6.7 letters learned. In the bottom row is the set of means for the four viewing groups. They increase from 6.4 to 9.3. In the bottom right-hand corner is the grand mean of 7.8. The entries in the cells are the cell means.

The set of cell means can also be graphed. The factor with the most levels is ordinarily placed on the $X$ axis. In this case that factor would be viewing. The cell means are then plotted on the graph, and one makes certain to place the mean in the appropriate place above the $X$ axis. The points are connected for each level of the second factor (the one not on the $X$ axis). So, as in Figure 15.1, the points for the advantaged groups are connected. To distinguish the two lines one can be solid and the other dashed, as in the figure.

**TABLE 15.2**   **Hypothetical Table of Means for Two-Way ANOVA**

| Socioeconomic Status | Never | Viewing Seldom | Moderate | Heavy | |
|---|---|---|---|---|---|
| Advantaged | 8.0 | 8.4 | 9.2 | 10.0 | 8.9 |
| Disadvantaged | 4.8 | 6.2 | 7.2 | 8.6 | 6.7 |
| | 6.4 | 7.3 | 8.2 | 9.3 | 7.8 |

*FIGURE 15.1*     **Graph of hypothetical means from the "Sesame Street" evaluation.**

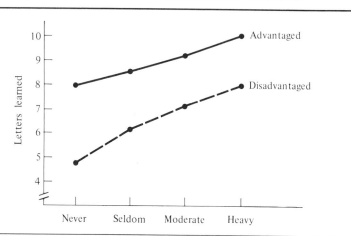

# The Concepts of Main Effect and Interaction

One purpose of conducting a two-way analysis of variance is to measure and test the effect of each of the two independent variables. So, with two-way ANOVA, two effects can be tested for the price of one. In two-way analysis of variance, however, there is more than one way to measure the effect of an independent variable. For instance, for the "Sesame Street" example, there is first the program's effect on advantaged children and second its effect on disadvantaged children. The *main effect* of a given independent variable is the effect of that variable averaged across all other levels of the other independent variable. One advantage of having equal numbers of subjects in each cell is that to compute the main effect of one independent variable, one adds the means across cells of the other independent variable and divides by the number of cells to compute the means of a main effect.

To interpret the main effect one examines the means for that factor. Returning to the table of means for the "Sesame Street" example, first the means are examined for the advantaged and disadvantaged subjects. They show that advantaged children learned more letters than disadvantaged children. To determine the main effect for viewing, the four means in the bottom row of Table 15.2 are examined. They show that the more often the children viewed "Sesame Street" the more letters they learned.

Alternatively, one can examine the main effects graphically. Returning to the "Sesame Street" example, it can be seen in Figure 15.1 that the line for advantaged children is always above the line for disadvantaged children. Thus, advantaged children outperform disadvantaged children at all four levels of viewing. Also the lines for both groups increase as the eye moves along the $X$ axis. So, children who view more "Sesame Street" learn more letters.

There are two major purposes in doing a two-way analysis of variance instead of doing separate studies, one for each independent variable. First, it is much more economical to have one study with about the same number of subjects and perform a two-way analysis of variance. One gets two studies for the effort of one. Second, with a two-way analysis of variance one gets information concerning the presence of interaction between the two variables. *Two variables are said to interact if the effect of one variable on the dependent variable varies as a function of the level of the other variable.* Consider the effect of "Sesame Street" on the learning of letters as shown in Table 15.2 and Figure 15.1. If the effect of the program is stronger for lower socioeconomic children than for higher socioeconomic children, it can be said that viewing "Sesame Street" and socioeconomic status *interacted* in causing the learning of letters. This is indicated in both the table and the figure. The effect of "Sesame Street" is greater for disadvantaged than for advantaged children.

The interaction between factor A and factor B is ordinarily referred to as the A by B interaction and it is usually symbolized by $A \times B$.

Predictions of interaction are very common in the social and behavioral sciences. For instance, one question of particular interest is the interaction between diagnostic category and form of therapy. If alcoholics were more helped by group therapy than traditional individual psychotherapy and neurotics were more helped by individual psychotherapy, it would be said that diagnostic category (alcoholic versus neurotic) interacts with mode of therapy (group versus individual).

Another example of interaction might be found when examining the effect of inhaling one milliliter of a toxic drug in the workplace and having a full versus an empty stomach. It might be found that inhaling a toxic drug with a full stomach has relatively little harmful effect, whereas the drug is quite toxic when inhaling it on an empty stomach. In this case, the drug (none versus inhaling one ml) interacts with having eaten (empty versus full stomach) to cause a toxic reaction.

In discussing an interaction, it is said that the effect of factor A on variable $X$ varies depending on the level of factor B. Alternatively, it must also be true that the effect of factor B on variable $X$ varies as a function of the level of factor A. Thus there is a choice in saying which variable's effect changes as a function of which other variable. If the interest is primarily in factor A, then the preference is to state that A's effect changes as a function of B. For instance, for the "Sesame Street" example, instead of saying that the effect of the program was greater for disadvantaged children, it could have been stated that the advantaged children outperformed the disadvantaged children least when both groups were heavy viewers of the program. Also, if A has more levels than B it is probably simpler to say that A's effect changes as function of B.

Interactions can also be represented graphically. The dependent variable is on the $Y$ axis and the independent variable with a larger number of levels is on the $X$ axis. The means of the dependent variable are graphed on the $X$ axis separately for each level of the other independent variable. If the distance between the lines on the graph varies, then an interaction is present. In Figure 15.2 are examples of graphs with interaction and with no interaction. (The diagrams in Figure 15.2 are idealized in that there is no sampling error; actually, graphs ordinarily do not show such clear patterns.)

In both graphs in the figure, six means from a 2 by 3 design are plotted. There are three levels of factor A and two levels of factor B. The $X$ axis is used to distinguish levels of A and two separate lines are drawn for the two levels of B. In the graph labeled I, the gap between the pair of B means increases as one moves along the $X$ axis. It is smallest for A1 and largest for A3. There is then an interaction between the two independent variables. The difference between the B means varies as a function of A. However, in the graph on the right labeled II, the gap remains the same. There is then no interaction between the independent variables. The effect of B is the same for the three levels of A.

To understand better the concept of interaction, examine the graph on the left of Figure 15.3 which is labeled as I. (Again, these are idealized patterns without sampling error.) The graph very clearly shows that A and B interact. At A1, there is no difference between B1 and B2. But as the eyes move to the right on the $X$ axis, the effect of B becomes larger. Although the graph on the left-hand side of Figure 15.3 shows clear interaction, it also shows clear main effects of A and B. For A, the A3 means are on average larger than the A2 and A1 means, and the A2 means are on average larger than the A1 means. For B, the average of the B1 means is larger than the average of the B2 means. So, the pattern of means in the left part of Figure 15.3 shows two main effects and an interaction.

**FIGURE 15.2**   **Illustration of interaction.**

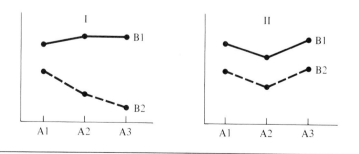

***FIGURE 15.3***    **Second illustration of interaction.**

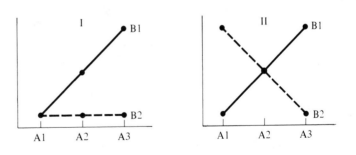

In Figure 15.3 the graph on the right shows pure interaction and no main effects. The distinctive feature about the graph is the crossover of the two B lines which must be present when there is interaction and no main effects.

Consider a final example based on actual data. West and Shults (1976) examined how much persons liked male and female names. They determined the commonness of the name by its frequency of occurrence in a college yearbook. They asked 148 persons to state how much they liked common male names such as David and John versus uncommon male names such as Jerome or Julius. They were also asked how much they liked common female names such as Mary and Carol versus uncommon names such as Melinda or Rosemary. There are two factors in this study. They are sex of the name, male or female, and commonness of the name, common or uncommon. The ratings were on a five-point scale, where a five is a favorable rating and a one is an unfavorable rating. The means are presented in Table 15.3.

The results show a main effect of the commonness of the name. Common

***TABLE 15.3***    **Favorableness of Rating**

| Commonness of Name | Sex of Name | | Average |
|---|---|---|---|
| | Male | Female | |
| Common | 3.540 | 3.240 | 3.390 |
| Uncommon | 2.420 | 2.980 | 2.700 |
| Average | 2.980 | 3.110 | 3.045 |

names are liked more than uncommon names (3.390 versus 2.700). Although male names are liked less than females (2.980 versus 3.110), this difference is too small to be statistically reliable. There is then no main effect for sex of name. But the two factors do interact. Overall, common names are liked better than uncommon names by .69 unit. This effect is strong for male names, a difference of 1.12 units, but the effect is relatively weak for females, a difference of .26 unit. The two factors clearly interact. The effect of commonness depends on gender.

# Estimation and Definitional Formulas

The complete model for two-way analysis of variance contains many terms. The dependent variable consists of the following terms:

1. the constant,
2. the main effect for factor A,
3. the main effect for factor B,
4. the interaction between A and B, and
5. the residual variable.

The estimates of these four terms are as follows:

1. The constant:   the grand mean or $\bar{X}_{...}$
2. The main effect for factor A at level $j$:  $\bar{X}_{.j.} - \bar{X}_{...}$
3. The main effect for factor B at level $k$:  $\bar{X}_{..k} - \bar{X}_{...}$
4. The interaction between A and B for cell $jk$:

$$\bar{X}_{.jk} - \bar{X}_{.j.} - \bar{X}_{..k} + \bar{X}_{...}$$

5. The residual score for person $i$ in cell $jk$: $X_{ijk} - \bar{X}_{.jk}$

The sum of squares for any effect involves the squares of all effects times the sample size that the effect is based on.

It is also necessary to define various totals, as follows:

$$T_{...} = \sum_i \sum_j \sum_k X_{ijk}$$

$$T_{.jk} = \sum_i X_{ijk}$$

$$T_{.j.} = \sum_i \sum_k X_{ijk}$$

$$T_{..k} = \sum_i \sum_j X_{ijk}$$

The $F$ distribution is used to evaluate the plausibility of the restricted models. There are three restricted models in two-way analysis of variance. In each, one of the effects (A, B, or A $\times$ B) is omitted.

# The Computation of Two-Way Analysis of Variance

A two-way analysis of variance amounts to little more than parts of three separate one-way analyses of variance. So, a sound understanding of one-way analysis of variance is essential for the understanding of two-way analysis of variance.

First, a one-way analysis of variance is computed for factor A ignoring factor B. Second, a one-way analysis of variance is computed for factor B ignoring factor A. Third, a big one-way analysis of variance is computed that treats the cells as levels of a single factor. This last ANOVA can be viewed as an analysis of the AB factor. The computation of two-way analysis of variance consists of the piecing together parts of these three different one-way analyses.

The sums of squares for the main effects of each of the factors ignoring the other are taken from the one-way analyses of variance. The sums of squares interaction is measured by taking the sum of squares from the big one-way analysis and subtracting the sum of squares for both of the main effects.

To compute the sum of squares for $A$, $B$, and $A \times B$ the correction term for the mean or $C$ is needed. It equals

$$C = \frac{T_{...}^{2}}{abn}$$

In words, it is simply the square of the sum of all scores, divided by the number of all the observations in the study.

To compute the sum of squares for factor A, or $SS_A$, each A total is squared, these squares are summed across the $a$ groups, this sum is divided by the number of observations that the totals are based on, and $C$ is subtracted. In terms of a formula,

$$SS_A = \frac{\sum_{j} T_{j.}^{2}}{bn} - C$$

The sum of squares for A using this formula results in the same sum of squares as would be obtained if B were ignored and a one-way analysis of variance sum of squares for A were computed.

To compute the sum of squares for factor B, or $SS_B$, each B total is squared, these squares are summed across the $b$ groups, this sum is divided by the number of observations that it is based on, and $C$ is subtracted. In terms of a formula,

$$SS_B = \frac{\sum_{k} T_{..k}^{2}}{an} - C$$

To compute the sum of squares for interaction, or $SS_{A \times B}$, the sum of squares AB, or $SS_{AB}$, is computed. This sum of squares is based on a one-way analysis of variance in which the cell means are treated as group means. Its formula is

$$SS_{AB} = \frac{\sum_j \sum_k T_{.jk}^2}{n} - C$$

The cell totals are divided by $n$ because each cell mean is based on $n$ observations. The formula for the sum of squares for interaction is

$$SS_{A \times B} = SS_{AB} - SS_A - SS_B$$

The sum of squares for subjects within the AB cells is computed indirectly. To compute it, first the sum of squares total, or $SS_{TOT}$, is computed. Its formula is

$$SS_{TOT} = \sum \sum \sum X_{ijk}^2 - C$$

The sum of squares for subjects within cells equals

$$SS_{S/AB} = SS_{TOT} - SS_{AB}$$

or, alternatively,

$$SS_{S/AB} = SS_{TOT} - SS_A - SS_B - SS_{A \times B}$$

There is a general formula for the sum of squares for a main effect. The formula for the main effect of D is

$$\begin{array}{l} \text{sum of} \\ \text{squares} \\ \text{for effect D} \end{array} = \text{sum} \begin{bmatrix} \text{each of the} \\ \text{D totals} \\ \text{squared} \end{bmatrix} \div \begin{array}{l} n \text{ each} \\ \text{D total} \\ \text{is based on} \end{array} - \begin{array}{l} \text{correction} \\ \text{term for the} \\ \text{grand mean} \end{array}$$

First, each D total is squared and summed across the set of D totals. This sum of totals squared is divided by the sample size of the totals. Finally, $C$, the correction term for the grand mean, is subtracted.

As was done with one-way analysis of variance, it is useful to diagram the partitioning of the sum of squares. In Figure 15.4 the large circle represents the total sum of squares. The area on inside the two overlapping circles represents the sum of squares AB. The area outside the two overlapping circles represents the $SS_{S/AB}$. It can then be graphically seen that the $SS_{S/AB}$ equals the $SS_{TOT}$ minus the $SS_{AB}$.

The area in the overlapping circles can be partitioned. The portion of the two smaller circles that does not overlap is the sum of squares for A (on the left) and the sum of squares for B (on the right). Note then that the $SS_A$ is not the entire circle on the left but only the nonoverlapping part.

The part of the two circles that overlaps is the sum of squares interaction. The diagram illustrates how the sum of squares AB equals the sum of squares for the two main effects plus the sum of squares for interaction. Thus the

**FIGURE 15.4**    **Circle diagram for two-way ANOVA.**

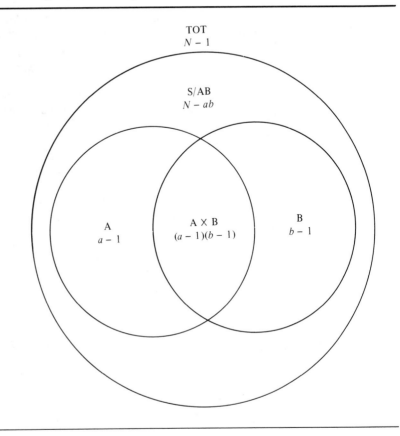

diagram illustrates the partitioning of the sum of squares, as well as their degrees of freedom.

After computing the sum of squares, the next step is to determine the degrees of freedom, as follows:

$$df_A = a - 1$$
$$df_B = b - 1$$

The degrees of freedom for main effect are as they were for one-way ANOVA: They each equal the number of groups less one.

The rule for determining the degrees of freedom for interaction is simple. The *degrees of freedom for interaction* equal the product of the degrees of freedom of its components; that is,

$$df_{A \times B} = (a - 1)(b - 1)$$

The degrees of freedom for subjects within cells equal

$$df_{S/AB} = N - ab$$

where $N$ is the total number of persons in the study. As with the sum of squares, the degrees of freedom partition in the same way:

$$df_{S/AB} = df_{TOT} - df_{AB}$$
$$df_{A \times B} = df_{AB} - df_A - df_B$$

The $df_{AB}$ equal $ab - 1$, the number of cells less one.

Each mean square is its sum of squares divided by its degrees of freedom. The individual mean squares are

$$MS_A = \frac{SS_A}{df_A}$$

$$MS_B = \frac{SS_B}{df_B}$$

$$MS_{A \times B} = \frac{SS_{A \times B}}{df_{A \times B}}$$

$$MS_{S/AB} = \frac{SS_{S/AB}}{df_{S/AB}}$$

For A, B, and A $\times$ B, the denominator of the $F$ ratio is $MS_{S/AB}$:

$$F(a-1, N-ab) = \frac{MS_A}{MS_{S/AB}}$$

$$F(b-1, N-ab) = \frac{MS_B}{MS_{S/AB}}$$

$$F((a-1)(b-1), N-ab) = \frac{MS_{A \times B}}{MS_{S/AB}}$$

These $F$ tests evaluate whether a restricted model, one that does not include the term in the numerator, is plausible. Note that the $df_n$ differs for these three $F$ tests if there are a different number of levels of A than B. So, in determining the statistical significance of the $F$ tests, different values from Appendix E must be used. A good rule of thumb is that $F$ must be about 4.0 or more to be significant at the .05 level of significance.

As with one-way analysis of variance, the basic results are summarized in an analysis of variance table (see Table 15.4). The column headings for two-way analysis of variance are the same as those for one way analysis of variance. They are source of variation, sum of squares, degrees of freedom, mean square, and $F$. The sources of variation are factor A, factor B, the A by B interaction, subjects within the AB cells, and total.

*TABLE 15.4*     **Analysis of Variance Table**

| Source of Variation | Sum of Squares | Degrees of Freedom | Mean Square | $F$ |
|---|---|---|---|---|
| Factor A | $SS_A$ | $df_A$ | $MS_A$ | $\dfrac{MS_A}{MS_{S/AB}}$ |
| Factor B | $SS_B$ | $df_B$ | $MS_B$ | $\dfrac{MS_B}{MS_{S/AB}}$ |
| A × B | $SS_{A \times B}$ | $df_{A \times B}$ | $MS_{A \times B}$ | $\dfrac{MS_{A \times B}}{MS_{S/AB}}$ |
| Subjects within cells (S/AB) | $SS_{S/AB}$ | $df_{S/AB}$ | $MS_{S/AB}$ | |
| Total (TOT) | $SS_{TOT}$ | $df_{TOT}$ | | |

The analysis of variance table neatly summarizes the computation and the results of the model testing.

## Assumptions and Power

The assumptions of two-way analysis of variance are identical to those of one-way analysis of variance and the two-sample $t$ test. They are (a) normally distributed residual variable, (b) equal variances in all the cells, and (c) independence of observations. The reader is referred to Chapter 13 for an extensive discussion of these assumptions.

The considerations for the power in the test of the main effects are the same as in one-way analysis of variance, presented in Chapter 14. The power of the test of interaction is ordinarily not as large as that of the main effect. A typical interaction is one in which the effect of one factor becomes weaker across levels of the other factor. Crossover interactions, as in graph II of Figure 15.3, are not common. Thus an interaction measures not some overall effect but the variation of an effect. Therefore main effects are tested with more power than interactions.

## Example

Imagine a researcher who wishes to measure the effect of a cigarette smoking on shortness of breath. She creates three groups of smokers: heavy, light, and none. She then divides these three groups by age: 30s, 40s, and 50s. Her raw data are

| *Smoking* | *Age* | | |
|---|---|---|---|
| *Level* | *30s* | *40s* | *50s* |
| Heavy | 4, 5, 6 | 5, 6, 10 | 7, 9, 11 |
| Light | 3, 3, 6 | 2, 4, 6 | 3, 5, 4 |
| None | 2, 2, 5 | 5, 3, 4 | 3, 6, 6 |

She has three persons in each cell of the design.

Some preliminary tables can simplify both the calculations and interpretation. The table of $n$'s for the study is as follows:

| | | Age | | | |
|---|---|---|---|---|---|
| | | 30s | 40s | 50s | Total |
| | Heavy | 3 | 3 | 3 | 9 |
| Smoking Level | Light | 3 | 3 | 3 | 9 |
| | None | 3 | 3 | 3 | 9 |
| | Total | 9 | 9 | 9 | 27 |

A table of total scores is as follows:

| | | Age | | | |
|---|---|---|---|---|---|
| | | 30s | 40s | 50s | Total |
| | Heavy | 15 | 21 | 27 | 63 |
| Smoking Level | Light | 12 | 12 | 12 | 36 |
| | None | 9 | 12 | 15 | 36 |
| | Total | 36 | 45 | 54 | 135 |

These tables of $n$'s and totals are useful in computing various sums of squares. Each total squared will be divided by its corresponding $n$. The means are as follows:

|  | Age | | | |
|--|------|------|------|---------|
|  | 30s | 40s | 50s | Average |

| Smoking Level | | 30s | 40s | 50s | Average |
|---------------|--------|-----|-----|-----|---------|
|  | Heavy | 5.0 | 7.0 | 9.0 | 7.0 |
|  | Light | 4.0 | 4.0 | 4.0 | 4.0 |
|  | None | 3.0 | 4.0 | 5.0 | 4.0 |
|  | Average | 4.0 | 5.0 | 6.0 | 5.0 |

The correction term for the mean is $135^2/27 = 675$. The sum of squares for smoking groups is

$$\frac{63^2 + 36^2 + 36^2}{9} - 675 = 54$$

The sum of squares for age is

$$\frac{36^2 + 45^2 + 54^2}{9} - 675 = 18$$

The sum of squares for cells is

$$\frac{15^2 + 21^2 + 27^2 + 12^2 + 12^2 + 12^2 + 9^2 + 12^2 + 15^2}{3} - 675 = 74$$

The sum of squares for interaction is

$$74 - 54 - 18 = 12$$

The sum of each squared observation is 813. The sum of squares for the total is then $813 - 675$ or 138. The sum of squares for persons within cells is $138 - 74 = 54$. The analysis of variance table is as follows:

| Source of Variation | Sum of Squares | Degrees of Freedom | Mean Square | F |
|---------------------|----------------|--------------------|-------------|-----|
| Smoke (S) | 54.0 | 2 | 27.0 | 9.0 |
| Age (A) | 18.0 | 2 | 9.0 | 3.0 |
| S × A | 12.0 | 4 | 3.0 | 1.0 |
| Persons within SA (P/SA) | 54.0 | 18 | 3.0 | |
| Total | 138.0 | 26 | | |

Only the main effect of smoking is statistically significant, and its significance level is the .001 level. The means show that heavy smokers have more difficulty breathing.

# Generalization to Higher-Order Analysis of Variance

The generalization to three- and four-way analysis of variance is straightforward. Again, the independent variables are denoted as A, B, and C. There are *a* levels within factor A, *b* levels within factor B, and *c* levels within factor C. There are *abc* cells, each with *n* persons. The total number of subjects in the study, *abcn*, is *N*.

The sums of squares for the main effects are computed exactly as they are computed in one- and two-way analysis of variance. For each level of a main effect, the total is squared and summed across levels, and this sum of squared totals is divided by the number of observations that each total is based on, and the correction term for the grand mean is subtracted.

To determine the interaction between two factors, three "one-way" analyses are done for the AB, AC, and BC means. These sums of squares are denoted as $SS_{AB}$, $SS_{AC}$, and $SS_{BC}$, respectively. The sums of squares for interaction equal

$$SS_{A \times B} = SS_{AB} - SS_A - SS_B$$
$$SS_{A \times C} = SS_{AC} - SS_A - SS_C$$
$$SS_{B \times C} = SS_{BC} - SS_B - SS_C$$

To determine the $SS_{A \times B \times C}$, first the sum of squares for ABC is computed. This is a one-way sum of squares in which the *abc* cells are treated as groups. The sum of squares for the A × B × C interaction is as follows:

$$SS_{A \times B \times C} = SS_{ABC} - SS_{A \times B} - SS_{A \times C} - SS_{B \times C} - SS_A - SS_B - SS_C$$

The sum of squares of subjects within cells equals

$$SS_{S/ABC} = SS_{TOT} - SS_{ABC}$$

where $SS_{TOT}$ equals the sum of each squared observation minus the correction term for the mean.

The degrees of freedom for the two-way interaction are computed in the usual way. They equal the product of the degrees of freedom of the components. They are then

$$df_{A \times B} = (a - 1)(b - 1)$$
$$df_{A \times C} = (a - 1)(c - 1)$$
$$df_{B \times C} = (b - 1)(c - 1)$$

The degrees of freedom for the A × B × C interaction are

$$df_{A \times B \times C} = (a - 1)(b - 1)(c - 1)$$

Like any interaction, its degrees of freedom are the product of the degrees of freedom of its components. The formula for $df_{S/ABC}$ is

$$df_{S/ABC} = N - abc$$

where $N$ is the total number of observations in the study.

As always, a mean square equals its sum of squares divided by its degrees of freedom. The $F$ test consists of each mean square divided by the mean square subjects within ABC cells.

As was done with two-way analysis of variance, it is useful to diagram the partitioning of the sum of squares. In Figure 15.5 the large circle represents the total sum of squares. The three overlapping circles inside of it represent the sum of squares ABC. The area outside of these three circles, but still within the large circle, represents the $SS_{S/ABC}$. It can then be graphically seen that the $SS_{S/ABC}$ equals the $SS_{TOT}$ minus the $SS_{ABC}$.

**FIGURE 15.5**    **Circle diagram for three-way ANOVA.**

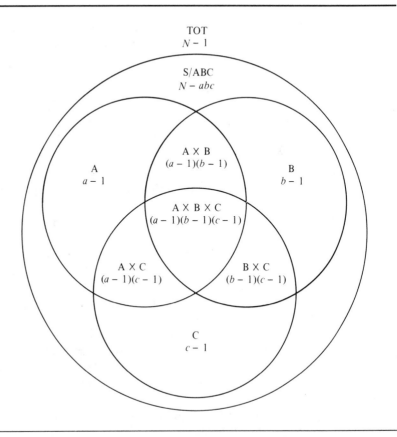

The parts of the three circles that do not overlap are the sums of square for A (on the left), for B (on the right), and for C (on the bottom). As with two-way ANOVA, the main effect for an effect is not the entire circle but only the nonoverlapping portion. The portion of the three smaller circles that completely overlaps is the sum of squares interaction of A by B by C. The part of the A and B circles that overlaps excluding C represents the A $\times$ B interaction. In a similar fashion the B $\times$ C and A $\times$ C interactions can be defined. So, the diagram illustrates the partitioning of the sum of squares.

Generalization to a four-way design follows along the same lines. Perhaps the most difficult aspect of four-way designs is that interactions can be very difficult to interpret.

# Repeated Measures Design

The estimation procedures for ANOVA have presumed that the groups are independent. In Chapter 13, a design is presented in which observations were matched, paired, or linked in some way. The most common way in which observations are linked is that they come from the same person. For example, consider a small study on the effects of psychotherapy on psychological adjustment. Six subjects were measured before and after psychotherapy on an adjustment scale. Higher scores indicate greater adjustment. The numbers are:

| Subject | Before | After |
|---------|--------|-------|
| 1 | 23 | 32 |
| 2 | 27 | 25 |
| 3 | 31 | 40 |
| 4 | 32 | 31 |
| 5 | 26 | 38 |
| 6 | 25 | 29 |

This is a paired design because one score in each group is linked to the same person.

It is also a two-way design. There are two independent variables. They are psychotherapy, before versus after, and person, 1 through 6. It is possible to compute a mean square subject, a mean square psychotherapy, and a mean square interaction. The two-way analysis of variance table is presented in Table 15.5. The main effect for subject refers to the extent to which subjects differ from one another across both time points. The subject by treatment interaction refers to whether the treatment is more effective for some subjects than others. There is no subjects within cells term that can be used as a denominator for the $F$ test. Instead the person by treatment interaction is used as the denominator for the $F$ test to test for the presence of treatment effects. The value of this $F$ exactly equals the value of $t^2$ that would be obtained from the paired $t$ test described in Chapter 13. The equivalence of repeated measures $F$ and paired $t^2$ occurs when the independent variable in the design has two levels.

*TABLE 15.4*    **Repeated Measures Example**

| Source | SS | *df* | MS | *F* |
|--------|-----|-----|-------|------|
| Treatment (T) | 80.08 | 1 | 80.08 | 4.80 |
| Person (S) | 135.42 | 5 | 27.08 | |
| S × T | 83.42 | 5 | 16.68 | |
| Total | 298.92 | 11 | | |

Wherever there is a series of observations for each person, a two-way analysis of variance can be computed. Such a design is commonly referred to as a *repeated measures design*. This notion of person as variable is a fundamental insight in understanding complicated analyses of variance. Subjects are a very special kind of independent variable, but they are a variable.

Repeated measures designs are very commonly employed in psychological research. In particular, most research in cognitive psychology uses repeated measures designs. Persons in these experiments receive stimuli that are arrayed to represent various levels of a given independent variable. It is not at all uncommon in these studies for a single person to provide data for as many as 25 experimental conditions.

There two major reasons for the popularity of repeated measures designs. First, with a repeated measures design the researcher needs fewer subjects to obtain the same number of observations than is the case with an independent groups design. Second, even if the number of observations is the same for both designs, a repeated measures design is still usually much more powerful than an independent groups design. The reason for this is that subject variation is removed from the residual variance.

Repeated measures designs do have their drawbacks. A complete discussion of these drawbacks can be found in advanced textbooks (Myers, 1979; Winer, 1971). These drawbacks are linked to the fact that measurements are almost always sequentially ordered.

# Summary

In two-way analysis of variance, two nominal variables affect a variable measured at the interval level. The creation of all possible combinations of two nominal variables is called *factorial design*. A particular combination in a factorial design is called a *cell*.

The *main effect* of a factor is the average effect of that factor across levels of the other variable. An *interaction* between two factors implies that the effect of one factor changes as a function of the other. An interaction can be assessed by an examination of a graph or table.

Sums of squares for the main effects are computed as in one-way analysis of variance. The interaction sum of squares equals the sum of squares for cells minus the sums of squares for main effects. The sum of squares for subjects within cells is the pooled sum of squares for each cell pooled across cells. This sum of squares is computed by subtracting the sum of squares cells from the sum of squares total. The degrees of freedom for interaction equal the product of the degrees of freedom for the main effects. The mean square for an effect equals its sum of squares divided by its degrees of freedom.

Hypotheses in two-way ANOVA are evaluated by an $F$ test. The denominator of the $F$ ratio is the mean square of subjects within cells. A significant $F$ ratio indicates that the term in the numerator must be included in the complete model.

*Repeated measures design* involves having each subject be in each level of the independent variable. In a repeated measures design the effect of a factor is tested by using the mean square interaction of subject by factor as the denominator of the $F$ ratio.

# Problems

1. Fill in the remainder of the source table from an equal $n$ study with ten subjects in each cell.

| Source of Variation | Sum of Squares | Degrees of Freedom | Mean Square | F |
|---|---|---|---|---|
| A | 6.3 | 3 | | |
| B | 4.3 | 2 | | |
| A × B | 6.0 | | | |
| Persons within AB (S/AB) | | | | |
| Total | 89.0 | | | |

State whether the effects are statistically significant.

2. A pig buyer wants to compare the fat content of bacon and ham in 20 different pigs, 10 from California and 10 from Nebraska. Do a two-way analysis of variance and interpret your results.

| | Birthplace | |
|---|---|---|
| | Nebraska | California |
| Ham | 30, 30, 25, 27, 26 | 26, 17, 37, 38, 34 |
| Bacon | 40, 28, 26, 29, 35 | 34, 29, 36, 37, 42 |

3. For the table below do a two-way analysis of variance and interpret your results.

| | B | | | | |
|---|---|---|---|---|---|
| | 1 | 2 | 3 | 4 | 5 |
| A 1 | 7, 9, 6, 5 | 3, 2, 1, 7 | 7, 8, 9, 3 | 1, 3, 8, 7 | 2, 4, 3, 1 |
| A 2 | 4, 3, 1, 6 | 4, 1, 3, 5 | 3, 1, 6, 5 | 1, 3, 4, 2 | 7, 6, 5, 9 |

4. A researcher is interested in studying the overjustification effect. Simply put, this effect states that people do not enjoy activities that they used to do solely for fun after they are paid for engaging in the behavior. A researcher wishes to investigate whether the effect is stronger for younger than for older children. To study the phenomenon the researcher has ten younger children (age four) play with a toy as well as ten older children (age seven). For each of these groups, five of the children were given candy as an incentive to play with the toy and five were not. The experimenter then measured the duration of time spent with the toy at a later period. The results are:

Older, rewarded: 44, 110, 12, 44, 59

Older, unrewarded: 79, 120, 112, 68, 39

Younger, rewarded: 64, 10, 34, 119, 78

Younger, unrewarded: 73, 10, 102, 49, 99

Conduct a two-way analysis of variance to see whether the groups differ.

5. Consider the following table of means.

| | B1 | B2 | B3 |
|---|---|---|---|
| A1 | 6.1 | 5.3 | 9.0 |
| A2 | 9.2 | 8.5 | 12.1 |

Interpret the main effects and interaction.

6. Five subjects are asked to learn material over a period of four days. Their data are

|        |    | Day |    |    |
|--------|----|-----|----|----|
| *Subject* | *1* | *2* | *3* | *4* |
| 1 | 10 | 15 | 18 | 20 |
| 2 | 15 | 18 | 18 | 22 |
| 3 | 10 | 24 | 22 | 25 |
| 4 | 12 | 19 | 20 | 25 |
| 5 | 16 | 31 | 40 | 42 |

Conduct a two-way ANOVA treating subject and day as factors. Test the effect of day.

7. Complete the following three-way ANOVA table, where there are two subjects in each cell.

| SV | SS | df | MS | F |
|----|----|----|----|----|
| A | 124 | 2 | | |
| B | 25 | 3 | | |
| C | 48 | 1 | | |
| A × B | 12 | | | |
| A × C | 19 | | | |
| B × C | 14 | | | |
| A × B × C | 12 | | | |
| S/ABC | | | | |
| Total | 318 | | | |

8. Construct a table of means for an experiment in which both independent variables have two levels. Designate the factors as A and B. Make the following three tables.

   a. a table with a main effect for B only
   b. a table with a main effect for A only
   c. a table with an interaction of A with B only

9. An experimenter wants to see whether the deleterious effects of alcohol are increased when one drinks on an empty stomach. He has 20 subjects learn nonsense symbols in one of four conditions: I: no alcohol, empty stomach; II: no alcohol, full stomach; III: alcohol, empty stomach; IV: alcohol, full stomach. He then measured the number correct out of 15 syllables, with the following results.

   I:  14, 12, 15, 14, 13

   II:  13, 15, 14, 12, 14

   III:  8, 7, 5, 9, 12

   IV:  10, 12, 14, 9, 11

Analyze the data by two-way ANOVA and interpret the results.

10. A researcher wants to see whether intelligence (IQ) is affected by birth order. He finds eight families with three children and measures the IQ of each member. The data are as follows

| | *Birth Order* | | |
| --- | --- | --- | --- |
| *Family* | *First* | *Second* | *Third* |
| 1 | 130 | 125 | 120 |
| 2 | 105 | 90 | 75 |
| 3 | 140 | 145 | 125 |
| 4 | 80 | 70 | 80 |
| 5 | 135 | 120 | 115 |
| 6 | 90 | 80 | 60 |
| 7 | 110 | 105 | 90 |
| 8 | 125 | 100 | 110 |

Test whether birth order affects IQ.

# 16 | *Testing Measures of Association*

In Chapters 6 and 7 the two most common measures of association were presented: regression and correlation coefficients. The regression coefficient, or $b$, measures the change in the criterion variable as a function of a one-unit change in the predictor variable. The correlation coefficient, or $r$, is a regression coefficient with both variables standardized (expressed as $Z$ scores); that is, both variables have means of zero and variances of one. It is a symmetric measure of association that varies from $-1$ to $+1$. For both measures of association, a value of zero indicates no linear association. In this chapter, methods are presented to test hypotheses about correlation and regression coefficients.

The use of the correlation coefficient or $r$ does not require a specification about the direction of the causal effect. That is, if a researcher correlates the degree to which a parent uses physical punishment and how aggressive the children are, there is no need to make any assumptions about which of the following causal patterns is true.

1. Physical punishment causes aggression.
2. Aggressive children make parents use physical aggression.
3. Physical punishment and aggression are two signs of a troubled family.

Correlations make no presumption about what is the independent variable and what is the dependent variable. There is then not a single complete model when one tests correlation coefficients because there are three distinct ways in which the correlation could come about. The complete model then presumes some unspecified causal network that brings about association between the variables. The restricted model is that there is no association between the variables.

A regression coefficient, when used in explanation and not in prediction or description, does make a clear statement about a causal ordering. The predictor variable is the independent variable and the criterion is the dependent variable. The complete model is

$$\begin{matrix} \text{dependent} \\ \text{variable} \end{matrix} = \text{constant} + \text{coefficient} \begin{bmatrix} \text{independent} \\ \text{variable} \end{bmatrix} + \begin{matrix} \text{residual} \\ \text{variable} \end{matrix}$$

The coefficient in the model is called the *regression coefficient*. Both variables are measured at the interval level of measurement. The complete model for regression is the same as one-way analysis of variance except that the independent variable is measured at the interval level. The restricted model is the same as the complete model, but the independent variable has no effect on the dependent variable.

In this chapter, first tests of correlation coefficients are presented because procedures to do so are relatively simpler than tests of regression coefficients. Then the somewhat more complicated tests of regression coefficients are described. In the last section of the chapter, rules for determining which type of test is most appropriate are presented.

# Tests of Correlation Coefficients

In this section, the following tests of correlation coefficients are presented:

1. How to test a single correlation coefficient.
2. How to test whether two correlation coefficients computed from different samples are equal. Correlations computed using different groups of persons are called *independent* correlations.
3. Testing more than two independent correlation coefficients.
4. How to test whether two correlations computed from the same sample are equal to each other.

## A Single Correlation Coefficient

Consider two variables: the number of times a person nods his or her head in a conversation and the degree to which the person likes his or her partner in the conversation. The two variables are nods and liking. The two variables are correlated across 30 pairs of persons and the correlation is .45. The correlation indicates that the more one nods during a conversation the more one likes one's partner. However, one might wonder whether the .45 value in the sample might have just occurred by chance. That is, if there were a thousand pairs of persons, would the correlation be zero? Is the .45 value due to sampling error or does it reflect a true positive correlation? A way is needed to evaluate whether a sample correlation coefficient is significantly different from zero.

It turns out that the distribution of *r* does not closely correspond to any of the major sampling distributions. However, for a population correlation of zero, $r$ divided by the square root of $1 - r^2$ is approximately normally distributed with a mean of zero and a variance of $n - 2$.

The test of the null hypothesis that a correlation coefficient equals zero is

$$t(n-2) = \frac{r\sqrt{n-2}}{\sqrt{1-r^2}}$$

In words, the correlation coefficient is divided by the square root of one minus the correlation squared, and then this quantity is multiplied by the square root of the sample size minus two. Under the null hypothesis that the population correlation equals zero, this quantity has a $t$ distribution with $n - 2$ degrees of freedom, where $n$ is the number of pairs of scores. So, one computes $r\sqrt{(n-2)/(1-r^2)}$ and determines whether it equals or exceeds the critical values for $t$ in Appendix D. The degrees of freedom are $n - 2$ and one rounds down to the closest value in the first column in Appendix D.

So for the nods and liking example if $r = .45$ and $n = 30$, then

$$t(28) = \frac{.45\sqrt{28}}{\sqrt{1-.45^2}} = 2.666$$

which is statistically significant at the .02 level. It would be concluded that the .45 correlation cannot be explained by sampling error and is significantly greater than zero.

If the correlation is significantly different from zero, the correlation can be either negative or positive. If the researcher wishes to allow the null hypothesis to be false in only one direction (e.g., he or she expects the correlation to be positive), then the $p$ value should be cut in half and the test is called a *one-tailed test*. Most researchers agree that a one-tailed test should not ordinarily be done because if the correlation is very large but in the unpredicted direction, it would still be deemed statistically significant.

*Assumptions.*    A correlation coefficient as a measure of association presumes that the relationship is linear. That is, a change in one unit in the $X$ variable results in the same amount of change in $Y$ regardless of the value of $X$. As explained in Chapter 7, other types of relationships are not adequately captured by a linear measure of association, and some are even totally missed. The reader is referred to Chapter 7 for a more extensive discussion of the linearity question.

The second assumption is that each pair of $(X, Y)$ scores is independent of all other pairs. Such an assumption presumes that person is the sampling unit. That is, each person provides one and only one pair of scores.

Both variables must be normally distributed. More technically, the two variables have a joint normal distribution. Research has indicated that $p$ values are not considerably altered by the violation of the normality assumption. However, the linearity and the independence assumptions cannot be violated with impunity.

***Interpretation.*** The discussion in Chapter 7 concerning the pitfalls in interpreting correlations is relevant. A significant correlation means that the variables are associated. It in no way indicates the direction of causation. Of course, if the researcher believes that the variables are causally related, a correlation is comforting; however, the correlation does not by itself indicate the direction of causation.

***Power.*** The power of a test is the probability of rejecting the null hypothesis when the null hypothesis is false. Tests of correlation have moderate levels of power. For a given value of the population $r$, a given $n$, and a given alpha, power can be determined. Table 16.1 gives the power for the correlation coefficient for a small, medium, and large effect sizes. As discussed in Chapter 7, a small correlation is a value of .1, a medium correlation is a value of .3, and a large correlation is .5. The values given in Table 16.1 are for the .05 level of significance. The $n$ in the table is the total sample size or the number of $(X, Y)$ pairs. The entry in the table is the power multiplied by 100. So if a researcher contemplates doing a study with 40 persons and the correlation is expected to be moderate in size, the probability of rejecting the null hypothesis is .48. This means that for every two times that the study is done, the null hypothesis will be rejected about once.

For a given $r$, alpha, and level of power desired, the $n$ that is needed for that power can be determined. These sample sizes are given in Table 16.2. As an example, consider the $n$ needed to achieve 70% power for a moderate

**TABLE 16.1**    **Power Table[a] for Correlation Coefficients, $\alpha = .05$**

| | Population Correlation | | |
|---|---|---|---|
| $n$ | .1 | .3 | .5 |
| 10 | 6 | 13 | 33 |
| 20 | 7 | 25 | 64 |
| 40 | 9 | 48 | 92 |
| 80 | 14 | 78 | 99 |
| 100 | 17 | 86 | 99 |
| 200 | 29 | 99 | 99 |

[a]Taken from Cohen (1977).
NOTE: Each entry in the table is the probability of rejecting the null hypothesis times 100 for a given population correlation and sample size.

*TABLE 16.2*    **Sample Size Required*ᵃ* for Correlation Coefficients, $\alpha = 0.5$**

| | Population Correlation | | |
|---|---|---|---|
| Power | .1 | .3 | .5 |
| .25 | 166 | 20 | 8 |
| .50 | 384 | 42 | 15 |
| .60 | 489 | 53 | 18 |
| .70 | 616 | 66 | 23 |
| .80 | 783 | 84 | 28 |
| .90 | 1046 | 112 | 37 |
| .95 | 1308 | 139 | 46 |
| .99 | 1828 | 194 | 63 |

[a]Taken from Cohen (1977).
NOTE: Each entry represents the sample size needed to achieve a given level of power for a given population correlation.

effect size of .3. The *n* would have to be 66 to have a 70% chance of being significant.

***Example.***    Manning and Wright (1983) investigated the degree to which learning pain control strategies would reduce the use of painkilling medication during labor and childbirth. For 52 women who were giving birth, Manning and Wright correlated how much time the women devoted to learning a pain control strategy with their use of painkilling drugs during labor. The correlation was found to be $-.22$, and hence the women who learned the pain control strategy used fewer drugs. The test of that correlation is

$$t(50) = \frac{-.22\sqrt{50}}{\sqrt{1 - (-.22)^2}} = -1.595$$

which is not significantly different from zero at the .05 level. It is concluded that the population correlation may be zero and that the $-.22$ correlation is within the limits of sampling error given a sample size of 52.

## Test of the Difference Between Two Independent Correlations

Assume that the correlation between nods and liking is computed separately for male and female pairs. For 15 male pairs the correlation is .13, and for 15 female pairs the correlation is .68. At issue is whether the correlation is significantly larger for females than it is for males. When the correlations are computed from two different samples (e.g., males and females) the two correlations are said to be *independent*.

To evaluate whether two correlations are significantly different from one another, one might be tempted to test this hypothesis by first testing whether the correlation is statistically significant for males and then testing whether the correlation is statistically significant for females. The $t$ for males is .473, which is not significant, and 3.344 for females, which is significant at the .01 level of significance. The fact that the correlation is significantly greater than zero for females and is not for males does not necessarily mean that the correlation is significantly larger for females than for males. Statistical logic does not follow ordinary logic. If the number $x$ is equal to zero and the number $y$ is greater than zero, then $y$ must be greater than $x$. This is simple logic. However, it is not necessarily true that if correlation $x$ is not significantly greater than zero and $y$ is significantly larger than zero, $y$ is a significantly larger correlation than $x$. One must explicitly test whether the two correlations differ and not rely on the significance tests of the correlations calculated individually. The example illustrates this. Although the .68 correlation is statistically significant and the .13 value is not, it will be seen that the difference between the correlations is not statistically significant.

To test whether these correlations are significantly different from one another, the correlations are transformed. Each correlation must be altered by what is called *Fisher's r to z* transformation. This $r$ to $z$ transformation is defined as

$$\frac{1}{2} \ln \left[ \frac{1 + r}{1 - r} \right]$$

Actually the transformation is not usually computed by hand or even by calculator, but rather the value is simply looked up in a table. Table 1 in Appendix F presents the Fisher $z$ transformation values for correlation coefficients. To find the Fisher $r$ to $z$ value in Appendix F, locate $r$ in the left column and then determine its $z$ in the right column. If $r$ is negative, follow the same procedure but give the $z$ value a negative sign. Table 2 in Appendix F contains a table for going from $z$ to $r$. First, round $z$ to two decimal places, and then locate the appropriate value of $z$ in the left column and top row of the table and find the appropriate value of $r$.

The $r$ to $z$ transformation has little or no effect on small correlations, but for large correlations the Fisher $z$ value is larger than $r$. Unlike $r$, the Fisher $z$ has no upper and lower limit. It is important not to confuse this transformation with the $Z$ or standardizing transformation. Fisher's $r$ to $z$ is for correlations and the $Z$-score transformation is for a sample of data. Also it should not be confused with the standard normal or $Z$ distribution. Fisher's $r$ to $z$ transformation is applied only to correlations.

The effect of this transformation is to make the sampling distribution of the transformed coefficient nearly normally distributed. When the population correlation is not equal to zero, the distribution of $r$ is somewhat skewed. Fisher's $z$ transformation also makes the variance of correlation coefficient approximately the same regardless of the value of the population correlation. For a given population correlation, the distribution of the Fisher's $z$ values for a sample of size $n$ has virtually a normal distribution with a variance of $1/(n-3)$. Thus, the standard error of a Fisher $z$ transformed correlation is $1/\sqrt{n-3}$.

If there are two correlations with sample sizes $n_1$ and $n_2$, respectively, they are each transformed into Fisher's $z$ values. These Fisher $z$ transformed values are denoted as $z_1$ and $z_2$. Under the null hypothesis that the population correlations are equal, the following has approximately a standard normal distribution.

$$Z = \frac{z_1 - z_2}{\sqrt{\dfrac{1}{n_1 - 3} + \dfrac{1}{n_2 - 3}}}$$

Therefore, if the above quantity is greater than or equal to 1.96 or less than or equal to –1.96, the two correlations are significantly different at the .05 level of significance. To determine the $p$ value, the value of $Z$ is located in Appendix C. The $p$ value equals twice the quantity of .5 minus the probability. So if $Z$ is 1.96 the $p$ value is two times .5000 – .4750 which equals .05.

For the .68 and .13 correlations for females and males, the $Z$ is only 1.71, which is not significant at the .05 level. Hence sampling error is a plausible explanation for the difference between the two correlations.

***Assumptions.***     The use of Fisher's $z$ to test the difference between correlations requires that the two correlations be independent. One condition required for independence is that the correlations are computed using two different sets of persons. If the same persons are used to compute both correlations, the correlations are called *correlated correlations*. This topic is discussed later in this chapter.

***Interpretation and Power.***     If the $Z$ is statistically significant, then it is concluded that the population correlations differ in the two groups. If the two are not significantly different, the null hypothesis that the correlations are equal is retained. However, the power of the test that compares the correlations between two samples is quite low. For instance, if $n_1 = n_2 = 50$ and population correlations are .10 and .40, which seems like a large difference, there is only a 35% chance of rejecting the null hypothesis. There must be fairly large sample sizes before having a reasonable chance of rejecting the null hypothesis that the two correlations are unequal.

This low power in showing that correlations differ in the two groups has made it very difficult to show that standardized tests, such as the Scholastic Aptitude Test or SAT, have differential validity across the races. Some have argued that standardized tests are less valid for minorities, particularly blacks. The validity of a standardized test is often measured by a correlation coefficient—for example, the correlation between SAT and college grades. Thus, critics of standardized tests have argued that the correlation is lower for blacks than for whites. However, because of low power, the null hypothesis of no difference is very difficult to show to be false, and so very rarely is the null hypothesis rejected.

***Example.*** Wheeler, Reis, and Nezlek (1983) correlated feelings of loneliness with the extent to which persons felt they had opportunities to disclose or discuss things about themselves with others. The correlations were computed separately for 43 men and 53 women. The correlation for the men was −.57 and for women it was −.21. So, persons who were lonely said they had few opportunities to discuss things about themselves. Using Table 1 of Appendix F, the $z$ value for the −.57 correlation is −.6475, and for the −.21 correlation the $z$ value is −.2132. The test that the coefficients differ is

$$Z = \frac{-.6475 - (-.2132)}{\sqrt{\dfrac{1}{43-3} + \dfrac{1}{53-3}}} = -2.05$$

which, from Appendix C, has a $p$ value of .0404—that is, two times (.5000 − .4798). Because the $p$ value is less than .05, the difference is judged statistically significant at the .05 level. So, loneliness and the absence of self disclosure correlate significantly more highly among men than women.

## More than Two Independent Correlations

Suppose the correlation between socioeconomic status and school achievement is computed for students for four schools. The null hypothesis to be tested is that the correlations do not vary across schools. The correlations are first transformed to Fisher $z$ values. The mean of the $z$ values, $\bar{z}$, is computed weighted by $n - 3$. So for the example of four schools,

$$\bar{z} = \frac{z_1(n_1 - 3) + z_2(n_2 - 3) + z_3(n_3 - 3) + z_4(n_4 - 3)}{n_1 - 3 + n_2 - 3 + n_3 - 3 + n_4 - 3}$$

This is the average of the four correlations weighted by sample size. In general to average correlations, the $r$'s are converted to Fisher $z$ values and are multiplied by the sample size less three, these values are summed, and this total is divided by the number of subjects in all groups less three times the number of correlations. This $\bar{z}$ can be converted back into a correlation by using Table 2 of Appendix F to obtain the average of the correlations.

To test whether the average correlation is significantly different from zero, the average $z$ is divided by

$$\sqrt{\frac{1}{n_1 - 3 + n_2 - 3 + n_3 - 3 + n_4 - 3}}$$

which is approximately distributed as $Z$, the standard normal distribution, given the null hypothesis of a zero correlation.

A researcher might also wish to test whether correlations, as a group, significantly differ from one another. Here the null hypothesis is not whether the average correlation is zero, but that the population correlations are the same in each group. To do so, the following is computed:

$$(n_1 - 3)(z_1 - \bar{z})^2 + (n_2 - 3)(z_2 - \bar{z})^2 + (n_3 - 3)(z_3 - \bar{z})^2 + (n_4 - 3)(z_4 - \bar{z})^2$$

where $\bar{z}$ is the Fisher's $z$ average of the correlations. In general, to evaluate whether correlations computed in $k$ groups are significantly different from one another, one first averages the Fisher $z$ values, weighting by sample size less three. One then computes the deviation of each Fisher $z$ from this average, squares, multiplies by sample size less three and sums. The resulting quantity is approximately distributed as chi square with $k - 1$ degrees of freedom, $k$ being the number of correlations. As described in Chapter 11, chi square (symbolized by $\chi^2$) is a positively skewed distribution with a lower limit of zero and an upper limit of positive infinity. If the chi square value exceeds the values tabled in Appendix G, it is deemed significant at the appropriate level. One rejects the null hypothesis that the correlation is the same for all groups.

***Example.*** Consider five hypothetical correlations between ability in mathematics and in reading in five different countries. The correlations and their Fisher $z$ values from Table 1 of Appendix F are as follows:

| Country | $r$ | $n$ | $z$ |
|---------|-----|-----|------|
| France  | .65 | 55  | .7753 |
| England | .53 | 76  | .5901 |
| Mexico  | .56 | 44  | .6328 |
| Italy   | .44 | 68  | .4722 |
| Canada  | .74 | 39  | .9505 |

The average $z$ is

$$\frac{52(.7753) + 73(.5901) + 41(.6328) + 65(.4722) + 36(.9505)}{52 + 73 + 41 + 65 + 36}$$

which equals .6526. Rounding this $z$ value to two decimal places and using Table 2 of Appendix F, it corresponds to an $r$ of .5717. The test that this correlation is zero is tested by

$$Z = \frac{.6526}{\sqrt{\dfrac{1}{52 + 73 + 41 + 65 + 36}}} = 10.66$$

which is statistically significant at the .001 level. So, across the five countries, the correlation between reading and mathematics ability is significantly different from zero at the .001 level of significance.

The test that the population correlations are all equal to each other is

$$52(.7753 - .6526)^2 + 73(.5901 - .6526)^2 + 41(.6328 - .6526)^2$$
$$+ 65(.4722 - .6526)^2 + 36(.9505 - .6526)^2 = 6.39$$

A $\chi^2$ with four degrees of freedom of 6.39 is not significant at the .05 level of significance. So, the correlation between mathematics and reading skill does not significantly differ from country to country.

## Correlated Correlations

The methods in the previous sections have assumed that when two or more correlations are being compared, different sets of subjects are being compared. Often there is one set of persons or one sample, and two correlations are computed from their data, and these two correlations are compared.

Ordinarily when two or more correlations are compared, the same two variables are correlated. But sometimes correlations involving different variables are compared. For instance, one might wish to compare the correlation of mother's education with child's verbal skill to the correlation of father's education with the child's verbal skill. So, the variables involved in the comparison of correlations need not be the same.

The Fisher $z$ transformation cannot be used for comparing these two correlations because the same persons are used. Consider the correlations between $X_1$ with $X_3$ and $X_2$ with $X_3$. If the correlation between $X_1$ and $X_2$ is 1.00, the correlation between $X_1$ and $X_3$ must be the same as the correlation between $X_2$ and $X_3$. This is a mathematical necessity. If the correlation between $X_1$ and $X_2$ is $-1.00$, the correlation between $X_1$ and $X_3$ must be the same as the correlation between $X_2$ and $X_3$ but with the opposite sign. Again this is a mathematical necessity. Thus, the size of the correlation between $X_1$ and $X_2$ influences how similar $r_{13}$ and $r_{23}$ are. The statistical test must take into account the degree to which $X_1$ correlates with $X_2$. This can be done by using a procedure known as the Williams modification of the Hotelling test.

For this test variable $X_3$ is correlated with two other variables, $X_1$ and $X_2$. There are then three correlations:

$r_{12}$: correlation between $X_1$ and $X_2$

$r_{13}$: correlation between $X_1$ and $X_3$

$r_{23}$: correlation between $X_2$ and $X_3$

The test of whether the population correlation between $X_1$ and $X_3$ equals the population correlation between $X_2$ and $X_3$ is

$$t(n-3) = \frac{(r_{13} - r_{23})\sqrt{(n-1)(1 + r_{12})}}{\sqrt{2K\dfrac{(n-1)}{(n-3)} + \dfrac{(r_{23} + r_{13})^2}{4}(1 - r_{12})^3}}$$

where

$$K = 1 - r_{12}^2 - r_{13}^2 - r_{23}^2 + 2r_{12}r_{13}r_{23}$$

The quantity is distributed as $t$ with $n - 3$ degrees of freedom given the null hypothesis that the population correlation between $X_1$ and $X_3$ is equal to the population correlation between $X_2$ and $X_3$. If the $t$ is statistically significant, it is concluded that the difference between the two correlations cannot be explained by sampling error.

The test just described has one variable ($X_3$) that is in both correlations. Consider a test of the difference between two correlation coefficients in one sample where none of the variables are the same. There are now four different variables: $X_1$, $X_2$, $X_3$, and $X_4$. They give rise to six different correlations:

$r_{12}$:   correlation between $X_1$ and $X_2$

$r_{34}$:   correlation between $X_3$ and $X_4$

$r_{13}$:   correlation between $X_1$ and $X_3$

$r_{14}$:   correlation between $X_1$ and $X_4$

$r_{23}$:   correlation between $X_2$ and $X_3$

$r_{24}$:   correlation between $X_2$ and $X_4$

At issue is the test that the population correlations between $X_1$ and $X_2$ and between $X_3$ and $X_4$ are equal to each other.

First, the correlations that are being compared are transformed into Fisher's $z$ values: $z_{12}$ and $z_{34}$. The test is

$$Z = \frac{\sqrt{(n-3)}(z_{12} - z_{34})}{\sqrt{2 - Q(1 - r^2)^2}}$$

where

$$Q = (r_{13} - r_{23}r)(r_{24} - r_{23}r) + (r_{14} - r_{13}r)(r_{23} - r_{13}r) + (r_{13} - r_{14}r)(r_{24} - r_{14}r)$$
$$+ (r_{14} - r_{24}r)(r_{23} - r_{24}r)$$

and

$$r = \frac{r_{12} + r_{34}}{2}$$

This test, called the *Pearson-Filon test,* is approximately distributed as $Z$, the standard normal distribution (*not* Fisher's $z$) under the null hypothesis that the population correlation between $X_1$ and $X_2$ is equal to the population correlation between $X_3$ and $X_4$. This standard normal approximation is quite good if $n$

is at least 20. This test was developed by Pearson and Filon (1898) and modified by Steiger (1980).

*Power.* Ordinarily tests of correlations are somewhat more powerful when computed from a single sample than from multiple samples. Nonetheless, the power of the test of the difference between independent correlations is so low that being more powerful still means the test of correlated correlations has relatively low power. It should be noted that in some special cases, the power in the one-sample case can actually be lower than in the two-sample case.

*Example.* The illustration is taken from Jacobson's (1977) research on the fear of peers among infants. He studied 23 infants and measured their cognitive development and their attachment to a parent when they are presented with a novel stimulus. Research in developmental psychology has shown that for very young children, intelligence and fear are positively associated. The child needs to have the intelligence to realize that a stimulus may be harmful. But as children mature, it is the intelligent children who are less afraid and feel less need to seek a parent for comfort. The older children realize that the novel stimulus is not harmful.

Two different correlations are to be compared. One correlation is between cognitive ability and fear for young infants and the second is between the same two variables for older infants. Jacobson's study confirms the theory. The correlation between cognitive ability and fear is .416 for infants of ten months, which indicates that the smarter children are more afraid. The correlation is $-.413$ between cognitive ability at ten months and fear at twelve months, which indicates that the smarter children are now less afraid. The correlation between fear over the two-month period is $-.343$. Fear at ten months is denoted as $X_1$, fear at twelve months as $X_2$, and cognitive ability as $X_3$. So the correlations are

$$r_{12} = -.343$$
$$r_{13} = \phantom{-}.416$$
$$r_{23} = -.413$$

These are correlated correlations that share one variable in common and that is the cognitive development measure. Using the Williams modification of the Hotelling test, the formula is

$$t(20) = \frac{(-.413 - .416)\sqrt{22[1 + (-.343)]}}{\sqrt{2K\dfrac{22}{20} + \dfrac{(-.413 + .416)^2}{4}[1 - (-.343)]^3}}$$

where

$$K = 1 - (-.343)^2 - .416^2 - (-.413)^2 + 2(-.343)(.416)(-.413)$$

The value of $K$ is .657 and the $t(20)$ is $-2.622$, which from Appendix D is statistically significant at the .02 level. Thus the correlations are significantly different at the .02 level of significance.

Jacobson also measured cognitive ability at twelve months. So, the correlation between cognitive ability and fear for infants at ten and twelve months can be compared. The correlation between ability and fear is .416 at ten months. At twelve months the correlation becomes $-.408$. These are two correlated correlations that share no variables in common. The variables are denoted as follows:

$X_1$:   ten-month cognitive ability

$X_2$:   ten-month fear

$X_3$:   twelve-month cognitive ability

$X_4$:   twelve-month fear

The six correlations between the four variables are:

$$r_{12} = \phantom{-}.416$$
$$r_{34} = -.408$$
$$r_{13} = \phantom{-}.556$$
$$r_{14} = -.413$$
$$r_{23} = -.075$$
$$r_{24} = -.343$$

The correlations that are to be compared are .416 and $-.408$. Their Fisher's $z$ values are .443 and $-.433$. (To increase accuracy, Appendix F is not used and the $z$'s are directly computed.) The average of $r_{12}$ and $r_{34}$ is

$$.004 = \frac{.416 + (-.408)}{2}$$

The value of $Q$ is

$$[.556 - (-.075)(.004)][-.343 - (-.075)(.004)]$$
$$+ [-.413 - (.556)(.004)][-.075 - (.556)(.004)]$$
$$+ [.556 - (-.413)(.004)][-.343 - (-.413)(.004)]$$
$$+ [-.413 - (-.343)(.004)][-.075 - (-.343)(.004)] = -.319$$

Now the Pearson-Filon test gives

$$Z = \frac{\sqrt{20}\,[.443 - (-.433)]}{\sqrt{2 - (-.319)(1.000)}} = 2.57$$

which from Appendix C has a $p$ value of .0102. Therefore the correlations between fear and cognitive ability are significantly different at ten and twelve months.

# Test of Regression Coefficients

In this section tests of regression coefficients are presented and the following cases discussed:

1. a single coefficient equal to zero,
2. two independent coefficients equal to each other, and
3. two correlated coefficients equal to each other.

As will be seen, tests of regression coefficients are more computationally complex than tests of correlation coefficients.

## Single Regression Coefficient

The regression equation in which the variable $X$ is the predictor and the variable $Y$ is the criterion is

$$Y = a + b_{YX}X + e$$

As an example, consider the number of packs of cigarettes smoked per day as the predictor variable and life expectancy in years as the criterion variable. Research has shown that the coefficient is about $-4.0$. That is, for every pack of cigarettes smoked per day, one lives on the average four fewer years.

The test that a regression coefficient $b_{YX}$ is not significantly different from zero is

$$t(n-2) = \frac{b_{YX}\sqrt{SS_X}}{s_{Y \cdot X}}$$

where $s_{Y \cdot X}$ is the standard deviation of the errors (see Chapter 6) and equals

$$s_{Y \cdot X} = \sqrt{\frac{SS_Y - b_{YX}{}^2 SS_X}{n - 2}}$$

The $SS_X$ and $SS_Y$ are the sum of squares for variables $X$ and $Y$, respectively. In analysis of variance terms, they are the sum of squares total. They equal

$$SS_X = \sum(X - \bar{X})^2$$

and

$$SS_Y = \sum(Y - \bar{Y})^2$$

Their computational formulas are

$$SS_X = \sum X^2 - \frac{(\sum X)^2}{n}$$

$$SS_Y = \sum Y^2 - \frac{(\sum Y)^2}{n}$$

An alternative and simpler way to test $b_{YX}$ is to convert $b_{YX}$ to $r_{XY}$ by the formula

$$r_{XY} = b_{YX}\left(\frac{s_X}{s_Y}\right)$$

The correlation coefficient can be tested for significance. The resulting $t$ value is the same as would be obtained by a direct test of $b_{YX}$. This fact can be used for determining the power of the test. One converts $b$ to $r$ and uses Table 16.1 to determine the power and Table 16.2 to determine the $n$ necessary to achieve a given level of power.

If $X$ and $Y$ are reversed by having $Y$ predict $X$, the test of a regression coefficient in which $Y$ predicts $X$ or $b_{XY}$ is

$$t(n\text{--}2) = \frac{b_{XY}\sqrt{SS_Y}}{s_{X \cdot Y}}$$

where

$$s_{X \cdot Y} = \sqrt{\frac{SS_X - b_{XY}^2 SS_Y}{n - 2}}$$

The value of $t$ will be the same regardless whether the test is of $b_{XY}$, $b_{YX}$, or $r_{XY}$.

As an example, assume that $b_{YX}$ is 1.5 and $SS_X$ is 33.5 and $SS_Y$ is 140.2 and $n$ is 131. The value of $s_{Y \cdot X}^2$ is

$$\frac{140.2 - 1.5^2(33.5)}{129} = .503$$

The test of the slope is

$$t(129) = \frac{1.5\sqrt{33.5}}{\sqrt{.503}} = 12.241$$

which is statistically significant at the .001 level of significance.

## Two Independent Regression Coefficients

In this case there is a pair of regression coefficients computed from two different groups of persons. For instance, there are regression coefficients for both males and females of the effect of cigarette smoking on life expectancy. The coefficient for males is –4.32 and for females the value is –3.93. That is, cigarette smoking reduces life expectancy more for males than for females.

The hypothesis of a difference between two regression coefficients is similar to the hypothesis of an interaction in two-way analysis of variance. Both evaluate whether the effect of one variable changes as a function of another. In two-way analysis of variance the two independent variables are measured at the nominal level of measurement. In the regression case one independent variable is at the interval level of measurement and its linear effect on the dependent variable is presumed to change as a function of a dichotomous independent variable.

In the restricted model there is a single slope, and in the complete model there is a slope for each group. The regression coefficient in each sample is computed. These coefficients are denoted as $b_1$ and $b_2$ for the two samples. At issue is whether the difference between the two coefficients can be explained by sampling error.

The test statistic is

$$t(n_1+n_2-4) = \frac{b_1 - b_2}{s_{Y \cdot X} \sqrt{\dfrac{1}{SS_{X_1}} + \dfrac{1}{SS_{X_2}}}}$$

where $SS_{X_1}$ and $SS_{X_2}$ are the sum of squares for $X$ for the first and second groups, respectively, and

$$s_{Y \cdot X} = \sqrt{\frac{SS_{Y_1} + SS_{Y_2} - b_1^2 SS_{X_1} - b_2^2 SS_{X_2}}{n_1 + n_2 - 4}}$$

or, alternatively,

$$s_{Y \cdot X} = \sqrt{\frac{(n_1 - 2)s_{Y \cdot X_1}^2 + (n_2 - 2)s_{Y \cdot X_2}^2}{n_1 + n_2 - 4}}$$

The formula for $s_{Y \cdot X}$ is the *pooled error standard deviation*. That is, it is a pooling or averaging of the two error variances, each weighted by its degrees of freeodm. This test seems to involve quite a bit of tedious computation. Actually it involves little more than computing and testing two regression coefficients.

As with the difference between two correlation coefficients, the power of the test of the difference between two regression coefficients is quite low. Even if the slopes are quite different, the sample sizes must be quite large before one has a reasonable chance of detecting that the slopes are indeed different.

If the null hypothesis that the regression coefficients are equal is not rejected, the two regression coefficients can be averaged by the following formula:

$$b_p = \frac{b_1 SS_{X_1} + b_2 SS_{X_2}}{SS_{X_1} + SS_{X_2}}$$

The term $b_p$ is called the *pooled regression coefficient*. The pooled coefficient can be viewed as a weighted average where the weights are the sum of squares of the predictor.

The pooled regression coefficient can be tested for significance using the formula

$$t(n_1+n_2-4) = \frac{b_p\sqrt{(SS_{X_1} + SS_{X_2})}}{s_{Y \cdot X}}$$

where $s_{Y \cdot X}$ is the pooled error standard deviation.

The above formula can be generalized to the case in which there is more than one regression coefficient. The generalization, which is found in advanced texts (Winer, 1971), is similar in its computation to a one-way analysis of variance.

As an example consider a researcher who investigates the effect of attitudes about wearing seat belts on behavior for those who heard a series of communications about the importance of wearing seat belts and a group who did not. The criterion is denoted as $B$ for behavior and the predictor as $A$ for attitude. The results are as follows:

|  | Communication | |
|---|---|---|
|  | *Heard* | *Did Not* |
| $b_{BA}$ | .58 | .23 |
| $SS_B$ | 68.1 | 56.3 |
| $SS_A$ | 25.0 | 19.1 |
| $n$ | 60 | 60 |

The pooled error standard deviation is

$$\sqrt{\frac{68.1 + 56.3 - .58^2(25.0) - .23^2(19.1)}{60 + 60 - 4}} = .996$$

The test that the slopes differ is

$$t(116) = \frac{.58 - .23}{.996\sqrt{\frac{1}{25.0} + \frac{1}{19.1}}} = 1.156$$

This value is not statistically significant at the .05 level. Therefore, the slopes do not significantly differ.

The pooled slope is

$$\frac{.58(25.0) + .23(19.1)}{25.0 + 19.1} = .428$$

The test that the pooled coefficient equals zero is

$$t(116) = \frac{.428\sqrt{(25.0 + 19.1)}}{.996} = 2.854$$

This value is statistically significant at the .001 level. So, attitude toward seat belts significantly predicts behavior across both communication groups.

## Two Correlated Regression Coefficients

This case is identical to the previous case except that the regression coefficients are computed from the same sample. For instance, does the number of siblings better predict fourth-grade vocabulary skill than fifth-grade vocabulary skill? If the same persons were measured at fourth and fifth grades, the regression coefficients are computed from the one sample and are correlated.

To test whether these coefficients are the same, the changes in vocabulary skill are computed by subtracting fourth-grade vocabulary scores from fifth-grade scores. This change score would be the criterion, and number of siblings would be the predictor variable in a regression equation. The test of the coefficient from this regression equation would evaluate the difference between regression coefficients. So if $X_1$ is used to predict $X_2$ and $X_3$, to evaluate the difference between $b_{21}$ and $b_{31}$ the regression coefficient $b_{(3-2)1}$ is computed. As with the paired $t$ test described in Chapter 13, difference scores can be used to test hypotheses with paired data.

# Choice of Test

Throughout the entire chapter an obvious question arises. Should the test of association be made using a correlation coefficient or regression coefficient? In the case of a single measure of association, the choice of the significance test does not matter. That is, the $t$ value is the same regardless of whether $r$ or $b$ is computed. However, when two or more measures of association are compared, the result from a test of the regression coefficients differs from the result from a test of the correlations. Which measure is to be preferred?

There are three important factors that can guide the decision. First, if there is a clear causal ordering of the two variables, then the regression coefficient is preferred. Because a regression coefficient assumes a causal ordering (the predictor causes the criterion) and a correlation coefficient does not, a regression coefficient is the coefficient of choice when the variables can be causally ordered.

Second, if the variances of the variables are not the same in both groups, the regression coefficient is preferred. As explained in Chapter 7, correlations are affected by variability. Variables with less variability tend to exhibit lower correlations. To evaluate whether two variances are equal the following statistical tests can be employed. For two independent groups, the ratio of sample variances is computed:

$$\frac{s_1^2}{s_2^2}$$

where $s_1^2$ is greater than $s_2^2$. If the variances are equal in the population, $s_1^2/s_2^2$ has an $F$ distribution with $n_1 - 1$ degrees of freedom on the numerator and $n_2 - 1$ degrees of freedom on the denominator. For this test, the $p$ value is doubled because the $F$ ratio is formed by always putting the larger variance on the numerator. If there is a single sample and the purpose is to test whether the variance of $X_1$ is different from the variance of $X_2$, begin by computing $X_1 - X_2$ and $X_1 + X_2$. Then correlate the difference, $X_1 - X_2$, with the sum, $X_1 + X_2$, and test whether it is significantly different from zero. The test of this correlation evaluates whether the two variances are equal.

Third, if the unit of measurement changes from group to group, the correlation coefficient is preferred. Thus, for example, if the groups are French and English children and the variables are vocabulary and intelligence, a correlation should be used because different tests would be used in the different countries. However, if males and females within a country were compared, the regression coefficient would be preferred.

# Summary

Tests of a regression coefficient or a correlation coefficient are accomplished by a $t$ test with degrees of freedom of sample size less two. Tests of two or more independent correlations is aided by the *Fisher's r to z* transformation. The Fisher's $r$ to $z$ transformation (not to be confused with the $Z$ or standard normal distribution) makes the distribution of the transformed correlation approximately normal. The Fisher's $z$ transformation can be used to pool correlations computed across different samples as well as test whether the correlations are equal.

Sometimes one seeks to compare two correlations that are computed from the same sample. These correlations are called *correlated correlations*. When correlations are themselves correlated, the tests are computationally complicated but straightforward. When the correlated correlations involve three variables, the Williams modification of the *Hotelling test* is used. When the correlated correlations involve four different variables the *Pearson-Filon* test is used.

A test of a single regression coefficient is identical to the $t$ test of a single correlation coefficient. Also, two regression coefficients from different samples can be tested for equality. If they are equal, they can be pooled and the pooled coefficient can be tested to determine whether it is different from zero.

The decision of whether to test either the correlation or the regression coefficient is aided by considerations of causal ordering, equal variances, and unit of measurement.

# Problems

1. According to Pulling et al. (1980) the correlation between age and susceptibility to glare is .742 for 148 subjects. Test whether the popula-

tion correlation between the two variables is different from zero. Interpret the result.

2. Convert the following correlations to Fisher $z$ values:

    a. $-.13$    b. $.07$    c. $.91$    d. $.73$

    e. $.41$    f. $-.32$    g. $-.21$    h. $.53$

3. Convert the following Fisher $z$ values to correlation coefficients:

    a. $-.86$    b. $-.43$    c. $.91$    d. $.06$

    e. $.19$    f. $.39$    g. $-.25$    h. $-1.03$

4. Given $b_{YX} = .31$, $n = 44$, $SS_X = 31.93$, and $SS_Y = 22.41$, test whether the population regression coefficient is significantly different from zero.

5. Evaluate whether the population correlations are equal if $r_1 = .23$, $r_2 = .48$, $n_1 = 212$, and $n_2 = 136$.

6. Given $r_{XY} = .39$ and $n = 84$, test whether the population correlation is significantly greater than zero.

7. Variables 1, 2, and 3 are measured on the same set of persons. Test whether the population correlation between variables 1 and 2 is equal to the population correlation between variables 1 and 3 if $r_{12} = .28$ and $r_{13} = .45$. The correlation between $X_2$ and $X_3$ is .63 and $n = 175$.

8. Given

$$r_{12} = .43 \qquad r_{14} = .10$$
$$r_{34} = .14 \qquad r_{23} = .16$$
$$r_{13} = .55 \qquad r_{24} = .49$$

and $n = 145$, test whether the difference between $r_{13}$ and $r_{23}$ is statistically significantly different from zero. Test also whether $r_{12}$ is significantly different from $r_{34}$.

9. Given the following,

| | Males | Females |
|---|---|---|
| $b_{YX}$ | .23 | .44 |
| $SS_Y$ | 44.5 | 38.2 |
| $SS_X$ | 63.2 | 58.9 |
| $n$ | 68 | 40 |

test whether the slopes are different from one another. Pool the two slopes and test whether the pooled slope is different from zero.

10. The following correlations are taken from Rasinski, Tyler, and Fridkin (1985):

| Sample | Correlation | n |
|--------|-------------|-----|
| 1 | .69 | 137 |
| 2 | .46 | 108 |
| 3 | .56 | 132 |
| 4 | .66 | 115 |

Average them using the Fisher $r$ to $z$ transformation. Test whether the pooled correlation is significantly different from zero. Also test whether the correlations significantly differ from each other.

11. Evaluate whether a correlation of .43 is statistically significant with a sample size of 68.

12. What is the power of the following tests?

|    | $r$ | $n$ |
|----|-----|-----|
| a. | .1 | 10 |
| b. | .5 | 40 |
| c. | .3 | 100 |
| d. | .1 | 200 |

13. For the following cases, how many subjects would be needed to achieve the desired level of power?

|    | $r$ | Power |
|----|-----|-------|
| a. | .1 | .50 |
| b. | .5 | .25 |
| c. | .3 | .90 |
| d. | .5 | .50 |

14. According to Holahan and Moos (1985), the correlation between seeing oneself as easy-going and feeling that one's family is supportive is .21 for 267 men. Test whether the correlation is statistically significant.

15. A sample consisting of 76 females was tested by Schifter and Ajzen (1985). These women's weight loss correlated .41 with the perceived control in losing weight and .25 with intention to lose weight. The correlation between intention and control is .36. Test whether perceived control correlates significantly higher with weight loss than perceived control correlates with intention.

16. According to Neff (1985), the relationships between education (E) and the reporting of depressive symptoms (D) are:

|          | Whites  | Blacks  |
|----------|---------|---------|
| $b_{DE}$ | −.0236  | −.0510  |
| $s_D$    | .47     | .59     |
| $s_E$    | 2.19    | 2.20    |
| $n$      | 658     | 171     |

a. Test each of the regression coefficients for statistical significance.
b. Test whether the coefficients are significantly different from each other.
c. Pool the coefficients and test whether the pooled coefficient is different from zero.

17. Given that $n = 148$ and

$$r_{12} = .67 \quad r_{13} = .40 \quad r_{23} = .26$$
$$r_{34} = .21 \quad r_{14} = .53 \quad r_{24} = .19$$

Test whether the correlation between variables one and two is significantly different from the correlation between variables three and four.

18. For the following correlations from three different groups of persons

| $r$ | $n$ |
|-----|-----|
| .61 | 96 |
| .23 | 39 |
| −.15 | 76 |

a. Average the correlations using Fisher $z$.
b. Test whether the average correlation is significantly different from zero.
c. Test whether the correlations differ from each other.

# 17 | *Models for Nominal Dependent Variables*

The preceding five chapters discussed models in which the dependent variable is assumed to be measured at the interval level of measurement. In this chapter models are considered in which the dependent variable is measured at the nominal level of measurement. And in the final chapter the dependent variable is assumed to be measured at the ordinal level of measurement. Techniques that do not assume an interval dependent variable are sometimes referred to as nonparametric or distribution-free methods. The term *distribution-free* is preferred because so-called nonparametric tests do test hypotheses about parameters. Methods that presume normality and homogeneity of variance such as the two-sample *t* test, analysis of variance, and regression will be called *distribution-tied* methods.

There are three major reasons for employing the methods described in this chapter and the next. The first reason is that sometimes the data are clearly not at the interval level of measurement. The dependent measure may be a set of ranks or a set of categories. In these cases it would be clearly inappropriate to use the methods described in the previous five chapters. So if the level of measurement of the dependent variable is clearly not at the interval level of measurement, the methods presented in this and the next chapter are appropriate.

The second use of distribution-free procedures is that one may be reasonably confident that the dependent variable is at the interval level of measurement, but one is worried about the assumptions made to perform a *t* or *F* test. In particular, one may be especially concerned that the assumption of a normal distribution for the residual variable is false. The dependent variable may be highly skewed or bimodal, and so it is quite likely that the residual variable does not have a normal distribution. One then desires to do a statistical test, but one is unwilling to make assumptions concerning the distribution of the residual variable. Because distribution-free methods make no assumptions concerning the distribution of any of the variables, they can be used with bimodal or highly skewed distributions. This is why these methods are called distribution-free.

If the residual variable does have a normal distribution and all the other assumptions are met, there is a cost in not doing an analysis that presumes the interval level of measurement. The techniques described in this and the next chapter have less power than the procedures described in the previous chapters when the assumptions made by distribution-tied tests are true. Thus analysis of variance and the two-sample $t$ test are more powerful statistical procedures than the distribution-free procedures described in this and the next chapter. However, if the classical assumptions of normality and homogeneity of residuals do not hold, the $p$ values obtained from analysis of variance are not correct and are usually too liberal, resulting in too many Type I errors. It can even happen that for some distributions, a distribution-free method is more powerful than a distribution-tied method.

A distribution-free test is ordinarily less powerful than its distribution-tied cousin because the distribution-free test ignores the interval information in the data. Consider the following pattern of numbers of two samples A and B.

A:   1, 2, 3, 6

B:   7, 8, 9, 12

There is no overlap in the numbers and the means (3.0 for sample A and 9.0 for sample B) differ by six units. Consider the pattern of the following numbers from two samples.

A:   1, 2, 3, 6

B:   107, 108, 109, 112

Again there is no overlap, but now the means (3.0 for sample A and 109.0 for sample B) differ by 106 units. A distribution-free test would see no difference between the two patterns, whereas a distribution-tied method would see the second pattern as more convincing evidence that the two groups differ.

However, distribution-free tests do have their advantages. Distribution-tied tests believe even the most anomalous aspect of the data. Consider again the first pattern of the numbers of two samples A and B:

A:   1, 2, 3, 6

B:   7, 8, 9, 12

There is no overlap in the numbers and the means differ by six units. Consider the following numbers from two samples.

A:   1, 2, 3, 6

B:   7, 8, 9, 120

The numbers are exactly the same except that the last number in the B sample is ten times larger in the second pattern. The distribution-free test would see no difference between the two patterns, whereas a distribution-tied method would see the second pattern as much more convincing evidence that the two groups differ even though the value of 120 would appear to be an outlier.

The third reason for choosing a distribution-free test is that it tests a different null hypothesis from the null hypothesis tested by the distribution-tied analog. For instance, consider the following two samples.

A:   1, 2, 3, 3, 4, 4, 4, 5, 5, 5, 6, 6, 7, 8

B:   1, 1, 1, 2, 2, 2, 2, 7, 7, 7, 7, 8, 8, 8

Both groups have means of 4.5, but clearly the groups differ. Sample B has more extreme scores than sample A. A distribution-free test can reveal such a difference, but a *t* test cannot.

Although there are clear-cut cases in which a distribution-free statistic is clearly superior to its distribution-tied cousin, the choice between the two may be more a matter of taste and custom than of right or wrong. For instance, researchers in medicine are much more likely to employ a distribution-free method than researchers in economics, even though data in medicine are no less likely to be normally distributed than in economics. Perhaps the preference is explainable by need to be somewhat more conservative when lives are at stake than when dollars are. I suspect, however, that the real reason has more to do with custom than anything else.

In cases in which the researcher is in doubt about the type of analysis, both types of tests might be employed. Most of the time the two sets of results agree. In such a case the distribution-tied tests are reported with mention that the distribution-free results are in essential agreement. In cases in which the analyses are in conflict, usually the distribution-free results are reported because they tend to be more conservative.

This chapter considers distribution-free tests in which the dependent variable is at the nominal level of measurement. Two basic types of models are considered. In the first, hypotheses concerning the distribution of a nominal dependent variable are tested. In the second, both the independent and dependent variable are at the nominal level of measurement. For this second model, either the scores can be independent across levels of the independent variable or they can be nonindependent. Different analysis strategies for models are needed when the groups are independent and when they are nonindependent.

First, this chapter shows how to test whether a nominal variable affects a second nominal variable in which the groups are independent. This test is commonly called a $\chi^2$ *test of independence*. Its distribution-tied analogs are the two-group *t* test, one-way ANOVA, and regression. The second test considered in this chapter is the *McNemar test*, which evaluates the effect of a dichotomous nominal variable on a dichotomous dependent variable in which the groups are nonindependent. Its distribution-tied analog is the paired *t* test which was presented in Chapter 13. The final test that is discussed evaluates the adequacy of an a priori prediction of a nominal variable's distribution. The test is commonly referred to as a $\chi^2$ *goodness of fit* test. Its distribution-tied analog is a *t* test of a constant which is presented in Chapter 12.

As was explained in Chapter 8, for a nominal variable the data can be converted into frequencies. A *frequency* of a category equals the total number of objects for the category of the nominal variable. For the statistical tests presented in this chapter, the $\chi^2$ distribution is the sampling distribution that is employed. In all cases the distribution of the test statistic is approximately $\chi^2$. The test statistic for these $\chi^2$ tests always compares the observed or actual frequencies to those frequencies expected under a restricted model.

# *Test of Independence of Two Nominal Variables*

In this case there are two nominal variables and the issue is whether the two variables are associated. One variable may be distinguished as independent and the other dependent or they may not be. Such a distinction does not affect the *p* value but it does affect the interpretation of the result.

As an example, consider a study by Brown (1981). He had a pair of persons stand in a mall talking to one another. Persons approaching the pair could either walk through the pair or walk around. Brown varied the racial composition of the pair. They were either both black, both white, or mixed race. So the independent variable is racial composition of the pair, and the dependent variable is the behavior of the subject: walking through versus around. A total of 508 subjects were observed, and the results are shown in Table 17.1.

The first row of the table consists of those who walked through. For instance, a total of 125 persons walked through the black pair. The second row consists of those who walked around. The final row is called the set of column margins and consists of the number of persons in the sample for type

**TABLE 17.1**  **Observed Frequencies for the Brown (1981) Study**

| Behavior | Racial Composition | | | Total |
|----------|-------|-------|-------|-------|
|          | Black | White | Mixed |       |
| Through  | 125   | 67    | 65    | 257   |
| Around   | 69    | 76    | 106   | 251   |
| Total    | 194   | 143   | 171   | 508   |

of pair. The final column contains the row margins. They give the total number of persons who walked through and around. The number in the bottom right-hand corner, 508, is the total number of persons in the study.

As discussed in Chapter 8, a table of frequencies is, by itself, not very interpretable. To increase interpretability the percentage of those who walked through for each racial composition is computed. Percentages are computed for each column because racial composition is the independent variable and behavior is the dependent variable. The result is shown in Table 17.2. The subjects are most likely to walk through the pair when the pair is black and least likely when the pair is mixed. Interestingly, the mixed-pair percentage does not fall halfway between the black and white pairs.

It might be asked whether these results could be explained by sampling error. Is it possible that, by chance, the subjects in the black condition just happened to be persons who would walk through any pair? Can the hypothesis that the racial composition does not affect behavior and that the observed differences are due to sampling error be ruled out?

If there is no association between the two nominal variables, then it is said that the two variables are independent. Thus, the complete model assumes that the two nominal variables are associated and the restricted model is that the two variables are independent.

To evaluate the restricted model, it is necessary to estimate the number of subjects who would walk through the black pair if the variables of racial composition and behavior were independent. The actual or observed number is compared with the *frequency expected* if the two variables were independent.

The expected frequency for a given cell equals the row margin times the column margin divided by the total number of persons. (Note that *frequencies,* not proportions, are used.) So, the expected number of persons who walk through the black pair is the row margin (257) times the column margin (194) divided by the total number of persons (508) or

$$\frac{(257)(194)}{508} = 98.15$$

TABLE 17.2    **Percentages by Column for the Brown (1981) Study**

| Behavior | Racial Composition | | |
|---|---|---|---|
|  | Black | White | Mixed |
| Through | 64 | 47 | 38 |
| Around | 36 | 53 | 62 |
| Total | 100 | 100 | 100 |

It is not at all unusual for the expected frequency to be a noninteger value. Normally the expected frequencies are computed to two decimal places.

The expected frequency is computed for every cell of the table. For the example, for the six cells of the table the expected frequencies are as shown in Table 17.3.

Note that the row and column margins of the table of expected frequencies are exactly the same as the observed frequencies. This mathematical necessity (within the limits of rounding error) can be used as a computational check to see whether the expected frequencies are computed correctly.

Now the observed frequency minus the expected frequency is computed for each cell. With these differences for each cell of the table the following is computed:

$$\frac{(\text{observed minus expected})^2}{\text{expected}}$$

and this quantity is added across all the cells of the table. This sum has approximately chi square distribution under the restricted model of independence. The degrees of freedom given $r$ rows and $c$ columns in the table are as follows:

$$\text{degrees of freedom} = (r - 1)(c - 1)$$

For the racial composition example, there are two rows and three columns. Thus, $(r - 1)(c - 1)$ equals 1 times 2, or 2. The chi square test of independence is

$$\chi^2[(r-1)(c-1)] = \text{sum} \frac{(\text{observed minus expected})^2}{\text{expected}}$$

The observed frequency is denoted as $o$ and the expected frequency is denoted as $e$. The mathematical formula for the chi square test of independence is

$$\chi^2[(r - 1)(c - 1)] = \sum \frac{(o - e)^2}{e}$$

*TABLE 17.3*   **Expected Frequencies for the Brown (1981) Study**

|  |  | Racial Composition | | |
| --- | --- | --- | --- | --- |
| Behavior | Black | White | Mixed | Total |
| Through | 98.15 | 72.34 | 86.51 | 257.00 |
| Around | 95.85 | 70.66 | 84.49 | 251.00 |
| Total | 194.00 | 143.00 | 171.00 | 508.00 |

The $\chi^2(2)$ for the example is 26.49. Using Appendix G, the $p$ value for the value of $\chi^2$ is less than .001, and so the null hypothesis that behavior and racial composition are unrelated is rejected. The differential probability of walking through racial pairs cannot be explained by chance.

For $2 \times 2$ tables (that is, a table with two rows and two columns), various measures of association were presented in Chapter 8. One such measure is the phi coefficient. As explained in Chapter 8, phi ($\phi$) is a correlation coefficient. If phi is known, $\chi^2$ can be computed directly

$$\chi^2(1) = N\phi^2$$

where $N$ is the total number of persons in the study. So $\chi^2$ equals the sample size times phi squared. This only applies to tests using $2 \times 2$ tables.

If the chi square is not significant, one concludes that the two variables are independent; that is, the variables are unrelated. If chi square is statistically significant, then one concludes that the variables are associated. To determine the direction of the association, one can compute percentages across rows or columns.

The fact that the degrees of freedom of the $\chi^2$ test are $(r-1)(c-1)$ is not as mysterious as might seem. Recall that the degrees of freedom for interaction in analysis of variance take on a similar form. They equal the product of the number of levels of the first independent variable less one times the number of levels of the second variable less one. The total number of cells in the table are $rc$, the number of rows times the number of columns. To test for independence, the row and column margins are used. The sum of the expected frequencies must equal these row and column margins. There are $r$ row margins and $c$ column margins. Because both the row column margins must sum to $N$, there is one constraint on the row and column margins. So the number of unconstrained frequencies is the total number of cells less the number of rows and columns plus one. In terms of symbols,

$$rc - r - c + 1$$

which equals

$$(r - 1)(c - 1)$$

This equals the degrees of freedom for the $\chi^2$ test of independence.

## Assumptions

One major assumption of the $\chi^2$ test is that observations are independent. To ensure that the assumption is met, the total $N$ must represent that many unique responses. *The same person must not enter the table more than once.* The number of persons must equal the number of observations.

The $\chi^2$ test of hypotheses of association between variables is only an approximate test. That is, the sum of the observed minus the expected squared divided by the expected has only approximately a $\chi^2$ distribution under the restricted model of independence. The $p$ values obtained are only approximate. How good the approximation is depends, in general, on the overall sample size. The larger the sample size, the better is the approximation. A good rule of thumb is that the total $N$ divided by the number of cells must be at least five before the approximation becomes quite good. In terms of symbols: $N$ must be greater than or equal to $5rc$; that is, $N \geq 5rc$.

# McNemar Test

The $\chi^2$ test of independence presumes that the observations are independent. It is not at all uncommon for observations to be linked. In Chapter 13, the paired $t$ test for scores that are linked or paired across two conditions is described. Described here is a similar procedure for linked scores in which both the independent and dependent variables are dichotomies.

Consider an election survey in which 100 persons are interviewed and 55 favor candidate A and 45 candidate B. These same 100 persons are interviewed again and asked who it is that they prefer. Now 49 prefer A and 51 prefer B. The issue is whether the percentage of those favoring the candidates has changed significantly over time. The independent variable is time, and the dependent variable is candidate preference. It would not be valid to employ a $\chi^2$ test of independence because the same persons were interviewed in both of the surveys.

To perform the McNemar test, one examines only those who have changed over time. So, the number who switched from candidate A to B is compared with the number who switched from B to A. If the independent variable had no effect on the dependent variable, within the limits of sampling error, these two numbers should be the same. The McNemar test evaluates the null hypothesis that the two types of changers are equal. If this null hypothesis is false, the null hypothesis that the independent variable has no effect on the dependent variable also is false.

There are two key frequencies that must be determined to compute the McNemar test. The persons who switch from one category to the other category for the dependent variable must be counted. The two frequencies are designated as $a$ and $d$. So, for the example, $a$ is the number who switched from candidate A to B and $d$ is the number who switched from B to A. The formula for the McNemar test is

$$\chi^2(1) = \frac{(|a - d| - 1.0)^2}{a + d}$$

(The expression $|a - d|$ is the absolute value of $a - d$. If $a - d$ is negative, the sign becomes positive.) The degrees of freedom for the McNemar test are one. The $-1.0$ term in the numerator is called the *correction for continuity*.

Such a correction improves the accuracy of the $\chi^2$ approximation. (A similar correction was proposed for the $\chi^2$ test of independence for $2 \times 2$ tables. Recent work has shown that the correction there is not necessary.)

## Assumptions

Even though the two groups are nonindependent, all other scores must be independent. Also, the $\chi^2$ distribution is used to approximate the sampling distribution of the McNemar test. If $a + d$ is small, the approximation is not very good. One rule of thumb is that $a + d$ must be at least ten before the test is performed. Even if the approximation were good for $a + d$ less than ten, the test would be of little use because its power would be so low.

## Example

Consider the following experiment. Mita, Dermer, and Knight (1978) took one picture of 33 persons, but for each person two different pictures were developed. One was a usual or regular picture. For the other, the negative was turned upside down before printing, causing the print to represent a mirror image of the person photographed. Each person and a person's friend were asked which of the two pictures they preferred. The normal print would show the way that others see the person, and the reversed print would show how the person would see him or herself as in a mirror. According to the social psychologist Robert Zajonc, individuals generally prefer the familiar, and so friends should prefer the regular photo and the persons themselves should prefer the reversed photo.

The independent variable from this experiment is friend versus self, and the dependent variable is picture chosen, regular or reversed. Although there are 66 persons in the study, only 33 of them are independent because there are actually 33 pairs of friends. To perform the McNemar test, it must be determined how many times the friend preferred the regular picture and the self preferred the reversed picture. According to Mita and his colleagues this number should be high relative to the number of times that the friend preferred the reversed picture and the self preferred the regular picture.

The results from the experiment by Mita, Dermer, and Knight are that 15 pairs operated as predicted and 7 pairs were in the opposite direction. The McNemar test result is

$$\frac{(\,|\,15 - 7\,|\, - 1.0)^2}{15 + 7}$$

The $\chi^2$ (1) value is 2.23. Using Appendix G, this value does not equal or exceed the value of 3.84 necessary for it to be statistically significant at the .05 level of significance. So although the results are in the predicted direction, they are not statistically significant. There is no statistically significant evidence that persons prefer the reversed picture of self and friends prefer the normal picture.

# $\chi^2$ *Goodness of Fit Test*

Sometimes a researcher has a hypothesis about the distribution of a nominal variable and wishes to evaluate it. Consider the following examples:

1. In a study of extrasensory perception, a researcher asks 40 supposed psychics whether a coin that is flipped is heads or tails. Of the 40 psychics, 24 are correct and 16 are incorrect. Is this significantly better than 20 correct and 20 incorrect expected by chance?
2. A computer scientist wants to test how random her random number generator is. She has a computer generate 1000 random integers from 1 to 10. If the generator is truly random, then each integer should appear 10% of the time.
3. A researcher seeks to compare whether enough women are called for jury duty in a given county of the United States. By using census data, the researcher determines that 52% of the adult population is female. Of 458 persons called for jury duty 212 are females.

In each of these cases, there is a nominal variable. For the first, it is heads or tails; for the second, it is integer from one to ten; and for the third, it is gender. The researcher has some way of predicting the percentage of cases for each category of the nominal variable. The *expected frequency* for a category equals the total $N$ times the proportion that is predicted for that category. So for each category of a nominal variable, there is an observed frequency and an expected frequency.

The observed frequency can be compared to the expected frequency. It turns out that the expression

$$\text{sum}\left[\frac{(\text{observed} - \text{expected})^2}{\text{expected}}\right]$$

has a $\chi^2$ distribution with $k - 1$ degrees of freedom, where $k$ is the number of categories of the nominal variable. If $\chi^2$ is significant, the model or theory that predicts the distribution is incorrect in some way. If $\chi^2$ is not significant, the frequencies are compatible with the theory.

Note that the formula for the $\chi^2$ goodness of fit test is identical to that for the $\chi^2$ test of independence. The difference between the two tests is in how the expected frequencies are computed.

## *Assumptions*

The $\chi^2$ goodness of fit test requires that observations be independent. One consequence of this assumption is that the same person may enter the table only once. A second assumption is that all expected values must be nonzero. If theory predicts that a category has no members, a $\chi^2$ test is not necessary. One need only see whether the category has any members. If it does the theory is falsified. For the $\chi^2$ approximation to be adequate, expected values should be at least five.

## Example

In 1866 the monk Gregor Mendel reported the results of his experiments on the inheritance of traits. Mendel took seeds that were pure strain yellow and pollinated them with pure strain green. A total of 529 plants were produced. If his theory of inheritance were correct, then 25% of the peas produced should be pure yellow, 25% pure green, and the remaining 50% should be a hybrid mixture of yellow and green.

What Mendel found was as follows:

|        |     |
|--------|-----|
| Yellow | 126 |
| Hybrid | 271 |
| Green  | 132 |
| Total  | 529 |

At issue is how well Mendel's theory of inheritance predicts the distribution of pea plant colors.

Because the theory predicts 25% yellow, 50% hybrid, and 25% green the expected frequency of plants are

$$\text{Yellow:} \quad .25 \times 529 = 132.25$$

$$\text{Hybrid:} \quad .50 \times 529 = 264.50$$

$$\text{Green:} \quad .25 \times 529 = 132.25$$

Each of these expected frequencies equals the proportion predicted by the theory times the total number of cases. The sum of these expected frequencies is 529, which is what it should be.

Now these expected frequencies are compared with the observed frequencies.

| Plant  | Observed | Expected | Observed–Expected |
|--------|----------|----------|-------------------|
| Yellow | 126      | 132.25   | –6.25             |
| Hybrid | 271      | 264.50   | 6.50              |
| Green  | 132      | 132.25   | –.25              |

Note that the sum of the observed minus expected is zero, which is a mathematical necessity. So, for Mendel's data, the $\chi^2$ is found to be

$$\chi^2(2) = \frac{(-6.25)^2}{132.25} + \frac{6.50^2}{264.50} + \frac{(-.25)^2}{132.25} = .46$$

Using Appendix G, a value of $\chi^2$ with two degrees of freedom requires a value of 5.99 to be significant at the .05 level of significance. So $\chi^2(2) = .46$

is not statistically significant. The degrees of freedom are two because there are three categories, making $k$ equal to three. Hence the difference between Mendel's obtained distribution of peas and the distribution expected by theory can be attributed to sampling error. The results are compatible with Mendel's theory.

# Other Models for Nominal Dependent Variables

There are many more complex models for nominal dependent variables than those considered in this chapter. For instance, more than one independent variable may be present and the effect of the interaction between the two independent variables may be of interest. Or one may wish to test the effect of a three-level nominal variable on a nominal variable with nonindependent groups. To estimate and test such models, a general method called *log linear analysis* can be used (Fienberg, 1977; Reynolds, 1977).

The model for log linear analysis is formally similar to an analysis of variance model. Like the methods presented in this chapter, log linear analysis produces a set of expected frequencies which are compared to the observed frequencies. However, for most log linear models the expected frequencies require extensive computation, and therefore computers must be used. The discrepancies between observed and expected frequencies are evaluated by the $\chi^2$ distribution. Log linear models are used primarily in survey research, but they could be applied to almost any area of research.

# Summary

The methods discussed in this chapter were developed for variables measured at the nominal level of measurement, whereas the methods discussed in the previous five chapters assume that the dependent variable is measured at the interval level of measurement. These methods, as well as those for ordinal dependent variables, are called *distribution-free* methods because no assumptions are made concerning the distribution of the residual variable. There are three reasons for using distribution-free methods. First, because the dependent variable may be clearly measured at the nominal or ordinal level of measurement, the procedures developed for interval data are inappropriate. Second, the dependent variable may be at the interval level of measurement, but the researcher may be unwilling to assume that the residual variable has a normal distribution. Third, the distribution-free test evaluates a different null hypothesis from that of the distribution-tied test.

The $\chi^2$ *test of independence* is used to evaluate association between a pair of nominally measured variables. It takes as the restricted model that there is no association between the two variables. The $\chi^2$ distribution is used as an approximation to evaluate the plausibility of the restricted model. It involves computing the frequencies expected given no association and comparing them with the observed frequencies. The expected frequency for a cell equals the cell's row margin times the cell's column margin divided by the total number of observations. The degrees of freedom of the test are the number of rows minus one, times the number of columns minus one.

The *McNemar test* evaluates whether a nominal independent variable affects a nominal dependent variable in which the groups are not independent. To use this test, the number of persons who switch from one category to the other is determined. The $\chi^2$ test has one degree of freedom.

For the $\chi^2$ *goodness of fit test* a theory predicts the relative frequencies for each category of the nominal variable. Like the $\chi^2$ test of independence, the goodness of fit test compares observed to expected frequencies. The number of degrees of freedom is the number of categories less one.

More complicated models for nominal dependent variables can be tested through the use of *log linear models*. Like the $\chi^2$ tests presented in this chapter, log linear models involve specifying a restricted model and making predictions concerning the expected frequencies. These expected frequencies are compared to the observed frequencies through the $\chi^2$ distribution.

# *Problems*

1. Locate in the $\chi^2$ table in Appendix G the minimal value of $\chi^2$ to achieve statistical significance.

   |     | df  | p level |
   | --- | --- | ------- |
   | a.  | 1   | .05     |
   | b.  | 5   | .01     |
   | c.  | 3   | .001    |
   | d.  | 2   | .05     |
   | e.  | 10  | .01     |
   | f.  | 19  | .10     |

2. For the following table compute and interpret the $\chi^2$ test of independence.

   |           | Male | Female |
   | --------- | ---- | ------ |
   | Yes       | 15   | 30     |
   | No        | 25   | 8      |
   | Undecided | 12   | 20     |

   Interpret the result.

3. A researcher seeks to compare how many women are called for jury duty in a given county of the United States, in relation to the number of men. By using census data, the researcher finds that 52% of the population

is female. Of 458 persons called for jury duty 212 are females. Are those called for jury duty representative of the general population?

4. The following table (Anderson, 1954) presents the relationship between seeing an ad and buying a product.

|  |  | See an Ad | |
|  |  | Yes | No |
|---|---|---|---|
| Buy the Product | Yes | 138 | 147 |
|  | No | 118 | 543 |

Compute a $\chi^2$ test of independence and interpret the result.

5. Below is a table of the preferences of blacks and whites to be stationed in a northern and southern camp during World War II (Stouffer, Suchman, Devinney, Star, & Williams, 1949).

|  |  | Blacks | Whites |
|---|---|---|---|
| Regional Preference | North | 2027 | 2024 |
|  | South | 2268 | 1717 |

Compute a $\chi^2$ and interpret the result.
    If one splits persons by where they were born, North versus South, one obtains the following pair of 2 × 2 tables.

| | Area of Birth | | | |
| Regional Preference | North | | South | |
| | Blacks | Whites | Blacks | Whites |
|---|---|---|---|---|
| North | 1263 | 1829 | 764 | 195 |
| South | 286 | 672 | 1982 | 1045 |

Compute the $\chi^2$ test of independence separately for those born in the North and those born in the South. For each group interpret the relationship.

6. In a study of extrasensory perception, a researcher asks 40 supposed psychics whether a coin that is flipped is heads or tails. Of the 40 psychics, 24 are correct and 16 are incorrect. Is this significantly better than the 20 correct and 20 incorrect expected by chance?

7. A computer scientist wants to test how random her random number generator is. She has the computer generate 1000 random integers from 1 to 10. If the generator is truly random, then each integer should appear about 10% of the time. She finds the following results.

| | | | | | *Integer* | | | | | |
|---|---|---|---|---|---|---|---|---|---|
| 1 | 2 | 3 | 4 | 5 | 6 | 7 | 8 | 9 | 10 |
| 105 | 99 | 101 | 111 | 85 | 103 | 101 | 96 | 101 | 98 |

Use a $\chi^2$ goodness of fit test to evaluate whether the ten numbers are equally likely.

8. A local politician wants to know if her popularity is improving. She had surveyed 112 persons and found that 40 thought that she was doing a good job and 72 did not. In a more recent survey, 50 thought that she was doing a good job and 62 did not. Given that the two groups are independent, test to see if her popularity is improving.

9. For problem 8, assume now that the same set of persons were interviewed at both times. The complete set of results are as follows:

| Time 1 | Time 2 | n |
|---|---|---|
| good | good | 35 |
| good | poor | 5 |
| poor | good | 15 |
| poor | poor | 57 |

Is her popularity significantly improving?

10. In problem 8 the candidate is rated as good by 40 and poor by 72 in her first survey. Test the hypothesis that as many persons like the candidate as dislike her.

11. An investigator has 27 mothers and fathers listen to recorded cries of their infant child and the cries of another child. Each parent is asked to identify the cries of their own child. The results are as follows:

| Father Correct | Mother Correct | n |
|---|---|---|
| yes | yes | 5 |
| yes | no | 1 |
| no | yes | 9 |
| no | no | 12 |

Are mothers better able than fathers to recognize the cries of their own infant?

12. Consider the variables of religion and support for or against abortion, where the entries represent observed frequencies.

| Abortion Attitude | Religion | | | |
|---|---|---|---|---|
| | *Protestant* | *Catholic* | *Jewish* | *Other* |
| Approve | 33 | 44 | 14 | 54 |
| Disapprove | 21 | 65 | 4 | 33 |

Compute a $\chi^2$ test of independence. Interpret the results.

# 18 | *Models for Ordinal Dependent Variables*

In the preceding chapter it was pointed out that statistical techniques that are used for variables measured at the interval level of measurement are not always appropriate. First, the dependent variable may clearly not be measured at the interval level of measurement. Second, the researcher may be unwilling to make the assumptions that are required for distribution-tied tests. For instance, the normality assumption may be clearly implausible. If either of these cases holds, a distribution-free test may be needed. In this chapter, the topic is the set of models for dependent variables measured at the ordinal level of measurement. As explained in Chapter 1, the ordinal level of measurement implies that the objects can only be rank ordered and that quantitative differences between pairs of objects cannot be assessed.

In this chapter all models have an ordinal dependent variable. The set of models to be considered are presented in Table 18.1. The Mann-Whitney $U$ test is the distribution-free analog of the two-sample $t$ test discussed in Chapter 13. The independent variable is a nominal variable with two levels. So for a Mann-Whitney test there are two groups of persons. Additionally, the dependent variable is measured at the ordinal level of measurement. The Kruskal-Wallis test is the distribution-free analog to one-way analysis of variance. There are multiple groups of persons with the Kruskal-Wallis test and so the independent variable is a nominally measured variable. Like the Mann-Whitney test, the dependent variable is measured at the ordinal level of measurement.

Both the Mann-Whitney and the Kruskal-Wallis presume that the groups are independent. If the groups are not independent, then different tests must be employed. If the independent variable is a dichotomy and the scores in each group are linked, the sign test is appropriate. The sign test's distribution-tied analog is the paired $t$ test described in Chapter 13. If there are more than two groups that are nonindependent, the appropriate test is Friedman two-way

**TABLE 18.1    Models for Ordinal Dependent Variables**

| Level of Measurement of the Independent Variable | Independent Groups | Test | Distribution-Tied Counterpart |
|---|---|---|---|
| Nominal (dichotomy) | Yes | Mann-Whitney | *t* test |
| Nominal (multilevel) | Yes | Kruskal-Wallis ANOVA | One-way ANOVA |
| Nominal (dichotomy) | No | Sign test | Paired *t* test |
| Nominal (multilevel) | No | Friedman two-way ANOVA | Repeated measures ANOVA |
| Ordinal | — | Rank-order coefficient | Correlation |

ANOVA. Its distribution-tied analog is the repeated measures ANOVA, which was presented in Chapter 15.

Finally, if both the independent and dependent variable are measured at the ordinal level of measurement, then the degree of association between the two variables is measured by the *rank-order coefficient*, sometimes called *Spearman's rho*. This coefficient is the distribution-free analog to the ordinary correlation coefficient. (Because the independent variable is not nominal, it is not relevant to refer to independent or nonindependent groups.)

It is important to realize that a distribution-free method evaluates different null hypotheses than the comparable distribution-tied method. If the different groups have the same distribution but different medians, the distribution-free tests evaluate whether the groups have equal medians. If, however, the groups have different distributions, then the null hypothesis becomes more complicated to state.

For the Mann-Whitney test and Kruskal-Wallis ANOVA, the general null hypothesis is that the groups, when considered as a single sample, all have mean percentile ranks of 50.0. For the sign test and Friedman two-way ANOVA, the null hypothesis is that, for each pair of conditions, persons are just as likely to have a higher score in one condition as they are in the other. When presenting these tests, for reasons of simplicity the null hypothesis will be stated that the groups have equal medians. It should be remembered that when the distributions are different, the null hypothesis is more complicated.

The rank-order coefficient measures any consistent positive or negative relationship between a pair of variables. The ordinary correlation coefficient, or $r$, measures only the linear association between a pair of variables. So, the null hypothesis for Spearman's rho is no positive or negative relationship between the variables.

The procedures that are to be presented in this chapter for ordinal data presume that there are no ties. If there are ties, then each score is given the mean of the tied rank. So, the following set of scores

$$5, 6, 6, 6, 9, 9, 12, 13, 13, 13, 13$$

would yield ranks of

$$1, 3, 3, 3, 5.5, 5.5, 7, 9.5, 9.5, 9.5, 9.5$$

Methods that correct for ties in the ranks for formulas described in this chapter are described in more advanced texts (Bradley, 1968; Siegel, 1956). However, not correcting for ties when there are not many seems to have little effect on the $p$ values.

When working with ranks there are two useful computational checks. The first is to make sure that the last rank (given that it is not tied) equals $n$, the sample size. If it does not, there is an error in the ranking. The second computational check is to compute the mean of the ranks. It should equal $(n + 1)/2$, even if there are tied ranks. If the mean of the ranks does not equal $(n + 1)/2$, there is an error in assigning ranks.

# Mann-Whitney U *Test*

The Mann-Whitney $U$ test is analogous to the two-sample $t$ test described in Chapter 13. However, the assumptions concerning normal distribution and homogeneity of variance are not made by the Mann-Whitney $U$ test. It is then a distribution-free "$t$ test." The test is fairly commonly used in medicine and the biological sciences, but it is relatively infrequently used by most social scientists. Nonetheless, it is an appropriate test when the assumption of normality seems totally implausible.

The Mann-Whitney $U$ test evaluates not the similarity of the means of two groups but rather any consistent difference in the mean percentile scores of the two groups. If the two groups have similar distributions, the Mann-Whitney evaluates whether the two groups have equal medians. Because the mean and median may not be the same, even in the population, the $t$ test and the Mann-Whitney test do not evaluate exactly the same null hypothesis.

The Mann-Whitney test begins with a ranking of all the scores ignoring the fact that the persons are in two different groups. Persons are therefore treated as if they were members of one large group. The ranks then are averaged for the persons in each of the groups, and the difference is computed. This difference between the ranks in the two groups will be denoted as $Q$. At issue is whether the difference between ranks is much larger than it would be if the ranks were assigned randomly.

One way to determine the unusualness of the value of $Q$ is through random assignment of a rank to each person. That is, the actual data are ignored and the subjects are rank ordered again, but this time the ranking is done randomly. Then with these random ranks, the value of $Q$ is computed. If this were done repeatedly, one would obtain a distribution for $Q$. One would then determine just how unusual the obtained value of $Q$ is relative to the values obtained for $Q$ by using a random procedure. This is the essence of the Mann-Whitney $U$ test. It essentially computes the difference between the average rank for the persons in the two groups and judges whether that difference between ranks could have occurred by chance. It does this by comparing the obtained value of $Q$ to what the value of $Q$ would be if persons were randomly assigned ranks.

Consider the following simple case: The number of persons in each group equals three. The data for the two groups are

Group 1:   12, 19, 18

Group 2:   25, 23, 30

The six scores are rank ordered from smallest to largest, as follows:

Group 1:   1, 3, 2

Group 2:   5, 4, 6

The mean or average rank is 2.0 for group 1 and 5.0 for group 2. The difference between the mean rank of group 1 from the mean rank of group 2 is 3.0. At issue is how unusual the value of 3.0 is. If ranks were randomly assigned to each of the six persons, the mean rank difference could be computed. If done enough times, the following mean rank differences with the following probabilities would be obtained.

| Difference in Mean Rank | Probability | Cumulative Probability |
|---|---|---|
| 3.00 | .05 | .05 |
| 2.33 | .05 | .10 |
| 1.67 | .10 | .20 |
| 1.00 | .15 | .35 |
| .33 | .15 | .50 |
| − .33 | .15 | .65 |
| −1.00 | .15 | .80 |
| −1.67 | .10 | .90 |
| −2.33 | .05 | .95 |
| −3.00 | .05 | 1.00 |

For instance, a difference between mean ranks of 1.00 or greater for two groups of size three would occur by chance 35% of the time. It can be seen that the obtained value of the difference between ranks of 3.0 is unusual and would occur by chance only 5% of the time. Because a value of −3.0 would also occur by chance 5%, the two-tailed $p$ value is .10.

For the Mann-Whitney test, the average difference between the ranks is not the statistic that is computed but rather a statistic that could be used to derive it. The statistic computed is the sum of the ranks of the group with the smaller sample size. To see that the sum of the ranks of one group yields the difference between mean ranks, consider the example of two groups of size three. If the sum of the ranks for one group is $R$, then it is a mathematical necessity that the mean difference between ranks must be $2R/3 − 7$. So if $R$ is six, the mean rank difference must be −3.0. The advantage of the sum of the ranks over the mean rank difference is that the sum of the ranks is always a positive integer, whereas this is not true of the mean rank difference. This fact makes it much easier to table the sum of the ranks rather than the mean difference.

To conduct a Mann-Whitney one proceeds as follows. All of the numbers are rank ordered from smallest to largest. Then the ranks in the smaller sized group are summed (not averaged). The sum of the ranks in the smaller sized group is denoted as $R$. If both groups have the same sample size, the sum of ranks of either group can be used. The sample size of the smaller group is denoted as $n_1$ and the sample size of the larger group as $n_2$. The Mann-Whitney test statistic, $U$ is:

$$U = n_1 n_2 + \frac{n_1(n_1 + 1)}{2.0} - R$$

where $R$ is the sum of the $n_1$ ranks. The value of $U$ ranges from zero to $n_1 n_2$. If the ranks are, on average, larger in the $n_1$ group, $U$ is small. If the ranks are larger, on average, in the $n_2$ group, then $U$ is large. So if the restricted model of equal medians is false, the value of $U$ is either very large or very small. To determine whether $U$ is unusually large or small depends on the sample sizes. If both $n_1$ and $n_2$ are less than or equal to 20, tables are used. If either is greater than 20, an approximation is used.

## Both $n_1$ and $n_2$ Are Less than or Equal to 20

First the value of $U$ is computed. Then the obtained value is compared to those values tabled in Appendix H. If the value of $U$ is greater than or equal to the larger value in the table or smaller than or equal to the smaller value in the table, then the restricted model that the medians of the two groups are equal is rejected. So for instance, if $n_1 = n_2 = 10$, the value of $U$ must exceed or equal 78 to be significant at the .05 level or be less than or equal to 28. In Appendix H, the smaller sample size $n_1$ is the first column, and $n_2$ is the second column. Four significance levels are given: .10, .05, .02, and .01.

## *Either* n*₁* *or* n*₂* *Is Greater than 20*

In this case, one does not use the tables in Appendix H, but rather relies on the fact that as the sample sizes increase, the distribution of $U$ approaches the normal distribution, with a mean of

$$\frac{n_1 n_2}{2}$$

and a standard deviation of

$$\sqrt{\frac{n_1 n_2 (n_1 + n_2 + 1)}{12}}$$

Using these facts, $U$ is converted into a variable that has approximately a standard normal distribution under the restricted model. This quantity is denoted as $Z_U$. That is, from $U$ its theoretical mean is subtracted and the difference is divided by its theoretical standard deviation. The complete formula is

$$Z_U = \frac{U - n_1 n_2 / 2}{\sqrt{n_1 n_2 (n_1 + n_2 + 1)/12}}$$

Although the formula looks complicated, it involves only the sum of the ranks and $n_1$ and $n_2$. The quantity $Z_U$ has a standard normal or $Z$ distribution. That is, given the restricted model, the statistic is approximately normally distributed, with a mean of zero and variance of one. Appendix C can be used to determine the $p$ value. The value closest to $Z_U$ (ignoring sign and rounding down) is located. Then take the probability for $Z$ and subtract it from .5, and multiply this difference by two. So for $Z_U = -2.51$, the probability is .4838. The $p$ value is $(.5000 - .4838) \times 2 = .0324$.

As was stated earlier the statistic is only approximately normally distributed. This means that the $p$ values are only approximate. How good the approximation is depends on $n_1$ and $n_2$. As they get larger, the approximation gets better.

## *Examples*

Consider the data in Table 18.2. Because the groups have the same sample size, either group's ranks can be summed. The sum of the ranks in group A is 71. Because $n_1$ and $n_2$ are both seven, the value of $U$ is

$$(7)(7) + \frac{7(7 + 1)}{2} - 71 = 6$$

Looking this value up in Appendix H, a value of $U$ of six with $n_1$ and $n_2$

*TABLE 18.2*    **Example of Mann-Whitney Test**

| | Group A | | Group B | |
| | Score | Rank | Score | Rank |
|---|---|---|---|---|
| | 23 | 5 | 34 | 8 |
| | 43 | 9 | 19 | 4 |
| | 53 | 11 | 11 | 1 |
| | 64 | 13 | 13 | 2 |
| | 27 | 7 | 25 | 6 |
| | 82 | 14 | 18 | 3 |
| | 63 | 12 | 51 | 10 |

equal to seven is statistically significant at the .05 level. This value indicates that the groups' distributions are significantly different.

Assume that there are two groups, $n_1 = 18$, $n_2 = 22$, and $U = 246$. Given $n_1 = 18$ and $n_2 = 22$, the expected mean is

$$\frac{(18)(22)}{2} = 198$$

The variance is

$$\frac{(18)(22)(18 + 22 + 1)}{12} = 1353$$

The square root of 1353 is 36.78. The test that $U$ does not differ from its population mean is

$$Z = \frac{246 - 198}{36.78} = 1.31$$

which is not significant. Therefore the distributions of the two groups do not significantly differ.

# Sign Test

Although the Mann-Whitney test does not presume normality or homogeneity of variance, it is still required that the scores be independent from one another. It may happen that scores are paired, as described for the paired $t$ test in Chapter 13. Each score in a given group is paired or linked to one and only one score in the other group. Scores can be paired because they come from the same person, come from a couple such as friends or littermates, or come from two persons who interact with each other.

A procedure called the *sign test* can be used to test hypotheses about the medians of two samples whose scores are linked. The sign test is very simple.

The two conditions are denoted as I and II, and it will be assumed that the same person is in both conditions. (As with any nonindependent group design, it need not be person that links together the pair of scores, but persons are used in the illustration.) If a person has exactly the same score in condition I as condition II, that person's scores are dropped from the analysis and the $n$ is reduced by one. Like the paired $t$ test, a difference is computed for each person. So the condition I score is subtracted from condition II score. The number of scores with positive *signs* is denoted as $c$. If $n$, the number of untied cases, is less than or equal to 25, then Appendix I is used to determine significance.

If $n$ is greater than 25, the following $Z$ approximation is used.

$$Z = \frac{|2c - n| - 1.0}{\sqrt{n}}$$

where $n$ is the number of persons who have different scores and $c$ is the number of persons whose difference score is positive. (The expression $|2c - n|$ is the absolute value, so the sign of $2c - n$ is always positive.) To determine the $p$ value, the probabilities in Appendix C are used. (See the discussion of $Z_U$ in the Mann-Whitney section.)

As an example, each of ten nine-year-old children work with a seven-year-old child on a task. Observers rate the degree of creativity for each child on the task. The hypothesis is that nine-year-olds are more creative than seven-year-olds. The data are as follows:

| Pair | Nine-Year-Old | Seven-Year-Old |
|------|---------------|----------------|
| 1 | 7 | 5 |
| 2 | 6 | 4 |
| 3 | 5 | 5 |
| 4 | 8 | 4 |
| 5 | 7 | 3 |
| 6 | 8 | 5 |
| 7 | 4 | 6 |
| 8 | 7 | 3 |
| 9 | 9 | 6 |
| 10 | 6 | 4 |

First, it is noted that because the two scores are the same for pair 3, that pair is dropped from the analysis. The $n$ now becomes nine. The difference scores are 2, 2, 4, 4, 3, –2, 4, 3, and 2. Of these nine differences, eight are positive. So $n$ is nine and $c$ is eight. Using Appendix I for these values, the result is statistically significant at the .05 level. So, it is concluded that the nine-year-olds are more creative than the seven-year-olds.

As a second example consider 45 persons who entered a smoking reduction program. One year later, 27 persons have reduced their amount of smoking but 18 increased. Because $n$ is greater than 25, the $Z$ method is used. The value of $Z$ is

$$Z = \frac{\left| (2)(27) - 45 \right| - 1.0}{\sqrt{45}}$$

which equals 1.19 with a $p$ value of .234, which is not statistically significant at the .05 level of significance. Thus the number of persons that reduced their smoking is not significantly greater than the number that increased.

# Kruskal-Wallis Analysis of Variance

The Mann-Whitney test is limited to a dichotomous independent variable, whereas the Kruskal-Wallis test allows for multilevel independent variables. Its distribution-tied cousin is one-way analysis of variance. Although Kruskal-Wallis and Mann-Whitney appear to be very different, it is a statistical fact that Mann-Whitney is a special case of Kruskal-Wallis.

Like the Mann-Whitney $U$ test, the Kruskal-Wallis test begins with a ranking of all of the data from smallest to largest. The ranks are summed in each group. It is the sum of these ranks that are analyzed. Also, like the Mann-Whitney test, the Kruskal-Wallis test evaluates whether the groups have any consistent differences in mean percentile rank.

The formula for the Kruskal-Wallis analysis of variance, called $H$, is

$$H = \left[ \frac{12}{N(N + 1)} \right] \left[ \sum \frac{R_j^2}{n_j} \right] - 3(N + 1)$$

where $N$ is the number of persons across groups, $k$ is the number of groups, $n_j$ is the sample size in the $j$th group, and $R_j$ is the sum of the ranks in the $j$th group. The quantity $H$ is approximately distributed as $\chi^2$ with $k - 1$ degrees of freedom under the restricted model that the medians of all the groups are equal. Hence a significant $\chi^2$ indicates that the groups differ in their medians. It has been found that this $\chi^2$ approximation is quite good if the sample size in every group is at least five.

The formula for the Kruskal-Wallis test looks bewildering. Actually its rationale, if not its derivation, is quite simple. Imagine that the scores are first ranked. Then using the ranks, a one-way ANOVA is computed. From this one-way ANOVA the mean squares for groups would be computed. Such a mean square would take the total of the ranks and square it. These terms are present in the Kruskal-Wallis formula. Also, the $3(N + 1)$ term in the formula is analogous to the correction term for the mean in ANOVA. This mean square for groups is not divided by the mean square for persons within groups but rather by a population variance for groups, and that is why the distribution is $\chi^2$ and not $F$ (see Chapter 11). The population variance can be determined because the scores are ranks, and the variance is therefore known.

Consider the data in Table 18.3. The sums of the ranks in the three groups are 63, 30, and 78. The Kruskal-Wallis statistic is

*TABLE 18.3*　**Kruskal-Wallis Analysis of Variance**

| | Group A | Rank | Group B | Rank | Group C | Rank |
|---|---|---|---|---|---|---|
| | 24 | 4 | 19 | 2 | 53 | 17 |
| | 29 | 7 | 21 | 3 | 46 | 14 |
| | 34 | 10 | 36 | 11 | 39 | 12 |
| | 47 | 15 | 17 | 1 | 42 | 13 |
| | 31 | 9 | 30 | 8 | 50 | 16 |
| | 55 | 18 | 25 | 5 | 28 | 6 |
| Sum | | 63 | | 30 | | 78 |

$$H = \left[\frac{12}{18(18 + 1)}\right]\left[\frac{63^2}{6} + \frac{30^2}{6} + \frac{78^2}{6}\right] - 3(19) = 7.05$$

Using the $\chi^2$ distribution with two degrees of freedom, the value of 7.05 is statistically significant at the .05 level of significance. So, the restricted model that the groups have the same population medians is rejected.

# Friedman Two-Way ANOVA

As described in Chapter 15, a design in which each person is at each level of a nominal independent variable is called repeated measures ANOVA. Here, the distribution-free analog to repeated measures ANOVA is presented. It is called the *Friedman two-way ANOVA.*

As in repeated measures ANOVA, each person is at every level of the independent variable or observations are linked across conditions in some way. However, with the Friedman test, the null hypothesis is that the groups' medians, as opposed to the groups' means, are equal.

To conduct a Friedman ANOVA, scores are separately ranked for each person. This is different from the Kruskal-Wallis, where the entire set of scores is ranked. The formula for the Friedman test is

$$\left[\frac{12}{nk(k + 1)}\right] \sum R_j^2 - 3n(k + 1)$$

where $n$ is the number of persons in the study, $k$ is the number of conditions, and $R_j$ is the sum of the ranks for condition $j$.

When there are two conditions and $k$ is two, the Friedman test is essentially identical to the sign test. Two minor adjustments should be made. First, the square root of the Friedman $\chi^2$ should be taken to make it comparable to the sign test $Z$ value. Second, the square root of $\chi^2$ is slightly larger than the $Z$ value of the sign test because the formula for sign test $Z$ has a 1.0 value subtracted, which is not the case in the Friedman test. If there are just two groups, the sign test should be preferred.

The data in Table 18.4 were presented earlier in this chapter. However, in this instance it is assumed that the data are from six subjects, each of whom is in every condition.

In the table, the three conditions are rank ordered for each subject. The sum of the ranks in the three conditions are 13, 7, and 16. The $n$ is six and $k$ is three. The Friedman statistic is

$$\left[\frac{12}{18(3 + 1)}\right](13^2 + 7^2 + 16^2) - 3(24) = 7.00$$

Using the $\chi^2$ distribution with two degrees of freedom, this value is statistically significant at the .05 level of significance.

# Spearman's Rank-Order Coefficient

In Chapter 8 Spearman's rank-order coefficient was described. It is a measure of association between two ordinally measured variables. Spearman's rank-order coefficient is denoted by $r_S$. Its formula is the standard correlation coefficient applied to ranks. For this measure the scores for each variable are separately rank ordered. Like $r$, $r_S$ can vary between $-1$ and $+1$, and zero indicates that there is no association between the two variables.

There are three major reasons for employing the rank-order coefficient instead of the distribution-tied test. First, the data may be truly ordinal, and not interval as assumed by the ordinary correlation coefficient. Second, it is useful in cases where it cannot be assumed that the variables have a normal distribution. Third, the relationship between the two variables may not be exactly linear. If as $X$ increases, $Y$ increases but in a nonlinear fashion, then the rank-order coefficient may be a more appropriate measure of association than the ordinary correlation coefficient.

*TABLE 18.4*    **Friedman Two-Way ANOVA**

| Person | Group A | Rank | Group B | Rank | Group C | Rank |
|--------|---------|------|---------|------|---------|------|
| 1 | 24 | 2 | 19 | 1 | 53 | 3 |
| 2 | 29 | 2 | 21 | 1 | 46 | 3 |
| 3 | 34 | 1 | 36 | 2 | 39 | 3 |
| 4 | 47 | 3 | 17 | 1 | 42 | 2 |
| 5 | 31 | 2 | 30 | 1 | 50 | 3 |
| 6 | 55 | 3 | 25 | 1 | 28 | 2 |
| Sum | | 13 | | 7 | | 16 |

As discussed in Chapter 8, the rank-order coefficient is an actual correlation between ranks. Besides actually correlating the ranks, there is a computationally simpler formula for the rank-order coefficient. It is based on the difference between each pair of ranks for all persons. The formula is

$$r_S = 1 - \frac{6 \sum D_i^2}{n(n^2 - 1)}$$

where $n$ is the sample size and $D_i$ is the difference between ranks for person $i$. This formula presumes that there are no ties. If there are ties, the ranks should be correlated using the regular formula for a correlation.

To evaluate whether the rank-order coefficient is significantly different from zero, the distribution of $r_S$ under the restricted model of no association can be obtained by randomly assigning the ranks to one of the two variables. Consider the following pairs of scores.

| Person | X | Y |
|--------|---|---|
| 1 | 5 | 3 |
| 2 | 8 | 9 |
| 3 | 6 | 4 |
| 4 | 4 | 1 |

If the scores are ranked separately for each variable, the following set of ranks would be obtained.

| | Rank of | |
|--------|---|---|
| Person | X | Y |
| 1 | 2 | 2 |
| 2 | 4 | 4 |
| 3 | 3 | 3 |
| 4 | 1 | 1 |

Thus, there is perfect correspondence in the ranks, and the rank order coefficient is 1.0.

To determine how unlikely a value of 1.0 is, the sampling distribution of $r_S$ for $n = 4$ is derived. The complete set of possible ranks is enumerated; there are a total 24 possible ranks. These 24 are listed by column, as follows:

```
1 1 1 1 1 1 2 2 2 2 2 2 3 3 3 3 3 3 4 4 4 4 4 4
2 2 3 3 4 4 1 1 3 3 4 4 1 1 2 2 4 4 1 1 2 2 3 3
3 4 2 4 2 3 3 4 1 4 1 3 2 4 1 4 1 2 2 3 1 3 1 2
4 3 4 2 3 2 4 3 4 1 3 1 4 2 4 1 2 1 3 2 3 1 2 1
```

Using these ranks for the X variable and the ranks 1, 2, 3, and 4 for the Y variable, these 24 pairs of ranks produce all the possible rank-order coefficients. There are eleven *different* rank-order coefficients with the following frequencies:

| Rank-Order Coefficient | Frequency | Probability |
|---|---|---|
| 1.0 | 1 | .04125 |
| .8 | 3 | .12375 |
| .6 | 1 | .04125 |
| .4 | 4 | .16500 |
| .2 | 2 | .08250 |
| .0 | 2 | .08250 |
| −.2 | 2 | .08250 |
| −.4 | 4 | .16500 |
| −.6 | 1 | .04125 |
| −.8 | 3 | .12375 |
| −1.0 | 1 | .04125 |

So, the obtained rank-order coefficient of 1.0 would occur by chance only 4.125% of the time. Allowing for a perfect negative rank-order coefficient, the exact $p$ value is .0825.

Fortunately, it is not necessary to do all this work. Tables and approximations are used to test $r_S$. The procedure used to test whether $r_S$ is equal to zero in the population depends on the sample size. If $n$ is less than or equal to 30, one uses the table in Appendix J. If the observed value of $r_S$ equals or exceeds the tabled value in Appendix J, the value of $r_S$ is statistically significantly different from zero at the appropriate level of significance. So for example, if $n$ is 15 and $r_S$ is .31, it does not exceed the critical values in Appendix J, and so the correlation is judged not to be statistically significant.

If $n$ is greater than 30, one uses the ordinary test of a correlation coefficient.

$$t(n-2) = \frac{r_S\sqrt{n-2}}{\sqrt{1-r_S^2}}$$

where $r_S$ is the Spearman rank-order coefficient and $n$ the sample size. So, if $n$ is greater than 30, the $t$ distribution is used as the test statistic. The use of the formula is only an approximation. How good the approximation is depends on $n$. As $n$ gets larger, the approximation gets better.

To illustrate the computations, consider the following example. A total of twelve countries are rank-ordered on their economic wealth and their rate of literacy. The results are

| Country: | A | B | C | D | E | F | G | H | I | J | K | L |
|---|---|---|---|---|---|---|---|---|---|---|---|---|
| Wealth: | 5 | 8 | 10 | 12 | 1 | 9 | 3 | 6 | 4 | 2 | 7 | 11 |
| Literacy: | 6 | 8 | 11 | 12 | 1 | 7 | 3 | 4 | 5 | 2 | 9 | 10 |

The sum of the discrepancies squared is 16 and the rank-order coefficient is

$$1 - \frac{6(16)}{12(12^2 - 1)} = .944$$

Using the table in Appendix J, it is found that a .944 coefficient with an *n* of 12 is statistically significant at the .002 level. Thus, the association between wealth and literacy cannot be explained by chance.

If 40 persons' intelligence and athletic ability are ranked, the rank-order coefficient might be .125. The test of this correlation is

$$t(38) = \frac{.125 \sqrt{40-2}}{\sqrt{1-.125^2}} = .777$$

This value is not statistically significant at the .05 level.

# Power Efficiency

To measure the relative power of a distribution-free and a distribution-tied method, statisticians have developed a measure called *power efficiency*. Assume that there are two tests A and B and A is the more powerful test. Let $n_a$ be the number of subjects needed to achieve a given level of power for test A and $n_b$ be the number of subjects needed for test B to achieve the same power as test A with $n_a$ observations. Because test A is more powerful than test B, $n_a$ must be less than $n_b$. The power efficiency of test B in relation to test A is defined as

$$100 \times \frac{n_a}{n_b} \text{ percent}$$

So, if the power efficiency of a given distribution-free test is 50%, one would need twice as many subjects for the distribution-free test to achieve the same power as with the distribution-tied test.

The power of the Mann-Whitney is comparable to the power of the standard two-sample *t* test. When the set of assumptions hold for the two-sample *t* test, the power efficiency of the Mann-Whitney test for moderate samples is about 95%. This value indicates that there is little loss of power in employing the Mann-Whitney *U* test instead of the *t* test when the distribution is normal. There exist certain types of distributions for which the Mann-Whitney *U* test has a power efficiency greater than 100%.

The power efficiency of the sign test compared to the paired *t* test depends on the sample size. For very small sample sizes ($n = 6$), the power efficiency of the sign test is 95.5%. For very large samples, the power efficiency drops to 63.7%.

The power of the Kruskal-Wallis test is measured in its efficiency versus the *F* test from an analysis of variance. When the set of assumptions hold for the *F* test, the power efficiency of the Kruskal-Wallis test is about 95%. There is little loss of power in employing the Kruskal-Wallis test even when the assumptions required by analysis of variance apply. This 95% figure refers to the normal distribution. There exist certain types of distributions for which the Kruskal-Wallis test has a power efficiency of greater than 100%.

The power efficiency of the Friedman two-way ANOVA depends on the number of conditions and the number of subjects. If there are only two conditions and many subjects, the power efficiency of the test can be as low as 63.7%. If there are either very few subjects or very many conditions, the power efficiency of the Friedman test can be as high as 95.5%.

The power efficiency of the rank-order coefficient is 91% compared to the ordinary correlation coefficient. So when the assumptions necessary for computing the ordinary correlation coefficient are true and $r$ is computed, one needs 91% of the subjects to have the same power to be able to reject the null hypothesis as one would need if Spearman's rho were computed. When the assumptions necessary for $r$ do not hold, the power efficiency of the rank-order coefficient may be almost as good as that of the Pearson $r$, and in some cases it is even better.

# Summary

When the dependent variable is a set of ranks or when one is unwilling to make the assumptions required in distribution-tied statistics, tests that require only variables at the ordinal level of measurement are useful.

The *Mann-Whitney test* is used to test whether a two-level independent variable affects an ordinally measured dependent variable. The test primarily evaluates whether the two groups have the same median. All the scores are ranked and the average rank of the two groups is compared. If the number of observations in both groups is less than or equal to 20, a table is used to determine statistical significance. If not, an approximation to the standard normal distribution is used.

When the independent variable is a dichotomy and the dependent variable is set of ranks and the two groups are nonindependent, the *sign test* is appropriate. The sign test involves determining which observation is larger. If the number of paired observations is less than or equal to 25, a table is used; and if greater than 25, a $\chi^2$ approximation is used.

The *Kruskal-Wallis test* is an extension of the Mann-Whitney test when there are more than two groups. Like the Mann-Whitney test, all the scores are initially ranked, and then analyzed. The test statistic, called $H$, is evaluated by a $\chi^2$ approximation. The degrees of freedom for the test are the number of groups less one.

When there are multiple groups that are nonindependent, *Friedman two-way ANOVA* can be employed. This test involves a ranking of the scores separately for each subject and then using a $\chi^2$ approximation. Its distribution-tied analog is a repeated measures ANOVA.

The *rank-order coefficient* is used to measure association between two ordinally measured variables. The scores for each variable are first rank ordered. The rank-order coefficient is a standard correlation of these ranks. With this measure the relationship between the variables need not be exactly linear. If the sample size is less than or equal to 30, a table is used to determine statistical significance. If $n$ is greater than 30, the standard $t$ test of a correlation can be used to approximate the $p$ value.

Distribution-free tests make weaker assumptions about the data. They do have somewhat less power than distribution-tied methods when the assumptions made by distribution-tied methods are true. However, the power efficiency of distribution-free tests is usually in the mid-90s. That is, the comparable distribution-tied test has the same power as the distribution-free test with about 95% of the subjects.

# Problems

1. For the following data compute a rank-order coefficient, test it, and interpret the results.

   | Person: | 1 | 2 | 3 | 4 | 5 | 6 | 7 | 8 |
   |---|---|---|---|---|---|---|---|---|
   | X: | 7 | 9 | 11 | 3 | 12 | 4 | 5 | 16 |
   | Y: | 10 | 7 | 6 | 12 | 4 | 5 | 8 | 3 |

2. Perform a Mann-Whitney $U$ test for the following data set

   A:  15, 21, 28, 17, 31, 24, 18

   B:  19,  7, 15,  8, 12, 19, 10

3. For the following data compute a rank-order coefficient, test it and interpret the results.

   | Person: | 1 | 2 | 3 | 4 | 5 | 6 |
   |---|---|---|---|---|---|---|
   | X: | 10 | 4 | 10 | 3 | 15 | 5 |
   | Y: | 6 | 7 | 6 | 11 | 3 | 8 |

4. Using a Kruskal-Wallis analysis of variance, test whether the groups' medians differ.

   I:  11, 18, 19, 24, 31

   II:  19, 27, 15,  8, 13

   III:  15, 12, 21, 29, 17

5. An experimenter investigated the success of three methods of lowering the level of cholesterol in the blood. Using a Kruskal-Wallis analysis of variance, test whether the groups' medians differ.

   I:  111, 128, 190, 214, 198

   II:  193, 207, 125, 88, 103, 176

   III:  150, 152, 221, 129, 171

6. Twelve subjects were measured before and after psychotherapy on an adjustment scale. Higher scores indicate greater adjustment. The numbers are as follows:

| Subject | Before | After |
|---|---|---|
| 1 | 23 | 32 |
| 2 | 27 | 25 |
| 3 | 31 | 40 |
| 4 | 32 | 31 |
| 5 | 26 | 38 |
| 6 | 25 | 29 |
| 7 | 25 | 31 |
| 8 | 24 | 24 |
| 9 | 33 | 40 |
| 10 | 22 | 34 |
| 11 | 36 | 38 |
| 12 | 29 | 25 |

Using a distribution-free test, evaluate whether persons improved after psychotherapy.

7. Subjects were asked to lift three weights and rank order them from lightest to heaviest. All three weights were identical in objective weight, but they differed in shape: spherical, conical, and cubical. For the 20 subjects the results were as follows.

| | Spherical | Conical | Cubical |
|---|---|---|---|
| Heaviest | 3 | 5 | 12 |
| Middle | 9 | 4 | 7 |
| Lightest | 8 | 11 | 1 |

The numbers in the table indicate the number of subjects who gave the object that rank. For example, 11 subjects felt that the conical object was the lightest. Do the three objects differ significantly in perceived weight?

8. The following scores are taken from a study that compared two different methods of increasing vocabulary. The scores of ten persons, five under each method, are

A:  16, 19, 20, 18, 24

B:  12, 15, 16, 15, 14

On the basis of a distribution-free test, is there any evidence that one method is superior to the other?

9. A program is developed to improve the intelligence (IQ), scores of preschool children. Two groups of children are randomly formed. Using a distribution-free test, test whether the program affects IQ score.

Treated group:  109, 123, 141, 119, 133, 117, 118, 120

Control group:  106, 103, 114, 120, 116, 107, 98

10. Twenty persons are randomly assigned to one of two treatments. In the treatment group, ten persons are taught a series of strategies to improve their memory. The control group learned none of the strategies. The scores on a memory test are

    Memory group:  88, 76, 83, 75, 64, 80, 76, 73, 84, 78

    Control group:  84, 73, 84, 78, 68, 78, 71, 70, 80, 79

    Using a distribution-free test, are the two groups different?

11. A psychologist studies the degree of happiness of people at various stages in life. His measure of general happiness varies from 0 to 60. In one study he compared the happiness of married and single men aged 25. Using a distribution-free test, is there a significant difference between the two groups?

    | Married | Single |
    |---------|--------|
    | 58 | 57 |
    | 45 | 44 |
    | 50 | 59 |
    | 54 | 44 |
    | 49 | 39 |
    | 39 | 60 |
    | 50 | 44 |
    | 51 | |

12. Nine persons were asked to rate the taste of cola A and cola B on a scale from one to ten. Using a distribution-free test, do persons significantly prefer one drink to the other?

    | Person | Cola A | Cola B |
    |--------|--------|--------|
    | 1 | 7 | 7 |
    | 2 | 8 | 9 |
    | 3 | 8 | 7 |
    | 4 | 9 | 5 |
    | 5 | 10 | 9 |
    | 6 | 9 | 7 |
    | 7 | 8 | 6 |
    | 8 | 8 | 10 |
    | 9 | 7 | 8 |

13. A psychologist is interested in the relationship between handedness and athletic ability. He measures the athletic ability of three groups of persons: left-handed, right-handed, and ambidextrous. His results are:

    Left-handed:    11, 13, 14, 13, 15

    Right-handed:   10, 8, 7, 10, 14

    Ambidextrous:   12, 8, 6, 11, 15

    Do a Kruskal-Wallis ANOVA to determine whether the groups significantly differ.

14. Problem 7 in Chapter 14 described a study of the effectiveness of three different treatments in relieving headache pain. The drugs studied were aspirin, acetaminophen, and a placebo. Ten different persons took one drug and rated their pain on a ten-point scale after three hours. The scores were

    Aspirin:            7, 6, 9, 5, 3, 5, 3, 2, 4, 2

    Acetaminophen:  5, 8, 6, 4, 7, 4, 6, 2, 3, 7

    Placebo:            9, 7, 8, 7, 5, 4, 6, 8, 3, 7

    Using a distribution-free test, evaluate whether the groups significantly differ.

15. A researcher seeks to compare the marital satisfaction of women who have been married for varying number of years. She finds the following (higher numbers, greater satisfaction).

    One Year:   56, 48, 57, 41

    Two Years:  63, 51, 65, 54

    Ten Years:  70, 61, 55, 58

    Using a distribution-free test, evaluate the effect of length of marriage on satisfaction.

16. The following data are taken from Diehl, Kluender, and Parker (1985).

    | Subject | I | II | III |
    |---------|-----|-----|-----|
    | DS | 20 | 19 | 21 |
    | MM | 21 | 18 | 20 |
    | JH | 28 | 24 | 31 |
    | JS | 17 | 6 | 10 |
    | MC | 15 | 13 | 10 |
    | CM | 20 | 13 | 18 |
    | LG | 28 | 22 | 22 |
    | LD | 30 | 24 | 26 |
    | RL | 23 | 18 | 17 |
    | TW | 29 | 19 | 21 |
    | TA | 30 | 16 | 25 |
    | VS | 34 | 29 | 32 |
    | CJ | 21 | 20 | 20 |

    Using a distribution-free test, test for an effect due to condition.

17. In a study involving 20 experimentals and 25 controls:

    a. The sum of the ranks of the 20 experimentals is 248. Do a Mann-Whitney test to determine if the groups' distributions differ.
    b. What would be your answer if the sum of the ranks was 342?

18. For the following values of the rank-order coefficient and $n$, state whether the correlation is significantly different from zero.

|     | $r_S$ | $n$ |
|-----|-------|-----|
| a.  | $-.21$ | 78  |
| b.  | $-.45$ | 42  |
| c.  | $.71$ | 12  |
| d.  | $.35$ | 20  |
| e.  | $.17$ | 99  |
| f.  | $.47$ | 29  |
| g.  | $.46$ | 33  |
| h.  | $-.19$ | 17  |

# Postscript

In Chapter 1 we began our journey. We have traveled through a sea of numbers, terms, formulas, and tables. Research in the social and behavioral sciences brings with it a bewildering array of symbols and terminology. If used properly, they can help us understand why human beings are the way they are. But even more important, they can help us understand how we can come to be more than what we are today.

# References

Anderson, T. W. (1954). Probability models for analyzing time changes in attitudes. In P. F. Lazarsfeld (Ed.), *Mathematical thinking in the social sciences*. Glencoe, Ill.: Free Press.

Ball, S., & Bogatz, G. A. (1970). *The first year of Sesame Street: An evaluation*. Princeton, N.J.: Educational Testing Service.

Ballard, K. D., & Crooks, T. J. (1984). Videotape modeling for preschool children with low levels of social interaction and low peer involvement in play. *Journal of Abnormal Child Psychology, 12,* 95–110.

Baxter, J. (1972). Clerical training for the mentally retarded on a college campus. *Education and Training of the Mentally Retarded, 7,* 135–140.

Benson, L., & Oslick, A. (1969). The uses and abuses of statistical methods in studies of legislative behavior: The 1836 Congressional Gag Rule decision as a test case. Paper presented to the Conference of Applications of Quantitative Methods to Political, Social, and Economic History, University of Chicago.

Boyer, R., & Savageau, D. (1985). *Places Rated Almanac*. New York: Rand McNally.

Bradley, J. V. (1968). *Distribution-free statistical tests*. Englewood Cliffs, N.J.: Prentice-Hall.

Brown, C. E. (1981). Shared space invasion and race. *Personality and Social Psychology Bulletin, 7,* 103–108.

Cohen, J. (1977). *Statistical power analysis for the behavioral sciences* (rev. ed.). New York: Academic Press.

Cutrona, C. E. (1982). Transition to college: Loneliness and the process of social adjustment. In L. A. Peplau & D. Perlman (Eds.), *Loneliness*. New York: Wiley.

DePaulo, B. M., & Rosenthal, R. (1979). Telling lies. *Journal of Personality and Social Psychology, 37,* 1713–1722.

Diehl, R. L., Kluender, K. R., & Parker, E. M. (1985). Are selective adaptation and contrast effects really distinct? *Journal of Experimental Psychology: Human Perception and Performance, 11,* 209–220.

Duncan, S. D., Jr., & Fiske, D. W. (1977). *Face-to-face interaction: Research, methods, and theory.* Hillsdale, N.J.: Erlbaum.

Fienberg, S. E. (1977). *The analysis of cross-classified categorical data.* Cambridge: MIT Press.

Hammersla, K. S. (1983). *Intrinsic motivation: A test of the perception of success hypothesis.* Unpublished doctoral dissertation, University of Connecticut, Storrs.

Harrison, C. L. (1980). *The attribution of intention for unfulfilled offers of aid: An application of indebtedness theory.* Unpublished master's thesis, University of Connecticut, Storrs.

Harrison, C. L. (1984). *Privacy, crowding, and loneliness.* Unpublished manuscript, University of Connecticut, Storrs.

Holahan, C. J., & Moos, R. H. (1985). Life stress and health: Personality, coping, and family support in stress resistance. *Journal of Personality and Social Psychology, 49,* 739–747.

Hughes, J. L., & McNamara, W. J. (1961). A comparative study of programmed and conventional instruction in industry. *Journal of Applied Psychology, 45,* 225–231.

Jacobson, J. (1977). *The determinants of early peer interaction.* Unpublished doctoral dissertation, Harvard University.

Jencks, C., Smith, M., Acland, H., Bane, M. J., Cohen, D., Gintis, H., Heynes, B., & Michelson, S. (1972). *Inequality.* New York: Basic Books.

Johnson, R. C., McClean, G. E., Yuen, S., Nagoshi, C. T., Ahern, F. M., & Cole, R. E. (1985). Galton's data a century later. *American Psychologist, 40,* 875–892.

Kenny, D. A., & La Voie, L. (1984). The social relations model. In L. Berkowitz (Ed.), *Advances in experimental social psychology,* vol. XVIII. New York: Academic Press.

Korytnyk, N. X., & Perkins, D. V. (1983). Effects of alcohol versus expectancy for alcohol on the incidence of graffiti following an experimental task. *Journal of Abnormal Psychology, 92,* 382–385.

Manning, M. M., & Wright, T. L. (1983). Self-efficacy expectancies, outcome expectancies, and persistence of pain control in childbirth. *Journal of Personality and Social Psychology, 45,* 421–431.

Milgram, S. (1963). Behavioral study of obedience. *Journal of Abnormal and Social Psychology, 67,* 371–378.

Mita, T. H., Dermer, M., & Knight, J. (1977). Reversed facial images and the mere-exposure hypothesis. *Journal of Personality and Social Psychology, 35,* 597–601.

Myers, J. L. (1979). *Fundamentals of experimental design* (3rd ed.). Boston: Allyn and Bacon.

Neff, J. A. (1985). Race and vulnerability to stress: An examination of differential vulnerability. *Journal of Personality and Social Psychology, 49,* 481–491.

Pearson, K., & Filon, L. N. G. (1898). Mathematical contributions to the theory of evolution. *Proceedings of the Royal Society of London, Series A, 191,* 259–262.

Pulling, N. H., Wolf, E., Sturgies, S. P., Vaillancourt, D. R., & Dolliver, J. J. (1980). Headlight glare resistance and driver age. *Human Factors, 22,* 103–112.

Rasinski, K., Tyler, T. R., & Fridkin, K. (1985). Exploring the function of legitimacy: Mediating effects of personal and institutional legitimacy on leadership endorsement and system support. *Journal of Personality and Social Psychology, 49,* 386–394.

Reynolds, H. T. (1977). *The analysis of cross-classifications.* New York: Free Press.

Robinson, L. B., & Hastie, R. (1985). Revision of beliefs when a hypothesis is eliminated from consideration. *Journal of Experimental Psychology: Human Perception and Performance, 11,* 443–456.

Rosenthal, R., & Jacobson, L. (1968). *Pygmalion in the classroom.* New York: Holt, Rinehart and Winston.

Rosenthal, R., & Rubin, D. B. (1979). A note on percent variance explained as a measure of the importance of effects. *Journal of Applied Social Psychology, 9,* 395–396.

Schifter, D. E., & Ajzen, I. (1985). Intention, perceived control, and weight loss: An application of the theory of planned behavior. *Journal of Personality and Social Psychology, 49,* 843–851.

Schwartz, R. G., & Leonard, L. B. (1984). Words, objects and actions in early lexical acquisition. *Journal of Speech and Hearing Research, 27,* 119–127.

Segal, M. W., (1974). Alphabet and attraction: An unobtrusive measure of the effect of propinquity in a field setting. *Journal of Personality and Social Psychology, 30,* 654–657.

Siegel, S. (1956). *Nonparametric statistics.* New York: McGraw-Hill.

Smith, M. (1980). Sex bias in counseling and psychotherapy. *Psychological Bulletin, 87,* 392–407.

Steiger, J. H. (1980). Tests for comparing elements of a correlation matrix. *Psychological Bulletin, 87,* 245–251.

Stoudt, H. W. (1981). Anthropometry of the elderly. *Human Factors, 23,* 29–37.

Stouffer, S. A., Suchman, E. A., Devinney, L. C., Star, S. A., & Williams, R. M., Jr. (1949). *The American soldier,* vol. I. Princeton, N.J.: Princeton University Press.

Taylor, R. B., & Ferguson, G. (1980). Solitude and intimacy: Linking territoriality and privacy experiences. *Journal of Nonverbal Behavior, 4,* 227–239.

Vinsel, A., Brown, B., Altman, I., & Foss, C. (1980). Privacy regulation, territorial displays, and effectiveness of individual functioning. *Journal of Personality and Social Psychology, 39,* 1104–1115.

West, S. G., & Shults, T. (1976). Liking for common and uncommon first names. *Personality and Social Psychology Bulletin, 2,* 299–302.

Wheeler, L., Reis, H., & Nezlek, J. (1983). Loneliness, social interaction, and sex roles. *Journal of Personality and Social Psychology, 45,* 943–953.

Winer, B. J. (1971). *Statistical principles in experimental design* (2nd ed.). New York: McGraw-Hill.

# Appendix A

# Two-Stretch Transformations of Proportions: Arcsin, Probit, and Logit

For proportions of .00 and 1.00, the values of .0025 and .9975 are used for probit and logit. Arcsin in radians equals $2\arcsin\sqrt{p}$, where $p$ is the proportion.

*Two-Stretch Transformations of Proportions*

| Proportion | Arcsin | Probit | Logit | Proportion | Arcsin | Probit | Logit |
|---|---|---|---|---|---|---|---|
| .00 | .000 | −2.807 | −5.989 | .51 | 1.591 | .025 | .040 |
| .01 | .200 | −2.326 | −4.595 | .52 | 1.611 | .050 | .080 |
| .02 | .284 | −2.054 | −3.892 | .53 | 1.631 | .075 | .120 |
| .03 | .348 | −1.881 | −3.476 | .54 | 1.651 | .100 | .160 |
| .04 | .403 | −1.751 | −3.178 | .55 | 1.671 | .126 | .201 |
| .05 | .451 | −1.645 | −2.944 | .56 | 1.691 | .151 | .241 |
| .06 | .495 | −1.555 | −2.752 | .57 | 1.711 | .176 | .282 |
| .07 | .536 | −1.476 | −2.587 | .58 | 1.731 | .202 | .323 |
| .08 | .574 | −1.405 | −2.442 | .59 | 1.752 | .228 | .364 |
| .09 | .609 | −1.341 | −2.314 | .60 | 1.772 | .253 | .405 |
| .10 | .644 | −1.282 | −2.197 | .61 | 1.793 | .279 | .447 |
| .11 | .676 | −1.227 | −2.091 | .62 | 1.813 | .305 | .490 |
| .12 | .707 | −1.175 | −1.992 | .63 | 1.834 | .332 | .532 |
| .13 | .738 | −1.126 | −1.901 | .64 | 1.855 | .358 | .575 |
| .14 | .767 | −1.080 | −1.815 | .65 | 1.875 | .385 | .619 |
| .15 | .795 | −1.036 | −1.735 | .66 | 1.897 | .412 | .663 |
| .16 | .823 | −.994 | −1.658 | .67 | 1.918 | .440 | .708 |
| .17 | .850 | −.954 | −1.586 | .68 | 1.939 | .468 | .754 |
| .18 | .876 | −.915 | −1.516 | .69 | 1.961 | .496 | .800 |
| .19 | .902 | −.878 | −1.450 | .70 | 1.982 | .524 | .847 |
| .20 | .927 | −.842 | −1.386 | .71 | 2.004 | .553 | .895 |
| .21 | .952 | −.806 | −1.325 | .72 | 2.026 | .583 | .944 |
| .22 | .976 | −.772 | −1.266 | .73 | 2.049 | .613 | .995 |
| .23 | 1.000 | −.739 | −1.208 | .74 | 2.071 | .643 | 1.046 |
| .24 | 1.024 | −.706 | −1.153 | .75 | 2.094 | .674 | 1.099 |
| .25 | 1.047 | −.674 | −1.099 | .76 | 2.118 | .706 | 1.153 |
| .26 | 1.070 | −.643 | −1.046 | .77 | 2.141 | .739 | 1.208 |
| .27 | 1.093 | −.613 | −.995 | .78 | 2.165 | .772 | 1.266 |
| .28 | 1.115 | −.583 | −.944 | .79 | 2.190 | .806 | 1.325 |
| .29 | 1.137 | −.553 | −.895 | .80 | 2.214 | .842 | 1.386 |
| .30 | 1.159 | −.524 | −.847 | .81 | 2.240 | .878 | 1.450 |
| .31 | 1.181 | −.496 | −.800 | .82 | 2.265 | .915 | 1.516 |
| .32 | 1.203 | −.468 | −.754 | .83 | 2.292 | .954 | 1.586 |
| .33 | 1.224 | −.440 | −.708 | .84 | 2.319 | .994 | 1.658 |
| .34 | 1.245 | −.412 | −.663 | .85 | 2.346 | 1.036 | 1.735 |
| .35 | 1.266 | −.385 | −.619 | .86 | 2.375 | 1.080 | 1.815 |
| .36 | 1.287 | −.358 | −.575 | .87 | 2.404 | 1.126 | 1.901 |
| .37 | 1.308 | −.332 | −.532 | .88 | 2.434 | 1.175 | 1.992 |
| .38 | 1.328 | −.305 | −.490 | .89 | 2.465 | 1.227 | 2.091 |
| .39 | 1.349 | −.279 | −.447 | .90 | 2.498 | 1.282 | 2.197 |
| .40 | 1.369 | −.253 | −.405 | .91 | 2.532 | 1.341 | 2.314 |
| .41 | 1.390 | −.228 | −.364 | .92 | 2.568 | 1.405 | 2.442 |
| .42 | 1.410 | −.202 | −.323 | .93 | 2.606 | 1.476 | 2.587 |
| .43 | 1.430 | −.176 | −.282 | .94 | 2.647 | 1.555 | 2.752 |
| .44 | 1.451 | −.151 | −.241 | .95 | 2.691 | 1.645 | 2.944 |
| .45 | 1.471 | −.126 | −.201 | .96 | 2.739 | 1.751 | 3.178 |
| .46 | 1.491 | −.100 | −.160 | .97 | 2.793 | 1.881 | 3.476 |
| .47 | 1.511 | −.075 | −.120 | .98 | 2.858 | 2.054 | 3.892 |
| .48 | 1.531 | −.050 | −.080 | .99 | 2.941 | 2.326 | 4.595 |
| .49 | 1.551 | −.025 | −.040 | 1.00 | 3.142 | 2.807 | 5.989 |
| .50 | 1.571 | .000 | .000 | | | | |

# *Random Number Table*

| | | | | | | | | | |
|---|---|---|---|---|---|---|---|---|---|
| 56307 | 81882 | 01267 | 60636 | 27616 | 94931 | 85877 | 33199 | 31923 | 04299 |
| 53170 | 66366 | 22597 | 69962 | 72660 | 36044 | 39661 | 46332 | 69063 | 69126 |
| 25441 | 24626 | 23769 | 44450 | 23392 | 55407 | 52835 | 80126 | 44220 | 21071 |
| 67767 | 75065 | 46060 | 52061 | 75922 | 75232 | 70485 | 02836 | 50285 | 76779 |
| 72158 | 36963 | 33973 | 61639 | 21384 | 11576 | 35060 | 83597 | 82196 | 57290 |
| 32677 | 29310 | 32886 | 42903 | 55303 | 84893 | 91062 | 92422 | 32258 | 79833 |
| 44262 | 70799 | 84371 | 61764 | 71740 | 12999 | 81527 | 95516 | 95997 | 63689 |
| 99218 | 78661 | 33220 | 90874 | 62120 | 15759 | 50368 | 63479 | 66303 | 27846 |
| 93070 | 74521 | 22764 | 95558 | 22262 | 09234 | 41209 | 65445 | 36943 | 90999 |
| 17536 | 73852 | 98382 | 45537 | 45349 | 16219 | 98549 | 69084 | 01392 | 71552 |
| 67557 | 97691 | 01644 | 11410 | 25441 | 52188 | 65424 | 42944 | 54006 | 88783 |
| 37779 | 64441 | 76173 | 59967 | 55136 | 37006 | 85750 | 56453 | 88846 | 60510 |
| 20213 | 92212 | 68812 | 60050 | 29080 | 89076 | 04321 | 78746 | 98507 | 06600 |
| 08921 | 99991 | 19084 | 80209 | 41627 | 27679 | 62120 | 77491 | 77637 | 44282 |
| 98842 | 07646 | 74416 | 91041 | 85667 | 71803 | 05700 | 21238 | 19419 | 94011 |
| 82280 | 31234 | 76089 | 59339 | 53797 | 38971 | 77804 | 80586 | 17913 | 73601 |
| 53589 | 92380 | 69774 | 55115 | 43007 | 49929 | 22053 | 22325 | 66889 | 02919 |
| 06830 | 07729 | 97336 | 46918 | 44137 | 20443 | 82949 | 16470 | 59820 | 99197 |
| 34600 | 97147 | 03860 | 01831 | 51246 | 73016 | 02354 | 31569 | 89891 | 81715 |
| 32719 | 79038 | 32970 | 91334 | 59276 | 22827 | 26529 | 34705 | 52333 | 68289 |
| 93195 | 72681 | 27993 | 81924 | 61702 | 90623 | 96750 | 97357 | 40916 | 79958 |
| 10092 | 95893 | 31966 | 43320 | 51706 | 95684 | 58690 | 87194 | 47942 | 33952 |
| 78348 | 68540 | 89975 | 44952 | 24521 | 54655 | 55386 | 71593 | 67767 | 49552 |
| 89096 | 09026 | 45642 | 53609 | 71573 | 88574 | 30753 | 21154 | 94450 | 44575 |
| 39034 | 34286 | 41125 | 40477 | 87507 | 14672 | 28411 | 03839 | 20589 | 38887 |
| 19878 | 65654 | 22974 | 71760 | 66679 | 51058 | 91689 | 88490 | 42003 | 50891 |
| 26570 | 21824 | 31589 | 18059 | 03149 | 19063 | 64797 | 43655 | 50702 | 98695 |
| 52585 | 18854 | 14107 | 48507 | 89515 | 35040 | 91648 | 38762 | 41920 | 02459 |
| 91104 | 78369 | 11514 | 29603 | 19251 | 18561 | 15864 | 41773 | 07080 | 79707 |
| 36189 | 17014 | 69188 | 32238 | 87884 | 87737 | 80774 | 21530 | 43175 | 48841 |
| 82782 | 54655 | 29874 | 23496 | 88302 | 63898 | 30585 | 47754 | 41042 | 30314 |
| 22012 | 10866 | 38364 | 73685 | 13103 | 62225 | 09214 | 87528 | 95914 | 40769 |
| 49406 | 02585 | 37988 | 50473 | 58106 | 03463 | 59109 | 11159 | 00389 | 00075 |
| 59025 | 49971 | 71781 | 22409 | 15320 | 06893 | 25943 | 86315 | 49113 | 27804 |
| 43467 | 70548 | 84830 | 60092 | 78808 | 89159 | 52752 | 82719 | 36441 | 42066 |
| 97377 | 83890 | 19586 | 30314 | 43509 | 71301 | 82279 | 67453 | 55261 | 60677 |
| 15111 | 08105 | 20548 | 01915 | 01727 | 70548 | 59318 | 10824 | 99343 | 03505 |
| 88386 | 63354 | 31924 | 44617 | 48988 | 06182 | 27197 | 81673 | 63334 | 88950 |
| 12142 | 67160 | 95286 | 26675 | 77010 | 97190 | 30125 | 48549 | 39243 | 35123 |
| 79854 | 37926 | 57771 | 17223 | 80732 | 61096 | 12016 | 78536 | 22305 | 62183 |
| 85876 | 94931 | 60364 | 85102 | 52459 | 18645 | 15320 | 43112 | 05115 | 87654 |
| 23099 | 58336 | 47273 | 98863 | 88888 | 86775 | 85709 | 08775 | 82321 | 29937 |
| 86462 | 44492 | 62413 | 25587 | 80565 | 85646 | 65048 | 45245 | 47105 | 09485 |
| 68269 | 72973 | 50577 | 36754 | 23601 | 32781 | 97252 | 20777 | 94701 | 44115 |
| 62873 | 63061 | 55094 | 00034 | 10301 | 47754 | 28285 | 05680 | 16533 | 52522 |
| 64337 | 33743 | 19000 | 82802 | 33847 | 48674 | 99134 | 66449 | 08419 | 53107 |
| 66428 | 03714 | 86211 | 55658 | 90644 | 55701 | 55178 | 73978 | 62371 | 65152 |
| 62622 | 90205 | 00347 | 86566 | 72116 | 85186 | 40331 | 95349 | 73622 | 07645 |
| 46102 | 13668 | 05909 | 45538 | 72911 | 57875 | 97043 | 99699 | 34768 | 99281 |
| 12100 | 90748 | 47106 | 98779 | 86210 | 91878 | 69816 | 55868 | 40456 | 57290 |
| 53212 | 43656 | 65508 | 16888 | 07248 | 06181 | 01685 | 33576 | 82697 | 02124 |
| 10050 | 97189 | 28076 | 53818 | 20213 | 90163 | 74082 | 41020 | 85291 | 23078 |
| 83659 | 51435 | 42463 | 86315 | 98088 | 31610 | 26863 | 57164 | 11054 | 19691 |
| 01225 | 72639 | 53630 | 27302 | 51664 | 45956 | 58607 | 38762 | 43969 | 96018 |
| 57101 | 80084 | 07248 | 44450 | 74416 | 52773 | 12936 | 52606 | 76633 | 71928 |
| 45349 | 92756 | 44011 | 07478 | 60699 | 71343 | 32007 | 67536 | 03693 | 64651 |
| 57394 | 28180 | 10176 | 85814 | 02229 | 19481 | 61200 | 54446 | 19502 | 93467 |
| 90226 | 07101 | 48193 | 46248 | 95119 | 17056 | 44429 | 80418 | 15780 | 78536 |
| 34684 | 47628 | 01392 | 58795 | 31673 | 82468 | 25316 | 14588 | 88637 | 36210 |

| | | | | | | | | | |
|---|---|---|---|---|---|---|---|---|---|
| 86001 | 68750 | 16324 | 92003 | 06244 | 32655 | 46018 | 25796 | 30377 | 87235 |
| 37946 | 87989 | 33388 | 15299 | 12894 | 90122 | 51915 | 82593 | 34182 | 49720 |
| 83994 | 24919 | 97377 | 96646 | 44220 | 70046 | 86922 | 54404 | 95287 | 41480 |
| 18499 | 97650 | 52793 | 04885 | 94868 | 92003 | 93488 | 09193 | 06286 | 83555 |
| 34350 | 25169 | 43844 | 08566 | 58021 | 76445 | 16114 | 15801 | 61828 | 92003 |
| 25316 | 27344 | 13271 | 75943 | 28913 | 41188 | 99009 | 42777 | 55094 | 87278 |
| 84872 | 22576 | 03525 | 78495 | 98967 | 03756 | 84914 | 59548 | 80146 | 87194 |
| 07373 | 27804 | 61451 | 81548 | 64002 | 83722 | 18624 | 34078 | 30753 | 10447 |
| 88051 | 64358 | 27741 | 55994 | 17202 | 01539 | 42296 | 36378 | 00389 | 76612 |
| 17160 | 48590 | 48653 | 68917 | 51455 | 72681 | 02480 | 33827 | 82237 | 04969 |
| 91689 | 01246 | 66638 | 90623 | 60531 | 18184 | 39577 | 95851 | 71531 | 13334 |
| 03651 | 40435 | 25023 | 89954 | 40247 | 57625 | 99552 | 90414 | 60866 | 19230 |
| 47105 | 83973 | 68017 | 34287 | 81443 | 06767 | 25734 | 87528 | 44889 | 43404 |
| 16993 | 51727 | 40707 | 93049 | 80523 | 86943 | 61158 | 55742 | 15612 | 03964 |
| 05617 | 62100 | 84370 | 34203 | 28912 | 15676 | 50912 | 63312 | 69440 | 19899 |
| 06161 | 33199 | 44680 | 28934 | 22430 | 11368 | 37445 | 78201 | 97796 | 09904 |
| 13856 | 85186 | 65843 | 43446 | 53087 | 93299 | 64086 | 71594 | 95328 | 92380 |
| 22472 | 35583 | 89431 | 47168 | 14943 | 81631 | 76215 | 09695 | 55219 | 85437 |
| 24312 | 03965 | 60239 | 10406 | 02940 | 92714 | 17746 | 61932 | 72702 | 09235 |
| 14902 | 08440 | 21593 | 99657 | 10552 | 47294 | 31129 | 97440 | 40372 | 81297 |
| 85959 | 67997 | 18875 | 84642 | 29791 | 62309 | 59695 | 85060 | 51706 | 21196 |
| 33597 | 01330 | 93655 | 33617 | 57060 | 57917 | 48821 | 93341 | 01058 | 47043 |
| 50117 | 26842 | 90727 | 16888 | 83785 | 51644 | 41251 | 91710 | 82488 | 78996 |
| 29875 | 63814 | 54592 | 00953 | 04613 | 63061 | 80606 | 48131 | 89766 | 33408 |
| 65759 | 31234 | 27114 | 56704 | 13898 | 11451 | 11389 | 31443 | 15194 | 31024 |
| 51539 | 46624 | 54550 | 51225 | 04530 | 14630 | 76633 | 10197 | 54300 | 91125 |
| 81861 | 30732 | 78180 | 54822 | 70276 | 91292 | 20255 | 52647 | 98799 | 29184 |
| 49908 | 00493 | 43676 | 33994 | 05784 | 61012 | 87047 | 29100 | 44805 | 67411 |
| 50912 | 76069 | 92902 | 59632 | 03065 | 41898 | 19168 | 08398 | 84621 | 23036 |
| 76299 | 74981 | 70945 | 99991 | 57352 | 52940 | 60824 | 56746 | 12602 | 15341 |
| 20715 | 16805 | 22598 | 23036 | 62413 | 12831 | 55930 | 45914 | 21635 | 62141 |
| 77344 | 19649 | 88846 | 98779 | 11723 | 41146 | 49281 | 42694 | 06663 | 83304 |
| 41418 | 02501 | 15069 | 94596 | 99636 | 54823 | 83032 | 15926 | 59987 | 96060 |
| 06705 | 84057 | 45182 | 79916 | 17871 | 72848 | 01393 | 35332 | 77135 | 20862 |
| 64337 | 82718 | 21635 | 23873 | 62539 | 16261 | 48277 | 69168 | 50996 | 75525 |
| 11012 | 18938 | 90100 | 94136 | 24772 | 01121 | 67306 | 86566 | 74165 | 78745 |
| 39620 | 09360 | 92442 | 60426 | 01267 | 47880 | 02982 | 55198 | 42463 | 50096 |
| 42756 | 73852 | 74919 | 90999 | 83743 | 74354 | 98340 | 44784 | 47900 | 07687 |
| 15069 | 30815 | 77637 | 54989 | 67139 | 99238 | 97294 | 24751 | 85709 | 70506 |
| 50912 | 99532 | 47440 | 21238 | 44931 | 43279 | 44296 | 61891 | 48486 | 57248 |
| 43174 | 97817 | 27365 | 58294 | 10301 | 22242 | 79017 | 26215 | 30879 | 86316 |
| 85625 | 45538 | 60155 | 33241 | 57311 | 56286 | 51665 | 86274 | 24897 | 75817 |
| 71739 | 59925 | 90602 | 17557 | 78557 | 65278 | 01811 | 10322 | 26947 | 45913 |
| 97043 | 61430 | 62037 | 78912 | 71029 | 65278 | 74250 | 68666 | 28452 | 15299 |
| 53254 | 67871 | 17494 | 85855 | 24395 | 77909 | 12309 | 27804 | 25232 | 02375 |
| 35312 | 97942 | 02062 | 07813 | 36232 | 19816 | 60196 | 58628 | 09298 | 25253 |
| 35521 | 70046 | 59360 | 11576 | 96792 | 10865 | 61827 | 28223 | 74709 | 04090 |
| 02773 | 95851 | 09799 | 86065 | 01769 | 22325 | 54132 | 78285 | 48277 | 07436 |
| 23308 | 06976 | 48862 | 42192 | 08754 | 77616 | 63040 | 24877 | 83868 | 74563 |
| 32300 | 55073 | 79351 | 25211 | 33889 | 74939 | 44680 | 54446 | 70527 | 90832 |
| 22639 | 57959 | 45475 | 03672 | 72702 | 34747 | 42379 | 35834 | 00556 | 74647 |
| 89264 | 35500 | 88803 | 47001 | 18080 | 73685 | 00347 | 37591 | 69482 | 44115 |
| 24480 | 04927 | 55303 | 23162 | 63793 | 11619 | 61326 | 29142 | 69021 | 19397 |
| 21468 | 85520 | 87130 | 02167 | 03316 | 66951 | 70109 | 79624 | 94115 | 18017 |
| 76633 | 97441 | 29665 | 50221 | 31004 | 58963 | 68854 | 56704 | 39410 | 59548 |
| 81401 | 57039 | 25651 | 39096 | 40916 | 05470 | 04112 | 28933 | 96918 | 63271 |
| 30418 | 83889 | 17536 | 35583 | 25651 | 26340 | 16282 | 65738 | 60699 | 45830 |
| 18038 | 98444 | 52167 | 09695 | 78682 | 39975 | 51330 | 36253 | 24521 | 27093 |
| 37152 | 17349 | 70234 | 28808 | 95537 | 64483 | 52585 | 08147 | 83032 | 26633 |

| | | | | | | | | | |
|---|---|---|---|---|---|---|---|---|---|
| 80355 | 59297 | 16825 | 63522 | 69015 | 18854 | 63083 | 51142 | 29414 | 62560 |
| 44639 | 45036 | 36734 | 80628 | 93153 | 22953 | 26737 | 33492 | 57729 | 52689 |
| 19669 | 91502 | 72116 | 48967 | 61159 | 94011 | 88344 | 24333 | 61744 | 18060 |
| 03024 | 17683 | 67181 | 38260 | 66302 | 51309 | 91230 | 91334 | 33764 | 74730 |
| 86127 | 43446 | 65843 | 17934 | 03818 | 13835 | 78432 | 04215 | 69314 | 43153 |
| 63626 | 86022 | 11723 | 55952 | 67474 | 00284 | 93864 | 32405 | 61284 | 41146 |
| 28703 | 65864 | 48151 | 19983 | 49574 | 29059 | 82321 | 42693 | 04613 | 88573 |
| 75420 | 00787 | 20506 | 79749 | 44471 | 19439 | 85960 | 06265 | 91606 | 03839 |
| 97545 | 08314 | 70360 | 04676 | 45056 | 90414 | 50159 | 89327 | 15445 | 05052 |
| 89724 | 09193 | 42505 | 61555 | 47440 | 59506 | 17662 | 62476 | 71363 | 11200 |
| 43969 | 21531 | 81443 | 21572 | 43927 | 46290 | 32091 | 93676 | 53128 | 42150 |
| 54341 | 64316 | 29038 | 52104 | 27700 | 68875 | 37947 | 51769 | 54216 | 13960 |
| 15780 | 55073 | 30376 | 21405 | 92819 | 47419 | 78264 | 05303 | 67808 | 48256 |
| 20589 | 13375 | 30251 | 99782 | 34224 | 99448 | 96959 | 23706 | 87967 | 62853 |
| 64295 | 58503 | 69648 | 54906 | 44220 | 44534 | 38825 | 74939 | 08461 | 75274 |
| 39787 | 08272 | 95119 | 55324 | 17160 | 99615 | 46019 | 27846 | 23936 | 06265 |
| 39661 | 82551 | 48235 | 70464 | 47106 | 86022 | 62748 | 52145 | 26403 | 72764 |
| 59737 | 98569 | 73789 | 68289 | 26654 | 21280 | 32928 | 16094 | 11096 | 94931 |
| 92024 | 76779 | 15111 | 18812 | 51623 | 23789 | 00179 | 87654 | 70611 | 90288 |
| 30585 | 34998 | 16407 | 90582 | 12852 | 11660 | 09004 | 36838 | 99594 | 78410 |
| 70360 | 91920 | 21594 | 50682 | 06746 | 06223 | 02438 | 31025 | 91230 | 78578 |
| 03651 | 65947 | 73120 | 69419 | 25901 | 23831 | 25567 | 39640 | 39578 | 08607 |
| 24814 | 76361 | 16658 | 14462 | 64965 | 84057 | 19670 | 31819 | 38406 | 87194 |
| 27908 | 42150 | 94073 | 55533 | 15947 | 05010 | 29247 | 75232 | 81192 | 32739 |
| 06704 | 82008 | 50451 | 62058 | 72911 | 06850 | 99678 | 40770 | 62288 | 65696 |
| 54676 | 13166 | 29414 | 00828 | 31966 | 07101 | 73706 | 94345 | 74584 | 02710 |
| 65006 | 57248 | 74291 | 41857 | 84245 | 25336 | 93780 | 07436 | 13020 | 64818 |
| 31756 | 30899 | 29289 | 52521 | 24103 | 79665 | 05575 | 46541 | 69900 | 68080 |
| 49866 | 52815 | 37152 | 04592 | 45600 | 89075 | 53296 | 81380 | 39578 | 34120 |
| 11640 | 42568 | 41501 | 63689 | 43677 | 72262 | 77344 | 81381 | 16115 | 79582 |
| 11431 | 19440 | 88009 | 99824 | 09465 | 48800 | 24855 | 13333 | 93279 | 23162 |
| 42129 | 00075 | 19712 | 81547 | 38490 | 35625 | 39160 | 48423 | 63375 | 85604 |
| 98674 | 05512 | 79352 | 78285 | 24814 | 52898 | 62120 | 54028 | 23099 | 83848 |
| 74542 | 52982 | 10552 | 58001 | 61033 | 19314 | 38825 | 98402 | 62998 | 36880 |
| 62915 | 62936 | 58984 | 89536 | 41794 | 52103 | 12894 | 51853 | 79184 | 63396 |
| 59026 | 52020 | 66512 | 41439 | 60280 | 72890 | 27658 | 80878 | 65132 | 82969 |
| 54299 | 40101 | 77051 | 83137 | 09381 | 98319 | 27323 | 56369 | 12852 | 13709 |
| 44555 | 45580 | 35395 | 82593 | 85207 | 47085 | 55805 | 46583 | 17579 | 75776 |
| 21803 | 86566 | 84872 | 09820 | 80063 | 37591 | 56725 | 20653 | 71907 | 85521 |
| 81276 | 84392 | 69691 | 32196 | 87131 | 91460 | 73413 | 45077 | 71656 | 62518 |
| 15571 | 54236 | 31422 | 19147 | 01644 | 24166 | 48904 | 91920 | 08837 | 26047 |
| 19335 | 18017 | 17202 | 39808 | 15027 | 55575 | 29456 | 27093 | 77512 | 95098 |
| 86922 | 92672 | 68018 | 61849 | 24270 | 05262 | 56349 | 20904 | 71447 | 87193 |
| 70318 | 67704 | 69607 | 81715 | 89599 | 34496 | 92986 | 36797 | 49866 | 78327 |
| 32426 | 56454 | 78139 | 30607 | 18289 | 22325 | 03108 | 80920 | 89348 | 34956 |
| 26027 | 49552 | 71279 | 24500 | 09632 | 22200 | 78264 | 28766 | 22346 | 09862 |
| 44889 | 66867 | 94116 | 32823 | 58859 | 51936 | 63835 | 46541 | 44387 | 19983 |
| 83660 | 27972 | 87925 | 25881 | 57395 | 30230 | 03735 | 03672 | 47189 | 85479 |
| 34350 | 48632 | 98381 | 69000 | 99887 | 76654 | 40414 | 69293 | 25692 | 25044 |
| 92192 | 51351 | 68519 | 34245 | 80690 | 09318 | 17202 | 12246 | 73371 | 57080 |
| 49782 | 51309 | 42254 | 88699 | 93865 | 47210 | 79477 | 99908 | 83408 | 00870 |
| 02940 | 69251 | 63208 | 00326 | 30836 | 07854 | 71155 | 02878 | 51037 | 74228 |
| 42505 | 23287 | 74709 | 40310 | 88595 | 11995 | 33513 | 87946 | 68854 | 97022 |
| 64881 | 33576 | 20966 | 74855 | 57980 | 77742 | 13396 | 26299 | 30335 | 86482 |
| 61493 | 16470 | 47064 | 74563 | 34224 | 22911 | 51497 | 84140 | 29833 | 63062 |
| 28160 | 92715 | 94283 | 07394 | 11095 | 67369 | 94074 | 32070 | 61410 | 44576 |
| 58064 | 79247 | 07122 | 42192 | 83242 | 29519 | 83576 | 38051 | 17662 | 49720 |
| 42923 | 96227 | 30962 | 47503 | 41501 | 27470 | 64504 | 72096 | 93237 | 96896 |

| | | | | | | | | | |
|---|---|---|---|---|---|---|---|---|---|
| 80021 | 62350 | 07373 | 91585 | 82279 | 80209 | 79895 | 00410 | 82489 | 06558 |
| 60489 | 96018 | 81150 | 44743 | 98172 | 07604 | 73664 | 93592 | 77135 | 95349 |
| 78683 | 16512 | 96792 | 74646 | 83827 | 26884 | 89431 | 19606 | 73287 | 83137 |
| 28662 | 41648 | 96165 | 51016 | 31255 | 58503 | 71698 | 48465 | 63250 | 88322 |
| 04195 | 90121 | 24354 | 40937 | 35688 | 19105 | 63500 | 47545 | 40205 | 30188 |
| 98758 | 81505 | 24981 | 13542 | 90895 | 28557 | 09925 | 86274 | 99385 | 27721 |
| 05533 | 35082 | 41376 | 38845 | 90351 | 06433 | 52250 | 32614 | 34559 | 98444 |
| 19753 | 67495 | 17745 | 85395 | 27239 | 69669 | 36148 | 56579 | 38030 | 61932 |
| 34015 | 99783 | 98005 | 20276 | 95621 | 38427 | 04655 | 25545 | 05324 | 36294 |
| 42965 | 48005 | 24604 | 13793 | 90435 | 31401 | 01685 | 10113 | 28159 | 41690 |
| 39787 | 33785 | 43216 | 34789 | 03986 | 66993 | 70861 | 77072 | 02648 | 94471 |
| 17369 | 74939 | 95704 | 51811 | 29456 | 63312 | 56684 | 96437 | 19920 | 15382 |
| 70652 | 88992 | 27156 | 31945 | 62078 | 39347 | 01016 | 35583 | 76675 | 23706 |
| 93111 | 96688 | 81192 | 45495 | 95621 | 14964 | 50117 | 65111 | 63459 | 36085 |
| 79101 | 14964 | 01142 | 61304 | 22388 | 08566 | 45265 | 51811 | 76382 | 72388 |
| 21510 | 63982 | 30041 | 47921 | 37904 | 38260 | 32133 | 66868 | 08921 | 51016 |
| 14107 | 61263 | 12977 | 74772 | 35061 | 21865 | 53755 | 76487 | 53086 | 91250 |
| 50368 | 25211 | 93572 | 08649 | 07625 | 80418 | 54048 | 51267 | 05282 | 12079 |
| 24646 | 77449 | 15153 | 19565 | 49072 | 31150 | 76633 | 59172 | 56934 | 31024 |
| 31004 | 33450 | 20757 | 77240 | 52584 | 93342 | 64839 | 67871 | 02689 | 67662 |
| 63208 | 13082 | 55470 | 48758 | 13396 | 85981 | 64045 | 48256 | 36901 | 41271 |
| 13563 | 35918 | 61744 | 03254 | 25274 | 26591 | 15822 | 68582 | 52459 | 69670 |
| 77595 | 20067 | 85249 | 09569 | 80523 | 35918 | 63793 | 96813 | 43132 | 71552 |
| 47022 | 84517 | 67850 | 36252 | 73496 | 30899 | 54802 | 01790 | 03567 | 66491 |
| 17955 | 46792 | 53463 | 52731 | 98255 | 30523 | 29540 | 52062 | 26947 | 71426 |
| 82697 | 27637 | 86880 | 28139 | 49741 | 54655 | 31923 | 17055 | 06160 | 07687 |
| 35604 | 45161 | 11430 | 30146 | 19084 | 20527 | 07917 | 65906 | 36148 | 92798 |
| 69565 | 17181 | 70150 | 27302 | 00639 | 48590 | 99678 | 66282 | 10385 | 45161 |
| 79519 | 62392 | 08126 | 89034 | 89640 | 56662 | 51414 | 86733 | 23225 | 84057 |
| 58942 | 00368 | 67808 | 84475 | 79854 | 63438 | 05868 | 97859 | 66386 | 27302 |
| 01017 | 50389 | 93697 | 81296 | 63585 | 87319 | 07833 | 66449 | 35981 | 95935 |
| 98632 | 30272 | 30000 | 49218 | 34015 | 48758 | 00640 | 61347 | 24312 | 06014 |
| 47900 | 58712 | 19461 | 56495 | 87549 | 64400 | 28494 | 52271 | 24563 | 76821 |
| 41125 | 55282 | 05700 | 72262 | 15612 | 52940 | 35311 | 08649 | 33137 | 28515 |
| 71489 | 38093 | 33220 | 65362 | 14023 | 35123 | 63542 | 97273 | 41460 | 79791 |
| 34224 | 24960 | 45056 | 03170 | 74793 | 29059 | 58858 | 88155 | 43007 | 47880 |
| 21928 | 36922 | 86294 | 55115 | 91983 | 52564 | 63124 | 49845 | 33304 | 52062 |
| 29247 | 90037 | 99385 | 91501 | 59360 | 24332 | 21426 | 50598 | 05240 | 11326 |
| 48193 | 05931 | 28829 | 78829 | 70401 | 65111 | 76215 | 60720 | 52585 | 44366 |
| 75169 | 01246 | 17662 | 87988 | 20632 | 90665 | 73162 | 46709 | 68812 | 70757 |

# Appendix C

# Probabilities for the Z Distribution

The value of $p$ is the probability that a sampled value is greater than or equal to zero and less than or equal to Z. To determine the probability that a sampled value is greater than or equal to Z, subtract $p$ from .5.

| Z | p | Z | p | Z | p |
|---|---|---|---|---|---|
| .00 | .0000 | .50 | .1915 | 1.00 | .3413 |
| .01 | .0040 | .51 | .1950 | 1.01 | .3438 |
| .02 | .0080 | .52 | .1985 | 1.02 | .3461 |
| .03 | .0120 | .53 | .2019 | 1.03 | .3485 |
| .04 | .0160 | .54 | .2054 | 1.04 | .3508 |
| .05 | .0199 | .55 | .2088 | 1.05 | .3531 |
| .06 | .0239 | .56 | .2123 | 1.06 | .3554 |
| .07 | .0279 | .57 | .2157 | 1.07 | .3577 |
| .08 | .0319 | .58 | .2190 | 1.08 | .3599 |
| .09 | .0359 | .59 | .2224 | 1.09 | .3621 |
| .10 | .0398 | .60 | .2257 | 1.10 | .3643 |
| .11 | .0438 | .61 | .2291 | 1.11 | .3665 |
| .12 | .0478 | .62 | .2324 | 1.12 | .3686 |
| .13 | .0517 | .63 | .2357 | 1.13 | .3708 |
| .14 | .0557 | .64 | .2389 | 1.14 | .3729 |
| .15 | .0596 | .65 | .2422 | 1.15 | .3749 |
| .16 | .0636 | .66 | .2454 | 1.16 | .3770 |
| .17 | .0675 | .67 | .2486 | 1.17 | .3790 |
| .18 | .0714 | .68 | .2517 | 1.18 | .3810 |
| .19 | .0753 | .69 | .2549 | 1.19 | .3830 |
| .20 | .0793 | .70 | .2580 | 1.20 | .3849 |
| .21 | .0832 | .71 | .2611 | 1.21 | .3869 |
| .22 | .0871 | .72 | .2642 | 1.22 | .3888 |
| .23 | .0910 | .73 | .2673 | 1.23 | .3907 |
| .24 | .0948 | .74 | .2704 | 1.24 | .3925 |
| .25 | .0987 | .75 | .2734 | 1.25 | .3944 |
| .26 | .1026 | .76 | .2764 | 1.26 | .3962 |
| .27 | .1064 | .77 | .2794 | 1.27 | .3980 |
| .28 | .1103 | .78 | .2823 | 1.28 | .3997 |
| .29 | .1141 | .79 | .2852 | 1.29 | .4015 |
| .30 | .1179 | .80 | .2881 | 1.30 | .4032 |
| .31 | .1217 | .81 | .2910 | 1.31 | .4049 |
| .32 | .1255 | .82 | .2939 | 1.32 | .4066 |
| .33 | .1293 | .83 | .2967 | 1.33 | .4082 |
| .34 | .1331 | .84 | .2995 | 1.34 | .4099 |
| .35 | .1368 | .85 | .3023 | 1.35 | .4115 |
| .36 | .1406 | .86 | .3051 | 1.36 | .4131 |
| .37 | .1443 | .87 | .3078 | 1.37 | .4147 |
| .38 | .1480 | .88 | .3106 | 1.38 | .4162 |
| .39 | .1517 | .89 | .3133 | 1.39 | .4177 |
| .40 | .1554 | .90 | .3159 | 1.40 | .4192 |
| .41 | .1591 | .91 | .3186 | 1.41 | .4207 |
| .42 | .1628 | .92 | .3212 | 1.42 | .4222 |
| .43 | .1664 | .93 | .3238 | 1.43 | .4236 |
| .44 | .1700 | .94 | .3264 | 1.44 | .4251 |
| .45 | .1736 | .95 | .3289 | 1.45 | .4265 |
| .46 | .1772 | .96 | .3315 | 1.46 | .4279 |
| .47 | .1808 | .97 | .3340 | 1.47 | .4292 |
| .48 | .1844 | .98 | .3365 | 1.48 | .4306 |
| .49 | .1879 | .99 | .3389 | 1.49 | .4319 |
| 1.50 | .4332 | 2.00 | .4772 | 2.50 | .4938 |
| 1.51 | .4345 | 2.01 | .4778 | 2.51 | .4940 |
| 1.52 | .4357 | 2.02 | .4783 | 2.52 | .4941 |
| 1.53 | .4370 | 2.03 | .4788 | 2.53 | .4943 |
| 1.54 | .4382 | 2.04 | .4793 | 2.54 | .4945 |
| 1.55 | .4394 | 2.05 | .4798 | 2.55 | .4946 |
| 1.56 | .4406 | 2.06 | .4803 | 2.56 | .4948 |

| Z | p | Z | p | Z | p |
|------|-------|------|-------|------|--------|
| 1.57 | .4418 | 2.07 | .4808 | 2.57 | .4949 |
| 1.58 | .4429 | 2.08 | .4812 | 2.58 | .4951 |
| 1.59 | .4441 | 2.09 | .4817 | 2.59 | .4952 |
| 1.60 | .4452 | 2.10 | .4821 | 2.60 | .4953 |
| 1.61 | .4463 | 2.11 | .4826 | 2.61 | .4955 |
| 1.62 | .4474 | 2.12 | .4830 | 2.62 | .4956 |
| 1.63 | .4484 | 2.13 | .4834 | 2.63 | .4957 |
| 1.64 | .4495 | 2.14 | .4838 | 2.64 | .4959 |
| 1.65 | .4505 | 2.15 | .4842 | 2.65 | .4960 |
| 1.66 | .4515 | 2.16 | .4846 | 2.66 | .4961 |
| 1.67 | .4525 | 2.17 | .4850 | 2.67 | .4962 |
| 1.68 | .4535 | 2.18 | .4854 | 2.68 | .4963 |
| 1.69 | .4545 | 2.19 | .4857 | 2.69 | .4964 |
| 1.70 | .4554 | 2.20 | .4861 | 2.70 | .4965 |
| 1.71 | .4564 | 2.21 | .4864 | 2.71 | .4966 |
| 1.72 | .4573 | 2.22 | .4868 | 2.72 | .4967 |
| 1.73 | .4582 | 2.23 | .4871 | 2.73 | .4968 |
| 1.74 | .4591 | 2.24 | .4875 | 2.74 | .4969 |
| 1.75 | .4599 | 2.25 | .4878 | 2.75 | .4970 |
| 1.76 | .4608 | 2.26 | .4881 | 2.80 | .4974 |
| 1.77 | .4616 | 2.27 | .4884 | 2.85 | .4978 |
| 1.78 | .4625 | 2.28 | .4887 | 2.90 | .4981 |
| 1.79 | .4633 | 2.29 | .4890 | 2.95 | .4984 |
| 1.80 | .4641 | 2.30 | .4893 | 3.00 | .4987 |
| 1.81 | .4649 | 2.31 | .4896 | 3.05 | .4989 |
| 1.82 | .4656 | 2.32 | .4898 | 3.10 | .4990 |
| 1.83 | .4664 | 2.33 | .4901 | 3.15 | .4992 |
| 1.84 | .4671 | 2.34 | .4904 | 3.20 | .4993 |
| 1.85 | .4678 | 2.35 | .4906 | 3.25 | .4994 |
| 1.86 | .4686 | 2.36 | .4909 | 3.30 | .4995 |
| 1.87 | .4693 | 2.37 | .4911 | 3.35 | .49960 |
| 1.88 | .4699 | 2.38 | .4913 | 3.40 | .49966 |
| 1.89 | .4706 | 2.39 | .4916 | 3.45 | .49972 |
| 1.90 | .4713 | 2.40 | .4918 | 3.50 | .49977 |
| 1.91 | .4719 | 2.41 | .4920 | 3.55 | .49981 |
| 1.92 | .4726 | 2.42 | .4922 | 3.60 | .49984 |
| 1.93 | .4732 | 2.43 | .4925 | 3.65 | .49987 |
| 1.94 | .4738 | 2.44 | .4927 | 3.70 | .49989 |
| 1.95 | .4744 | 2.45 | .4929 | 3.75 | .49991 |
| 1.96 | .4750 | 2.46 | .4931 | 3.80 | .49993 |
| 1.97 | .4756 | 2.47 | .4932 | 3.85 | .49994 |
| 1.98 | .4761 | 2.48 | .4934 | 3.90 | .49995 |
| 1.99 | .4767 | 2.49 | .4936 | 3.95 | .49996 |

# Appendix D

# *Two-Tailed Critical Values of* t

To determine the approximate *p* value, first round down the degrees of freedom *(df)*. Then locate the largest value that the *t* test statistic equals or exceeds. Finally, read the column heading for the approximate *p* value.

## Critical Values for t

| df | .20 | .10 | .05 | .02 | .01 | .002 | .001 |
|----|-----|-----|-----|-----|-----|------|------|
| 1 | 3.078 | 6.314 | 12.706 | 31.821 | 63.657 | 318.309 | 636.619 |
| 2 | 1.886 | 2.920 | 4.303 | 6.965 | 9.925 | 22.327 | 31.598 |
| 3 | 1.638 | 2.353 | 3.182 | 4.541 | 5.841 | 10.214 | 12.924 |
| 4 | 1.533 | 2.132 | 2.776 | 3.747 | 4.604 | 7.173 | 8.610 |
| 5 | 1.476 | 2.015 | 2.571 | 3.365 | 4.032 | 5.893 | 6.869 |
| 6 | 1.440 | 1.943 | 2.447 | 3.143 | 3.707 | 5.208 | 5.959 |
| 7 | 1.415 | 1.895 | 2.365 | 2.998 | 3.499 | 4.785 | 5.408 |
| 8 | 1.397 | 1.860 | 2.306 | 2.896 | 3.355 | 4.501 | 5.041 |
| 9 | 1.383 | 1.833 | 2.262 | 2.821 | 3.250 | 4.297 | 4.781 |
| 10 | 1.372 | 1.812 | 2.228 | 2.764 | 3.169 | 4.144 | 4.587 |
| 11 | 1.363 | 1.796 | 2.201 | 2.718 | 3.106 | 4.025 | 4.437 |
| 12 | 1.356 | 1.782 | 2.179 | 2.681 | 3.055 | 3.930 | 4.318 |
| 13 | 1.350 | 1.771 | 2.160 | 2.650 | 3.012 | 3.852 | 4.221 |
| 14 | 1.345 | 1.761 | 2.145 | 2.624 | 2.977 | 3.787 | 4.140 |
| 15 | 1.341 | 1.753 | 2.131 | 2.602 | 2.947 | 3.733 | 4.073 |
| 16 | 1.337 | 1.746 | 2.120 | 2.583 | 2.921 | 3.686 | 4.015 |
| 17 | 1.333 | 1.740 | 2.110 | 2.567 | 2.898 | 3.646 | 3.965 |
| 18 | 1.330 | 1.734 | 2.101 | 2.552 | 2.878 | 3.610 | 3.922 |
| 19 | 1.328 | 1.729 | 2.093 | 2.539 | 2.861 | 3.579 | 3.883 |
| 20 | 1.325 | 1.725 | 2.086 | 2.528 | 2.845 | 3.552 | 3.850 |
| 21 | 1.323 | 1.721 | 2.080 | 2.518 | 2.831 | 3.527 | 3.819 |
| 22 | 1.321 | 1.717 | 2.074 | 2.508 | 2.819 | 3.505 | 3.792 |
| 23 | 1.319 | 1.714 | 2.069 | 2.500 | 2.807 | 3.485 | 3.768 |
| 24 | 1.318 | 1.711 | 2.064 | 2.492 | 2.797 | 3.467 | 3.745 |
| 25 | 1.316 | 1.708 | 2.060 | 2.485 | 2.787 | 3.450 | 3.725 |
| 26 | 1.315 | 1.706 | 2.056 | 2.479 | 2.779 | 3.435 | 3.707 |
| 27 | 1.314 | 1.703 | 2.052 | 2.473 | 2.771 | 3.421 | 3.690 |
| 28 | 1.313 | 1.701 | 2.048 | 2.467 | 2.763 | 3.408 | 3.674 |
| 29 | 1.311 | 1.699 | 2.045 | 2.462 | 2.756 | 3.396 | 3.659 |
| 30 | 1.310 | 1.697 | 2.042 | 2.457 | 2.750 | 3.385 | 3.646 |
| 35 | 1.306 | 1.690 | 2.030 | 2.438 | 2.724 | 3.340 | 3.591 |
| 40 | 1.303 | 1.684 | 2.021 | 2.423 | 2.704 | 3.307 | 3.551 |
| 45 | 1.301 | 1.679 | 2.014 | 2.412 | 2.690 | 3.281 | 3.520 |
| 50 | 1.299 | 1.676 | 2.009 | 2.403 | 2.678 | 3.261 | 3.496 |
| 55 | 1.297 | 1.673 | 2.004 | 2.396 | 2.668 | 3.245 | 3.476 |
| 60 | 1.296 | 1.671 | 2.000 | 2.390 | 2.660 | 3.232 | 3.460 |
| 70 | 1.294 | 1.667 | 1.994 | 2.381 | 2.648 | 3.211 | 3.435 |
| 80 | 1.292 | 1.664 | 1.990 | 2.374 | 2.639 | 3.195 | 3.416 |
| 90 | 1.291 | 1.662 | 1.987 | 2.368 | 2.632 | 3.183 | 3.402 |
| 100 | 1.290 | 1.660 | 1.984 | 2.364 | 2.626 | 3.174 | 3.390 |
| 120 | 1.289 | 1.658 | 1.980 | 2.358 | 2.617 | 3.153 | 3.373 |
| 200 | 1.286 | 1.652 | 1.972 | 2.345 | 2.601 | 3.131 | 3.340 |
| 500 | 1.283 | 1.648 | 1.965 | 2.334 | 2.586 | 3.107 | 3.310 |
| ∞ | 1.282 | 1.645 | 1.960 | 2.326 | 2.576 | 3.090 | 3.291 |

# Appendix E

# Critical Values for the F Distribution

To determine the approximate $p$ value, round down the degrees of freedom on the numerator ($df_n$) and the degrees of freedom on the denominator ($df_d$). Locate $df_n$ in the top row and $df_d$ in the first column. Using the row labels (.10, .05, .01, and .001), determine the approximate $p$ value for the intersection of $df_n$ and $df_d$.

| $df_n$ | | 1 | 2 | 3 | 4 | 5 | 6 | 7 | 8 | 9 | 10 |
|---|---|---|---|---|---|---|---|---|---|---|---|
| $df_d$ | | | | | | | | | | | |
| 1 | .10 | 39.86 | 49.50 | 53.59 | 55.83 | 57.24 | 58.20 | 58.91 | 59.44 | 59.86 | 60.19 |
| | .05 | 161.4 | 199.5 | 215.7 | 224.6 | 230.2 | 234.0 | 236.8 | 238.9 | 240.5 | 241.9 |
| 2 | .10 | 8.53 | 9.00 | 9.16 | 9.24 | 9.29 | 9.33 | 9.35 | 9.37 | 9.38 | 9.39 |
| | .05 | 18.51 | 19.00 | 19.16 | 19.25 | 19.30 | 19.33 | 19.35 | 19.37 | 19.38 | 19.40 |
| | .01 | 98.50 | 99.00 | 99.17 | 99.25 | 99.30 | 99.33 | 99.36 | 99.37 | 99.39 | 99.40 |
| | .001 | 998.5 | 999.0 | 999.2 | 999.2 | 999.3 | 999.3 | 999.4 | 999.4 | 999.4 | 999.4 |
| 3 | .10 | 5.54 | 5.46 | 5.39 | 5.34 | 5.31 | 5.28 | 5.27 | 5.25 | 5.24 | 5.23 |
| | .05 | 10.13 | 9.55 | 9.28 | 9.12 | 9.01 | 8.94 | 8.89 | 8.85 | 8.81 | 8.79 |
| | .01 | 34.12 | 30.82 | 29.46 | 28.71 | 28.24 | 27.91 | 27.67 | 27.49 | 27.35 | 27.23 |
| | .001 | 167.0 | 148.5 | 141.1 | 137.1 | 134.6 | 132.8 | 131.6 | 130.6 | 129.9 | 129.2 |
| 4 | .10 | 4.54 | 4.32 | 4.19 | 4.11 | 4.05 | 4.01 | 3.98 | 3.95 | 3.94 | 3.92 |
| | .05 | 7.71 | 6.94 | 6.59 | 6.39 | 6.26 | 6.16 | 6.09 | 6.04 | 6.00 | 5.96 |
| | .01 | 21.20 | 18.00 | 16.69 | 15.98 | 15.52 | 15.21 | 14.98 | 14.80 | 14.66 | 14.55 |
| | .001 | 74.14 | 61.25 | 56.18 | 53.44 | 51.71 | 50.53 | 49.66 | 49.00 | 48.47 | 48.05 |
| 5 | .10 | 4.06 | 3.78 | 3.62 | 3.52 | 3.45 | 3.40 | 3.37 | 3.34 | 3.32 | 3.30 |
| | .05 | 6.61 | 5.79 | 5.41 | 5.19 | 5.05 | 4.95 | 4.88 | 4.82 | 4.77 | 4.74 |
| | .01 | 16.26 | 13.27 | 12.06 | 11.39 | 10.97 | 10.67 | 10.46 | 10.29 | 10.16 | 10.05 |
| | .001 | 47.18 | 37.12 | 33.20 | 31.09 | 29.75 | 28.84 | 28.16 | 27.64 | 27.24 | 26.92 |
| 6 | .10 | 3.78 | 3.46 | 3.29 | 3.18 | 3.11 | 3.05 | 3.01 | 2.98 | 2.96 | 2.94 |
| | .05 | 5.99 | 5.14 | 4.76 | 4.53 | 4.39 | 4.28 | 4.21 | 4.15 | 4.10 | 4.06 |
| | .01 | 13.75 | 10.92 | 9.78 | 9.15 | 8.75 | 8.47 | 8.26 | 8.10 | 7.98 | 7.87 |
| | .001 | 35.51 | 27.00 | 23.70 | 21.92 | 20.81 | 20.03 | 19.46 | 19.03 | 18.69 | 18.41 |
| 7 | .10 | 3.59 | 3.26 | 3.07 | 2.96 | 2.88 | 2.83 | 2.78 | 2.75 | 2.72 | 2.70 |
| | .05 | 5.59 | 4.74 | 4.35 | 4.12 | 3.97 | 3.87 | 3.79 | 3.73 | 3.68 | 3.64 |
| | .01 | 12.25 | 9.55 | 8.45 | 7.85 | 7.46 | 7.19 | 6.99 | 6.84 | 6.72 | 6.62 |
| | .001 | 29.25 | 21.69 | 18.77 | 17.19 | 16.21 | 15.52 | 15.02 | 14.63 | 14.33 | 14.08 |
| 8 | .10 | 3.46 | 3.11 | 2.92 | 2.81 | 2.73 | 2.67 | 2.62 | 2.59 | 2.56 | 2.54 |
| | .05 | 5.32 | 4.46 | 4.07 | 3.84 | 3.69 | 3.58 | 3.50 | 3.44 | 3.39 | 3.35 |
| | .01 | 11.26 | 8.65 | 7.59 | 7.01 | 6.63 | 6.37 | 6.18 | 6.03 | 5.91 | 5.81 |
| | .001 | 25.42 | 18.49 | 15.83 | 14.39 | 13.49 | 12.86 | 12.40 | 12.04 | 11.77 | 11.54 |
| 9 | .10 | 3.36 | 3.01 | 2.81 | 2.69 | 2.61 | 2.55 | 2.51 | 2.47 | 2.44 | 2.42 |
| | .05 | 5.12 | 4.26 | 3.86 | 3.63 | 3.48 | 3.37 | 3.29 | 3.23 | 3.18 | 3.14 |
| | .01 | 10.56 | 8.02 | 6.99 | 6.42 | 6.06 | 5.80 | 5.61 | 5.47 | 5.35 | 5.26 |
| | .001 | 22.86 | 16.39 | 13.90 | 12.56 | 11.71 | 11.13 | 10.70 | 10.37 | 10.11 | 9.89 |
| 10 | .10 | 3.29 | 2.92 | 2.73 | 2.61 | 2.52 | 2.46 | 2.41 | 2.38 | 2.35 | 2.32 |
| | .05 | 4.96 | 4.10 | 3.71 | 3.48 | 3.33 | 3.22 | 3.14 | 3.07 | 3.02 | 2.98 |
| | .01 | 10.04 | 7.56 | 6.55 | 5.99 | 5.64 | 5.39 | 5.20 | 5.06 | 4.94 | 4.85 |
| | .001 | 21.04 | 14.91 | 12.55 | 11.28 | 10.48 | 9.92 | 9.52 | 9.20 | 8.96 | 8.75 |
| 11 | .10 | 3.23 | 2.86 | 2.66 | 2.54 | 2.45 | 2.39 | 2.34 | 2.30 | 2.27 | 2.25 |
| | .05 | 4.84 | 3.98 | 3.59 | 3.36 | 3.20 | 3.09 | 3.01 | 2.95 | 2.90 | 2.85 |
| | .01 | 9.65 | 7.21 | 6.22 | 5.67 | 5.32 | 5.07 | 4.89 | 4.74 | 4.63 | 4.54 |
| | .001 | 19.69 | 13.81 | 11.56 | 10.35 | 9.58 | 9.05 | 8.66 | 8.35 | 8.12 | 7.92 |
| 12 | .10 | 3.18 | 2.81 | 2.61 | 2.48 | 2.39 | 2.33 | 2.28 | 2.24 | 2.21 | 2.19 |
| | .05 | 4.75 | 3.89 | 3.49 | 3.26 | 3.11 | 3.00 | 2.91 | 2.85 | 2.80 | 2.75 |
| | .01 | 9.33 | 6.93 | 5.95 | 5.41 | 5.06 | 4.82 | 4.64 | 4.50 | 4.39 | 4.30 |
| | .001 | 18.64 | 12.97 | 10.80 | 9.63 | 8.89 | 8.38 | 8.00 | 7.71 | 7.48 | 7.29 |

| $df_n$ | | 12 | 15 | 20 | 24 | 30 | 40 | 60 | 120 | ∞ |
|---|---|---|---|---|---|---|---|---|---|---|
| $df_d$ | | | | | | | | | | |
| 1 | .10 | 60.71 | 61.22 | 61.74 | 62.00 | 62.26 | 62.53 | 62.79 | 63.06 | 63.33 |
|   | .05 | 243.9 | 245.9 | 248.0 | 249.1 | 250.1 | 251.1 | 252.2 | 253.3 | 254.3 |
| 2 | .10 | 9.41 | 9.42 | 9.44 | 9.45 | 9.46 | 9.47 | 9.47 | 9.48 | 9.49 |
|   | .05 | 19.41 | 19.43 | 19.45 | 19.45 | 19.46 | 19.47 | 19.48 | 19.49 | 19.50 |
|   | .01 | 99.42 | 99.43 | 99.45 | 99.46 | 99.47 | 99.47 | 99.48 | 99.49 | 99.50 |
|   | .001 | 999.4 | 999.4 | 999.4 | 999.5 | 999.5 | 999.5 | 999.5 | 999.5 | 999.5 |
| 3 | .10 | 5.22 | 5.20 | 5.18 | 5.18 | 5.17 | 5.16 | 5.15 | 5.14 | 5.13 |
|   | .05 | 8.74 | 8.70 | 8.66 | 8.64 | 8.62 | 8.59 | 8.57 | 8.55 | 8.53 |
|   | .01 | 27.05 | 26.87 | 26.69 | 26.60 | 26.50 | 26.41 | 26.32 | 26.22 | 26.13 |
|   | .001 | 128.3 | 127.4 | 126.4 | 125.9 | 125.4 | 125.0 | 124.5 | 124.0 | 123.5 |
| 4 | .10 | 3.90 | 3.87 | 3.84 | 3.83 | 3.82 | 3.80 | 3.79 | 3.78 | 3.76 |
|   | .05 | 5.91 | 5.86 | 5.80 | 5.77 | 5.75 | 5.72 | 5.69 | 5.66 | 5.63 |
|   | .01 | 14.37 | 14.20 | 14.02 | 13.93 | 13.84 | 13.75 | 13.65 | 13.56 | 13.46 |
|   | .001 | 47.41 | 46.76 | 46.10 | 45.77 | 45.43 | 45.09 | 44.75 | 44.40 | 44.05 |
| 5 | .10 | 3.27 | 3.24 | 3.21 | 3.19 | 3.17 | 3.16 | 3.14 | 3.12 | 3.10 |
|   | .05 | 4.68 | 4.62 | 4.56 | 4.53 | 4.50 | 4.46 | 4.43 | 4.40 | 4.36 |
|   | .01 | 9.89 | 9.72 | 9.55 | 9.47 | 9.38 | 9.29 | 9.20 | 9.11 | 9.02 |
|   | .001 | 26.42 | 25.91 | 25.39 | 25.14 | 24.87 | 24.60 | 24.33 | 24.06 | 23.79 |
| 6 | .10 | 2.90 | 2.87 | 2.84 | 2.82 | 2.80 | 2.78 | 2.76 | 2.74 | 2.72 |
|   | .05 | 4.00 | 3.94 | 3.87 | 3.84 | 3.81 | 3.77 | 3.74 | 3.70 | 3.67 |
|   | .01 | 7.72 | 7.56 | 7.40 | 7.31 | 7.23 | 7.14 | 7.06 | 6.97 | 6.88 |
|   | .001 | 17.99 | 17.56 | 17.12 | 16.89 | 16.67 | 16.44 | 16.21 | 15.99 | 15.75 |
| 7 | .10 | 2.67 | 2.63 | 2.59 | 2.58 | 2.56 | 2.54 | 2.51 | 2.49 | 2.47 |
|   | .05 | 3.57 | 3.51 | 3.44 | 3.41 | 3.38 | 3.34 | 3.30 | 3.27 | 3.23 |
|   | .01 | 6.47 | 6.31 | 6.16 | 6.07 | 5.99 | 5.91 | 5.82 | 5.74 | 5.65 |
|   | .001 | 13.71 | 13.32 | 12.93 | 12.73 | 12.53 | 12.33 | 12.12 | 11.91 | 11.70 |
| 8 | .10 | 2.50 | 2.46 | 2.42 | 2.40 | 2.38 | 2.36 | 2.34 | 2.32 | 2.29 |
|   | .05 | 3.28 | 3.22 | 3.15 | 3.12 | 3.08 | 3.04 | 3.01 | 2.97 | 2.93 |
|   | .01 | 5.67 | 5.52 | 5.36 | 5.28 | 5.20 | 5.12 | 5.03 | 4.95 | 4.86 |
|   | .001 | 11.19 | 10.84 | 10.48 | 10.30 | 10.11 | 9.92 | 9.73 | 9.53 | 9.33 |
| 9 | .10 | 2.38 | 2.34 | 2.30 | 2.28 | 2.25 | 2.23 | 2.21 | 2.18 | 2.16 |
|   | .05 | 3.07 | 3.01 | 2.94 | 2.90 | 2.86 | 2.83 | 2.79 | 2.75 | 2.71 |
|   | .01 | 5.11 | 4.96 | 4.81 | 4.73 | 4.65 | 4.57 | 4.48 | 4.40 | 4.31 |
|   | .001 | 9.57 | 9.24 | 8.90 | 8.72 | 8.55 | 8.37 | 8.19 | 8.00 | 7.81 |
| 10 | .10 | 2.28 | 2.24 | 2.20 | 2.18 | 2.16 | 2.13 | 2.11 | 2.08 | 2.06 |
|   | .05 | 2.91 | 2.85 | 2.77 | 2.74 | 2.70 | 2.66 | 2.62 | 2.58 | 2.54 |
|   | .01 | 4.71 | 4.56 | 4.41 | 4.33 | 4.25 | 4.17 | 4.08 | 4.00 | 3.91 |
|   | .001 | 8.45 | 8.13 | 7.80 | 7.64 | 7.47 | 7.30 | 7.12 | 6.94 | 6.76 |
| 11 | .10 | 2.21 | 2.17 | 2.12 | 2.10 | 2.08 | 2.05 | 2.03 | 2.00 | 1.97 |
|   | .05 | 2.79 | 2.72 | 2.65 | 2.61 | 2.57 | 2.53 | 2.49 | 2.45 | 2.40 |
|   | .01 | 4.40 | 4.25 | 4.10 | 4.02 | 3.94 | 3.86 | 3.78 | 3.69 | 3.60 |
|   | .001 | 7.63 | 7.32 | 7.01 | 6.85 | 6.68 | 6.52 | 6.35 | 6.17 | 6.00 |
| 12 | .10 | 2.15 | 2.10 | 2.06 | 2.04 | 2.01 | 1.99 | 1.96 | 1.93 | 1.90 |
|   | .05 | 2.69 | 2.62 | 2.54 | 2.51 | 2.47 | 2.43 | 2.38 | 2.34 | 2.30 |
|   | .01 | 4.16 | 4.01 | 3.86 | 3.78 | 3.70 | 3.62 | 3.54 | 3.45 | 3.36 |
|   | .001 | 7.00 | 6.71 | 6.40 | 6.25 | 6.09 | 5.93 | 5.76 | 5.59 | 5.42 |

| | $df_n$ | 1 | 2 | 3 | 4 | 5 | 6 | 7 | 8 | 9 | 10 |
|---|---|---|---|---|---|---|---|---|---|---|---|
| $df_d$ | | | | | | | | | | | |
| | .10 | 3.14 | 2.76 | 2.56 | 2.43 | 2.35 | 2.28 | 2.23 | 2.20 | 2.16 | 2.14 |
| 13 | .05 | 4.67 | 3.81 | 3.41 | 3.18 | 3.03 | 2.92 | 2.83 | 2.77 | 2.71 | 2.67 |
| | .01 | 9.07 | 6.70 | 5.74 | 5.21 | 4.86 | 4.62 | 4.44 | 4.30 | 4.19 | 4.10 |
| | .001 | 17.81 | 12.31 | 10.21 | 9.07 | 8.35 | 7.86 | 7.49 | 7.21 | 6.98 | 6.80 |
| | .10 | 3.10 | 2.73 | 2.52 | 2.39 | 2.31 | 2.24 | 2.19 | 2.15 | 2.12 | 2.10 |
| 14 | .05 | 4.60 | 3.74 | 3.34 | 3.11 | 2.96 | 2.85 | 2.76 | 2.70 | 2.65 | 2.60 |
| | .01 | 8.86 | 6.51 | 5.56 | 5.04 | 4.69 | 4.46 | 4.28 | 4.14 | 4.03 | 3.94 |
| | .001 | 17.14 | 11.78 | 9.73 | 8.62 | 7.92 | 7.43 | 7.08 | 6.80 | 6.58 | 6.40 |
| | .10 | 3.07 | 2.70 | 2.49 | 2.36 | 2.27 | 2.21 | 2.16 | 2.12 | 2.09 | 2.06 |
| 15 | .05 | 4.54 | 3.68 | 3.29 | 3.06 | 2.90 | 2.79 | 2.71 | 2.64 | 2.59 | 2.54 |
| | .01 | 8.68 | 6.36 | 5.42 | 4.89 | 4.56 | 4.32 | 4.14 | 4.00 | 3.89 | 3.80 |
| | .001 | 16.59 | 11.34 | 9.34 | 8.25 | 7.57 | 7.09 | 6.74 | 6.47 | 6.26 | 6.08 |
| | .10 | 3.05 | 2.67 | 2.46 | 2.33 | 2.24 | 2.18 | 2.13 | 2.09 | 2.06 | 2.03 |
| 16 | .05 | 4.49 | 3.63 | 3.24 | 3.01 | 2.85 | 2.74 | 2.66 | 2.59 | 2.54 | 2.49 |
| | .01 | 8.53 | 6.23 | 5.29 | 4.77 | 4.44 | 4.20 | 4.03 | 3.89 | 3.78 | 3.69 |
| | .001 | 16.12 | 10.97 | 9.00 | 7.94 | 7.27 | 6.81 | 6.46 | 6.19 | 5.98 | 5.81 |
| | .10 | 3.03 | 2.64 | 2.44 | 2.31 | 2.22 | 2.15 | 2.10 | 2.06 | 2.03 | 2.00 |
| 17 | .05 | 4.45 | 3.59 | 3.20 | 2.96 | 2.81 | 2.70 | 2.61 | 2.55 | 2.49 | 2.45 |
| | .01 | 8.40 | 6.11 | 5.18 | 4.67 | 4.34 | 4.10 | 3.93 | 3.79 | 3.68 | 3.59 |
| | .001 | 15.72 | 10.66 | 8.73 | 7.68 | 7.02 | 6.56 | 6.22 | 5.96 | 5.75 | 5.58 |
| | .10 | 3.01 | 2.62 | 2.42 | 2.29 | 2.20 | 2.13 | 2.08 | 2.04 | 2.00 | 1.98 |
| 18 | .05 | 4.41 | 3.55 | 3.16 | 2.93 | 2.77 | 2.66 | 2.58 | 2.51 | 2.46 | 2.41 |
| | .01 | 8.29 | 6.01 | 5.09 | 4.58 | 4.25 | 4.01 | 3.84 | 3.71 | 3.60 | 3.51 |
| | .001 | 15.38 | 10.39 | 8.49 | 7.46 | 6.81 | 6.35 | 6.02 | 5.76 | 5.56 | 5.39 |
| | .10 | 2.99 | 2.61 | 2.40 | 2.27 | 2.18 | 2.11 | 2.06 | 2.02 | 1.98 | 1.96 |
| 19 | .05 | 4.38 | 3.52 | 3.13 | 2.90 | 2.74 | 2.63 | 2.54 | 2.48 | 2.42 | 2.38 |
| | .01 | 8.18 | 5.93 | 5.01 | 4.50 | 4.17 | 3.94 | 3.77 | 3.63 | 3.52 | 3.43 |
| | .001 | 15.08 | 10.16 | 8.28 | 7.26 | 6.62 | 6.18 | 5.85 | 5.59 | 5.39 | 5.22 |
| | .10 | 2.97 | 2.59 | 2.38 | 2.25 | 2.16 | 2.09 | 2.04 | 2.00 | 1.96 | 1.94 |
| 20 | .05 | 4.35 | 3.49 | 3.10 | 2.87 | 2.71 | 2.60 | 2.51 | 2.45 | 2.39 | 2.35 |
| | .01 | 8.10 | 5.85 | 4.94 | 4.43 | 4.10 | 3.87 | 3.70 | 3.56 | 3.46 | 3.37 |
| | .001 | 14.82 | 9.95 | 8.10 | 7.10 | 6.46 | 6.02 | 5.69 | 5.44 | 5.24 | 5.08 |
| | .10 | 2.96 | 2.57 | 2.36 | 2.23 | 2.14 | 2.08 | 2.02 | 1.98 | 1.95 | 1.92 |
| 21 | .05 | 4.32 | 3.47 | 3.07 | 2.84 | 2.68 | 2.57 | 2.49 | 2.42 | 2.37 | 2.32 |
| | .01 | 8.02 | 5.78 | 4.87 | 4.37 | 4.04 | 3.81 | 3.64 | 3.51 | 3.40 | 3.31 |
| | .001 | 14.59 | 9.77 | 7.94 | 6.95 | 6.32 | 5.88 | 5.56 | 5.31 | 5.11 | 4.95 |
| | .10 | 2.95 | 2.56 | 2.35 | 2.22 | 2.13 | 2.06 | 2.01 | 1.97 | 1.93 | 1.90 |
| 22 | .05 | 4.30 | 3.44 | 3.05 | 2.82 | 2.66 | 2.55 | 2.46 | 2.40 | 2.34 | 2.30 |
| | .01 | 7.95 | 5.72 | 4.82 | 4.31 | 3.99 | 3.76 | 3.59 | 3.45 | 3.35 | 3.26 |
| | .001 | 14.38 | 9.61 | 7.80 | 6.81 | 6.19 | 5.76 | 5.44 | 5.19 | 4.99 | 4.83 |
| | .10 | 2.94 | 2.55 | 2.34 | 2.21 | 2.11 | 2.05 | 1.99 | 1.95 | 1.92 | 1.89 |
| 23 | .05 | 4.28 | 3.42 | 3.03 | 2.80 | 2.64 | 2.53 | 2.44 | 2.37 | 2.32 | 2.27 |
| | .01 | 7.88 | 5.66 | 4.76 | 4.26 | 3.94 | 3.71 | 3.54 | 3.41 | 3.30 | 3.21 |
| | .001 | 14.19 | 9.47 | 7.67 | 6.69 | 6.08 | 5.65 | 5.33 | 5.09 | 4.89 | 4.73 |

| | $df_n$ | 12 | 15 | 20 | 24 | 30 | 40 | 60 | 120 | ∞ |
|---|---|---|---|---|---|---|---|---|---|---|
| $df_d$ | | | | | | | | | | |
| | .10 | 2.10 | 2.05 | 2.01 | 1.98 | 1.96 | 1.93 | 1.90 | 1.88 | 1.85 |
| 13 | .05 | 2.60 | 2.53 | 2.46 | 2.42 | 2.38 | 2.34 | 2.30 | 2.25 | 2.21 |
| | .01 | 3.96 | 3.82 | 3.66 | 3.59 | 3.51 | 3.43 | 3.34 | 3.25 | 3.17 |
| | .001 | 6.52 | 6.23 | 5.93 | 5.78 | 5.63 | 5.47 | 5.30 | 5.14 | 4.97 |
| | .10 | 2.05 | 2.01 | 1.96 | 1.94 | 1.91 | 1.89 | 1.86 | 1.83 | 1.80 |
| 14 | .05 | 2.53 | 2.46 | 2.39 | 2.35 | 2.31 | 2.27 | 2.22 | 2.18 | 2.13 |
| | .01 | 3.80 | 3.66 | 3.51 | 3.43 | 3.35 | 3.27 | 3.18 | 3.09 | 3.00 |
| | .001 | 6.13 | 5.85 | 5.56 | 5.41 | 5.25 | 5.10 | 4.94 | 4.77 | 4.60 |
| | .10 | 2.02 | 1.97 | 1.92 | 1.90 | 1.87 | 1.85 | 1.82 | 1.79 | 1.76 |
| 15 | .05 | 2.48 | 2.40 | 2.33 | 2.29 | 2.25 | 2.20 | 2.16 | 2.11 | 2.07 |
| | .01 | 3.67 | 3.52 | 3.37 | 3.29 | 3.21 | 3.13 | 3.05 | 2.96 | 2.87 |
| | .001 | 5.81 | 5.54 | 5.25 | 5.10 | 4.95 | 4.80 | 4.64 | 4.47 | 4.31 |
| | .10 | 1.99 | 1.94 | 1.89 | 1.87 | 1.84 | 1.81 | 1.78 | 1.75 | 1.72 |
| 16 | .05 | 2.42 | 2.35 | 2.28 | 2.24 | 2.19 | 2.15 | 2.11 | 2.06 | 2.01 |
| | .01 | 3.55 | 3.41 | 3.26 | 3.18 | 3.10 | 3.02 | 2.93 | 2.84 | 2.75 |
| | .001 | 5.55 | 5.27 | 4.99 | 4.85 | 4.70 | 4.54 | 4.39 | 4.23 | 4.06 |
| | .10 | 1.96 | 1.91 | 1.86 | 1.84 | 1.81 | 1.78 | 1.75 | 1.72 | 1.69 |
| 17 | .05 | 2.38 | 2.31 | 2.23 | 2.19 | 2.15 | 2.10 | 2.06 | 2.01 | 1.96 |
| | .01 | 3.46 | 3.31 | 3.16 | 3.08 | 3.00 | 2.92 | 2.83 | 2.75 | 2.65 |
| | .001 | 5.32 | 5.05 | 4.78 | 4.63 | 4.48 | 4.33 | 4.18 | 4.02 | 3.85 |
| | .10 | 1.93 | 1.89 | 1.84 | 1.81 | 1.78 | 1.75 | 1.72 | 1.69 | 1.66 |
| 18 | .05 | 2.34 | 2.27 | 2.19 | 2.15 | 2.11 | 2.06 | 2.02 | 1.97 | 1.92 |
| | .01 | 3.37 | 3.23 | 3.08 | 3.00 | 2.92 | 2.84 | 2.75 | 2.66 | 2.57 |
| | .001 | 5.13 | 4.87 | 4.59 | 4.45 | 4.30 | 4.15 | 4.00 | 3.84 | 3.67 |
| | .10 | 1.91 | 1.86 | 1.81 | 1.79 | 1.76 | 1.73 | 1.70 | 1.67 | 1.63 |
| 19 | .05 | 2.31 | 2.23 | 2.16 | 2.11 | 2.07 | 2.03 | 1.98 | 1.93 | 1.88 |
| | .01 | 3.30 | 3.15 | 3.00 | 2.92 | 2.84 | 2.76 | 2.67 | 2.58 | 2.49 |
| | .001 | 4.97 | 4.70 | 4.43 | 4.29 | 4.14 | 3.99 | 3.84 | 3.68 | 3.51 |
| | .10 | 1.89 | 1.84 | 1.79 | 1.77 | 1.74 | 1.71 | 1.68 | 1.64 | 1.61 |
| 20 | .05 | 2.28 | 2.20 | 2.12 | 2.08 | 2.04 | 1.99 | 1.95 | 1.90 | 1.84 |
| | .01 | 3.23 | 3.09 | 2.94 | 2.86 | 2.78 | 2.69 | 2.61 | 2.52 | 2.42 |
| | .001 | 4.82 | 4.56 | 4.29 | 4.15 | 4.00 | 3.86 | 3.70 | 3.54 | 3.38 |
| | .10 | 1.87 | 1.83 | 1.78 | 1.75 | 1.72 | 1.69 | 1.66 | 1.62 | 1.59 |
| 21 | .05 | 2.25 | 2.18 | 2.10 | 2.05 | 2.01 | 1.96 | 1.92 | 1.87 | 1.81 |
| | .01 | 3.17 | 3.03 | 2.88 | 2.80 | 2.72 | 2.64 | 2.55 | 2.46 | 2.36 |
| | .001 | 4.70 | 4.44 | 4.17 | 4.03 | 3.88 | 3.74 | 3.58 | 3.42 | 3.26 |
| | .10 | 1.86 | 1.81 | 1.76 | 1.73 | 1.70 | 1.67 | 1.64 | 1.60 | 1.57 |
| 22 | .05 | 2.23 | 2.15 | 2.07 | 2.03 | 1.98 | 1.94 | 1.89 | 1.84 | 1.78 |
| | .01 | 3.12 | 2.98 | 2.83 | 2.75 | 2.67 | 2.58 | 2.50 | 2.40 | 2.31 |
| | .001 | 4.58 | 4.33 | 4.06 | 3.92 | 3.78 | 3.63 | 3.48 | 3.32 | 3.15 |
| | .10 | 1.84 | 1.80 | 1.74 | 1.72 | 1.69 | 1.66 | 1.62 | 1.59 | 1.55 |
| 23 | .05 | 2.20 | 2.13 | 2.05 | 2.01 | 1.96 | 1.91 | 1.86 | 1.81 | 1.76 |
| | .01 | 3.07 | 2.93 | 2.78 | 2.70 | 2.62 | 2.54 | 2.45 | 2.35 | 2.26 |
| | .001 | 4.48 | 4.23 | 3.96 | 3.82 | 3.68 | 3.53 | 3.38 | 3.22 | 3.05 |

| $df_n$ | | 1 | 2 | 3 | 4 | 5 | 6 | 7 | 8 | 9 | 10 |
|---|---|---|---|---|---|---|---|---|---|---|---|
| $df_d$ | | | | | | | | | | | |
| 24 | .10 | 2.93 | 2.54 | 2.33 | 2.19 | 2.10 | 2.04 | 1.98 | 1.94 | 1.91 | 1.88 |
| | .05 | 4.26 | 3.40 | 3.01 | 2.78 | 2.62 | 2.51 | 2.42 | 2.36 | 2.30 | 2.25 |
| | .01 | 7.82 | 5.61 | 4.72 | 4.22 | 3.90 | 3.67 | 3.50 | 3.36 | 3.26 | 3.17 |
| | .001 | 14.03 | 9.34 | 7.55 | 6.59 | 5.98 | 5.55 | 5.23 | 4.99 | 4.80 | 4.64 |
| 25 | .10 | 2.92 | 2.53 | 2.32 | 2.18 | 2.09 | 2.02 | 1.97 | 1.93 | 1.89 | 1.87 |
| | .05 | 4.24 | 3.39 | 2.99 | 2.76 | 2.60 | 2.49 | 2.40 | 2.34 | 2.28 | 2.24 |
| | .01 | 7.77 | 5.57 | 4.68 | 4.18 | 3.85 | 3.63 | 3.46 | 3.32 | 3.22 | 3.13 |
| | .001 | 13.88 | 9.22 | 7.45 | 6.49 | 5.88 | 5.46 | 5.15 | 4.91 | 4.71 | 4.56 |
| 26 | .10 | 2.91 | 2.52 | 2.31 | 2.17 | 2.08 | 2.01 | 1.96 | 1.92 | 1.88 | 1.86 |
| | .05 | 4.23 | 3.37 | 2.98 | 2.74 | 2.59 | 2.47 | 2.39 | 2.32 | 2.27 | 2.22 |
| | .01 | 7.72 | 5.53 | 4.64 | 4.14 | 3.82 | 3.59 | 3.42 | 3.29 | 3.18 | 3.09 |
| | .001 | 13.74 | 9.12 | 7.36 | 6.41 | 5.80 | 5.38 | 5.07 | 4.83 | 4.64 | 4.48 |
| 27 | .10 | 2.90 | 2.51 | 2.30 | 2.17 | 2.07 | 2.00 | 1.95 | 1.91 | 1.87 | 1.85 |
| | .05 | 4.21 | 3.35 | 2.96 | 2.73 | 2.57 | 2.46 | 2.37 | 2.31 | 2.25 | 2.20 |
| | .01 | 7.68 | 5.49 | 4.60 | 4.11 | 3.78 | 3.56 | 3.39 | 3.26 | 3.15 | 3.06 |
| | .001 | 13.61 | 9.02 | 7.27 | 6.33 | 5.73 | 5.31 | 5.00 | 4.76 | 4.57 | 4.41 |
| 28 | .10 | 2.89 | 2.50 | 2.29 | 2.16 | 2.06 | 2.00 | 1.94 | 1.90 | 1.87 | 1.84 |
| | .05 | 4.20 | 3.34 | 2.95 | 2.71 | 2.56 | 2.45 | 2.36 | 2.29 | 2.24 | 2.19 |
| | .01 | 7.64 | 5.45 | 4.57 | 4.07 | 3.75 | 3.53 | 3.36 | 3.23 | 3.12 | 3.03 |
| | .001 | 13.50 | 8.93 | 7.19 | 6.25 | 5.66 | 5.24 | 4.93 | 4.69 | 4.50 | 4.35 |
| 29 | .10 | 2.89 | 2.50 | 2.28 | 2.15 | 2.06 | 1.99 | 1.93 | 1.89 | 1.86 | 1.83 |
| | .05 | 4.18 | 3.33 | 2.93 | 2.70 | 2.55 | 2.43 | 2.35 | 2.28 | 2.22 | 2.18 |
| | .01 | 7.60 | 5.42 | 4.54 | 4.04 | 3.73 | 3.50 | 3.33 | 3.20 | 3.09 | 3.00 |
| | .001 | 13.39 | 8.85 | 7.12 | 6.19 | 5.59 | 5.18 | 4.87 | 4.64 | 4.45 | 4.29 |
| 30 | .10 | 2.88 | 2.49 | 2.28 | 2.14 | 2.05 | 1.98 | 1.93 | 1.88 | 1.85 | 1.82 |
| | .05 | 4.17 | 3.32 | 2.92 | 2.69 | 2.53 | 2.42 | 2.33 | 2.27 | 2.21 | 2.16 |
| | .01 | 7.56 | 5.39 | 4.51 | 4.02 | 3.70 | 3.47 | 3.30 | 3.17 | 3.07 | 2.98 |
| | .001 | 13.29 | 8.77 | 7.05 | 6.12 | 5.53 | 5.12 | 4.82 | 4.58 | 4.39 | 4.24 |
| 40 | .10 | 2.84 | 2.44 | 2.23 | 2.09 | 2.00 | 1.93 | 1.87 | 1.83 | 1.79 | 1.76 |
| | .05 | 4.08 | 3.23 | 2.84 | 2.61 | 2.45 | 2.34 | 2.25 | 2.18 | 2.12 | 2.08 |
| | .01 | 7.31 | 5.18 | 4.31 | 3.83 | 3.51 | 3.29 | 3.12 | 2.99 | 2.89 | 2.80 |
| | .001 | 12.61 | 8.25 | 6.60 | 5.70 | 5.13 | 4.73 | 4.44 | 4.21 | 4.02 | 3.87 |
| 60 | .10 | 2.79 | 2.39 | 2.18 | 2.04 | 1.95 | 1.87 | 1.82 | 1.77 | 1.74 | 1.71 |
| | .05 | 4.00 | 3.15 | 2.76 | 2.53 | 2.37 | 2.25 | 2.17 | 2.10 | 2.04 | 1.99 |
| | .01 | 7.08 | 4.98 | 4.13 | 3.65 | 3.34 | 3.12 | 2.95 | 2.82 | 2.72 | 2.63 |
| | .001 | 11.97 | 7.76 | 6.17 | 5.31 | 4.76 | 4.37 | 4.09 | 3.87 | 3.69 | 3.54 |
| 120 | .10 | 2.75 | 2.35 | 2.13 | 1.99 | 1.90 | 1.82 | 1.77 | 1.72 | 1.68 | 1.65 |
| | .05 | 3.92 | 3.07 | 2.68 | 2.45 | 2.29 | 2.17 | 2.09 | 2.02 | 1.96 | 1.91 |
| | .01 | 6.85 | 4.79 | 3.95 | 3.48 | 3.17 | 2.96 | 2.79 | 2.66 | 2.56 | 2.47 |
| | .001 | 11.38 | 7.32 | 5.79 | 4.95 | 4.42 | 4.04 | 3.77 | 3.55 | 3.38 | 3.24 |
| ∞ | .10 | 2.71 | 2.30 | 2.08 | 1.94 | 1.85 | 1.77 | 1.72 | 1.67 | 1.63 | 1.60 |
| | .05 | 3.84 | 3.00 | 2.60 | 2.37 | 2.21 | 2.10 | 2.01 | 1.94 | 1.88 | 1.83 |
| | .01 | 6.63 | 4.61 | 3.78 | 3.32 | 3.02 | 2.80 | 2.64 | 2.51 | 2.41 | 2.32 |
| | .001 | 10.83 | 6.91 | 5.42 | 4.62 | 4.10 | 3.74 | 3.47 | 3.27 | 3.10 | 2.96 |

| $df_d$ | | $df_n$ 12 | 15 | 20 | 24 | 30 | 40 | 60 | 120 | ∞ |
|---|---|---|---|---|---|---|---|---|---|---|
| 24 | .10 | 1.83 | 1.78 | 1.73 | 1.70 | 1.67 | 1.64 | 1.61 | 1.57 | 1.53 |
| | .05 | 2.18 | 2.11 | 2.03 | 1.98 | 1.94 | 1.89 | 1.84 | 1.79 | 1.73 |
| | .01 | 3.03 | 2.89 | 2.74 | 2.66 | 2.58 | 2.49 | 2.40 | 2.31 | 2.21 |
| | .001 | 4.39 | 4.14 | 3.87 | 3.74 | 3.59 | 3.45 | 3.29 | 3.14 | 2.97 |
| 25 | .10 | 1.82 | 1.77 | 1.72 | 1.69 | 1.66 | 1.63 | 1.59 | 1.56 | 1.52 |
| | .05 | 2.16 | 2.09 | 2.01 | 1.96 | 1.92 | 1.87 | 1.82 | 1.77 | 1.71 |
| | .01 | 2.99 | 2.85 | 2.70 | 2.62 | 2.54 | 2.45 | 2.36 | 2.27 | 2.17 |
| | .001 | 4.31 | 4.06 | 3.79 | 3.66 | 3.52 | 3.37 | 3.22 | 3.06 | 2.89 |
| 26 | .10 | 1.81 | 1.76 | 1.71 | 1.68 | 1.65 | 1.61 | 1.58 | 1.54 | 1.50 |
| | .05 | 2.15 | 2.07 | 1.99 | 1.95 | 1.90 | 1.85 | 1.80 | 1.75 | 1.69 |
| | .01 | 2.96 | 2.81 | 2.66 | 2.58 | 2.50 | 2.42 | 2.33 | 2.23 | 2.13 |
| | .001 | 4.24 | 3.99 | 3.72 | 3.59 | 3.44 | 3.30 | 3.15 | 2.99 | 2.82 |
| 27 | .10 | 1.80 | 1.75 | 1.70 | 1.67 | 1.64 | 1.60 | 1.57 | 1.53 | 1.49 |
| | .05 | 2.13 | 2.06 | 1.97 | 1.93 | 1.88 | 1.84 | 1.79 | 1.73 | 1.67 |
| | .01 | 2.93 | 2.78 | 2.63 | 2.55 | 2.47 | 2.38 | 2.29 | 2.20 | 2.10 |
| | .001 | 4.17 | 3.92 | 3.66 | 3.52 | 3.38 | 3.23 | 3.08 | 2.92 | 2.75 |
| 28 | .10 | 1.79 | 1.74 | 1.69 | 1.66 | 1.63 | 1.59 | 1.56 | 1.52 | 1.48 |
| | .05 | 2.12 | 2.04 | 1.96 | 1.91 | 1.87 | 1.82 | 1.77 | 1.71 | 1.65 |
| | .01 | 2.90 | 2.75 | 2.60 | 2.52 | 2.44 | 2.35 | 2.26 | 2.17 | 2.06 |
| | .001 | 4.11 | 3.86 | 3.60 | 3.46 | 3.32 | 3.18 | 3.02 | 2.86 | 2.69 |
| 29 | .10 | 1.78 | 1.73 | 1.68 | 1.65 | 1.62 | 1.58 | 1.55 | 1.51 | 1.47 |
| | .05 | 2.10 | 2.03 | 1.94 | 1.90 | 1.85 | 1.81 | 1.75 | 1.70 | 1.64 |
| | .01 | 2.87 | 2.73 | 2.57 | 2.49 | 2.41 | 2.33 | 2.23 | 2.14 | 2.03 |
| | .001 | 4.05 | 3.80 | 3.54 | 3.41 | 3.27 | 3.12 | 2.97 | 2.81 | 2.64 |
| 30 | .10 | 1.77 | 1.72 | 1.67 | 1.64 | 1.61 | 1.57 | 1.54 | 1.50 | 1.46 |
| | .05 | 2.09 | 2.01 | 1.93 | 1.89 | 1.84 | 1.79 | 1.74 | 1.68 | 1.62 |
| | .01 | 2.84 | 2.70 | 2.55 | 2.47 | 2.39 | 2.30 | 2.21 | 2.11 | 2.01 |
| | .001 | 4.00 | 3.75 | 3.49 | 3.36 | 3.22 | 3.07 | 2.92 | 2.76 | 2.59 |
| 40 | .10 | 1.71 | 1.66 | 1.61 | 1.57 | 1.54 | 1.51 | 1.47 | 1.42 | 1.38 |
| | .05 | 2.00 | 1.92 | 1.84 | 1.79 | 1.74 | 1.69 | 1.64 | 1.58 | 1.51 |
| | .01 | 2.66 | 2.52 | 2.37 | 2.29 | 2.20 | 2.11 | 2.02 | 1.92 | 1.80 |
| | .001 | 3.64 | 3.40 | 3.15 | 3.01 | 2.87 | 2.73 | 2.57 | 2.41 | 2.23 |
| 60 | .10 | 1.66 | 1.60 | 1.54 | 1.51 | 1.48 | 1.44 | 1.40 | 1.35 | 1.29 |
| | .05 | 1.92 | 1.84 | 1.75 | 1.70 | 1.65 | 1.59 | 1.53 | 1.47 | 1.39 |
| | .01 | 2.50 | 2.35 | 2.20 | 2.12 | 2.03 | 1.94 | 1.84 | 1.73 | 1.60 |
| | .001 | 3.31 | 3.08 | 2.83 | 2.69 | 2.55 | 2.41 | 2.25 | 2.08 | 1.89 |
| 120 | .10 | 1.60 | 1.55 | 1.48 | 1.45 | 1.41 | 1.37 | 1.32 | 1.26 | 1.19 |
| | .05 | 1.83 | 1.75 | 1.66 | 1.61 | 1.55 | 1.50 | 1.43 | 1.35 | 1.25 |
| | .01 | 2.34 | 2.19 | 2.03 | 1.95 | 1.86 | 1.76 | 1.66 | 1.53 | 1.38 |
| | .001 | 3.02 | 2.78 | 2.53 | 2.40 | 2.26 | 2.11 | 1.95 | 1.76 | 1.54 |
| ∞ | .10 | 1.55 | 1.49 | 1.42 | 1.38 | 1.34 | 1.30 | 1.24 | 1.17 | 1.00 |
| | .05 | 1.75 | 1.67 | 1.57 | 1.52 | 1.46 | 1.39 | 1.32 | 1.22 | 1.00 |
| | .01 | 2.18 | 2.04 | 1.88 | 1.79 | 1.70 | 1.59 | 1.47 | 1.32 | 1.00 |
| | .001 | 2.74 | 2.51 | 2.27 | 2.13 | 1.99 | 1.84 | 1.66 | 1.45 | 1.00 |

# I. Transformation of r to Fisher's z

# II. Transformation of Fisher's z to r

## *I. Transformation of* r *to Fisher's* z

| r | z | r | z | r | z |
|---|---|---|---|---|---|
| .01 | .0100 | .34 | .3541 | .67 | .8107 |
| .02 | .0200 | .35 | .3654 | .68 | .8291 |
| .03 | .0300 | .36 | .3769 | .69 | .8480 |
| .04 | .0400 | .37 | .3884 | .70 | .8673 |
| .05 | .0500 | .38 | .4001 | .71 | .8872 |
| .06 | .0601 | .39 | .4118 | .72 | .9076 |
| .07 | .0701 | .40 | .4236 | .73 | .9287 |
| .08 | .0802 | .41 | .4356 | .74 | .9505 |
| .09 | .0902 | .42 | .4477 | .75 | .9730 |
| .10 | .1003 | .43 | .4599 | .76 | .9962 |
| .11 | .1104 | .44 | .4722 | .77 | 1.0203 |
| .12 | .1206 | .45 | .4847 | .78 | 1.0454 |
| .13 | .1307 | .46 | .4973 | .79 | 1.0714 |
| .14 | .1409 | .47 | .5101 | .80 | 1.0986 |
| .15 | .1511 | .48 | .5230 | .81 | 1.1270 |
| .16 | .1614 | .49 | .5361 | .82 | 1.1568 |
| .17 | .1717 | .50 | .5493 | .83 | 1.1881 |
| .18 | .1820 | .51 | .5627 | .84 | 1.2212 |
| .19 | .1923 | .52 | .5763 | .85 | 1.2562 |
| .20 | .2027 | .53 | .5901 | .86 | 1.2933 |
| .21 | .2132 | .54 | .6042 | .87 | 1.3331 |
| .22 | .2237 | .55 | .6184 | .88 | 1.3758 |
| .23 | .2342 | .56 | .6328 | .89 | 1.4219 |
| .24 | .2448 | .57 | .6475 | .90 | 1.4722 |
| .25 | .2554 | .58 | .6625 | .91 | 1.5275 |
| .26 | .2661 | .59 | .6777 | .92 | 1.5890 |
| .27 | .2769 | .60 | .6931 | .93 | 1.6584 |
| .28 | .2877 | .61 | .7089 | .94 | 1.7380 |
| .29 | .2986 | .62 | .7250 | .95 | 1.8318 |
| .30 | .3095 | .63 | .7414 | .96 | 1.9459 |
| .31 | .3205 | .64 | .7582 | .97 | 2.0923 |
| .32 | .3316 | .65 | .7753 | .98 | 2.2976 |
| .33 | .3428 | .66 | .7928 | .99 | 2.6467 |

NOTE: For a given correlation, $r$, its corresponding Fisher's $z$ is presented.

## *II. Transformation of Fisher's z to r*

| z | .00 | .01 | .02 | .03 | .04 | .05 | .06 | .07 | .08 | .09 |
|---|-----|-----|-----|-----|-----|-----|-----|-----|-----|-----|
| .0 | .0000 | .0100 | .0200 | .0300 | .0400 | .0500 | .0599 | .0699 | .0798 | .0898 |
| .1 | .0997 | .1096 | .1194 | .1293 | .1391 | .1489 | .1586 | .1684 | .1781 | .1877 |
| .2 | .1974 | .2070 | .2165 | .2260 | .2355 | .2449 | .2543 | .2636 | .2729 | .2821 |
| .3 | .2913 | .3004 | .3095 | .3185 | .3275 | .3364 | .3452 | .3540 | .3627 | .3714 |
| .4 | .3800 | .3885 | .3969 | .4053 | .4136 | .4219 | .4301 | .4382 | .4462 | .4542 |
| .5 | .4621 | .4699 | .4777 | .4854 | .4930 | .5005 | .5080 | .5154 | .5227 | .5299 |
| .6 | .5370 | .5441 | .5511 | .5581 | .5649 | .5717 | .5784 | .5850 | .5915 | .5980 |
| .7 | .6044 | .6107 | .6169 | .6231 | .6291 | .6351 | .6411 | .6469 | .6527 | .6584 |
| .8 | .6640 | .6696 | .6751 | .6805 | .6858 | .6911 | .6963 | .7014 | .7064 | .7114 |
| .9 | .7163 | .7211 | .7259 | .7306 | .7352 | .7398 | .7443 | .7487 | .7531 | .7574 |
| 1.0 | .7616 | .7658 | .7699 | .7739 | .7779 | .7818 | .7857 | .7895 | .7932 | .7969 |
| 1.1 | .8005 | .8041 | .8076 | .8110 | .8144 | .8178 | .8210 | .8243 | .8275 | .8306 |
| 1.2 | .8337 | .8367 | .8397 | .8426 | .8455 | .8483 | .8511 | .8538 | .8565 | .8591 |
| 1.3 | .8617 | .8643 | .8668 | .8692 | .8717 | .8741 | .8764 | .8787 | .8810 | .8832 |
| 1.4 | .8854 | .8875 | .8896 | .8917 | .8937 | .8957 | .8977 | .8996 | .9015 | .9033 |
| 1.5 | .9051 | .9069 | .9087 | .9104 | .9121 | .9138 | .9154 | .9170 | .9186 | .9201 |
| 1.6 | .9217 | .9232 | .9246 | .9261 | .9275 | .9289 | .9302 | .9316 | .9329 | .9341 |
| 1.7 | .9354 | .9366 | .9379 | .9391 | .9402 | .9414 | .9425 | .9436 | .9447 | .9458 |
| 1.8 | .9468 | .9478 | .9488 | .9498 | .9508 | .9517 | .9527 | .9536 | .9545 | .9554 |
| 1.9 | .9562 | .9571 | .9579 | .9587 | .9595 | .9603 | .9611 | .9618 | .9626 | .9633 |
| 2.0 | .9640 | .9647 | .9654 | .9661 | .9667 | .9674 | .9680 | .9687 | .9693 | .9699 |
| 2.1 | .9705 | .9710 | .9716 | .9721 | .9727 | .9732 | .9737 | .9743 | .9748 | .9753 |
| 2.2 | .9757 | .9762 | .9767 | .9771 | .9776 | .9780 | .9785 | .9789 | .9793 | .9797 |
| 2.3 | .9801 | .9905 | .9809 | .9812 | .9816 | .9820 | .9823 | .9827 | .9830 | .9833 |
| 2.4 | .9837 | .9840 | .9843 | .9846 | .9849 | .9852 | .9855 | .9858 | .9861 | .9863 |
| 2.5 | .9866 | .9869 | .9871 | .9874 | .9876 | .9879 | .9881 | .9884 | .9886 | .9888 |
| 2.6 | .9890 | .9892 | .9895 | .9897 | .9899 | .9901 | .9903 | .9905 | .9906 | .9908 |
| 2.7 | .9910 | .9912 | .9914 | .9915 | .9917 | .9919 | .9920 | .9922 | .9923 | .9925 |
| 2.8 | .9926 | .9928 | .9929 | .9931 | .9932 | .9933 | .9935 | .9936 | ,9937 | .9938 |
| 2.9 | .9940 | .9941 | .9942 | .9943 | .9944 | .9945 | .9946 | .9947 | .9949 | .9950 |

NOTE: To determine *r* locate the digits of *z* to the left and right of the decimal place in the first column. Locate the second decimal place in the top column. The intersection of the row and column gives *r*.

# Appendix G

# Critical Values for Chi Square

To determine the approximate $p$ value, locate the degrees of freedom ($df$) in the first column. Then find the largest value that the chi square test statistic equals or exceeds, and read the column heading for the approximate $p$ value.

*Critical Values for Chi Square*

| df | .20 | .10 | .05 | .02 | .01 | .001 |
|----|-----|-----|-----|-----|-----|------|
| 1 | 1.64 | 2.71 | 3.84 | 5.41 | 6.63 | 10.83 |
| 2 | 3.22 | 4.61 | 5.99 | 7.82 | 9.21 | 13.82 |
| 3 | 4.64 | 6.25 | 7.81 | 9.84 | 11.34 | 16.27 |
| 4 | 5.99 | 7.78 | 9.49 | 11.67 | 13.28 | 18.47 |
| 5 | 7.29 | 9.24 | 11.07 | 13.39 | 15.09 | 20.52 |
| 6 | 8.56 | 10.64 | 12.59 | 15.03 | 16.81 | 22.46 |
| 7 | 9.80 | 12.02 | 14.07 | 16.62 | 18.48 | 24.32 |
| 8 | 11.03 | 13.36 | 15.51 | 18.17 | 20.09 | 26.12 |
| 9 | 12.24 | 14.68 | 16.92 | 19.68 | 21.67 | 27.88 |
| 10 | 13.44 | 15.99 | 18.31 | 21.16 | 23.21 | 29.59 |
| 11 | 14.63 | 17.28 | 19.68 | 22.62 | 24.72 | 31.26 |
| 12 | 15.81 | 18.55 | 21.03 | 24.05 | 26.22 | 32.91 |
| 13 | 16.98 | 19.81 | 22.36 | 25.47 | 27.69 | 34.53 |
| 14 | 18.15 | 21.06 | 23.68 | 26.87 | 29.14 | 36.12 |
| 15 | 19.31 | 22.31 | 25.00 | 28.26 | 30.58 | 37.70 |
| 16 | 20.46 | 23.54 | 26.30 | 29.63 | 32.00 | 39.25 |
| 17 | 21.62 | 24.77 | 27.59 | 31.00 | 33.41 | 40.79 |
| 18 | 22.76 | 25.99 | 28.87 | 32.35 | 34.81 | 42.31 |
| 19 | 23.90 | 27.20 | 30.14 | 33.69 | 36.19 | 43.82 |
| 20 | 25.04 | 28.41 | 31.41 | 35.02 | 37.57 | 45.32 |
| 21 | 26.17 | 29.62 | 32.67 | 36.34 | 38.93 | 46.80 |
| 22 | 27.30 | 30.81 | 33.92 | 37.66 | 40.29 | 48.27 |
| 23 | 28.43 | 32.01 | 35.17 | 38.97 | 41.64 | 49.73 |
| 24 | 29.55 | 33.20 | 36.42 | 40.27 | 42.98 | 51.18 |
| 25 | 30.68 | 34.38 | 37.65 | 41.57 | 44.31 | 52.62 |
| 26 | 31.80 | 35.56 | 38.89 | 42.86 | 45.64 | 54.05 |
| 27 | 32.91 | 36.74 | 40.11 | 44.14 | 46.96 | 55.48 |
| 28 | 34.03 | 37.92 | 41.34 | 45.42 | 48.28 | 56.89 |
| 29 | 35.14 | 39.09 | 42.56 | 46.69 | 49.59 | 58.30 |
| 30 | 36.25 | 40.26 | 43.77 | 47.96 | 50.89 | 59.70 |

# Appendix H

# Two-Tailed Critical Values for the Mann-Whitney U Test

To determine the approximate $p$ value, locate $n_1$, the number of scores in the smaller group, in the first column and $n_2$ in the second column. The value of $U$ must be smaller than or equal to the first number, or larger than or equal to the second number, to be significant at the level given by the column heading.

| $n_1$ | $n_2$ | .10 | .05 | .02 | .01 |
|---|---|---|---|---|---|
| 3 | 3 | 0–9 | — | — | — |
| 3 | 4 | 0–12 | — | — | — |
| 4 | 4 | 1–15 | 0–16 | — | — |
| 2 | 5 | 0–10 | — | — | — |
| 3 | 5 | 1–14 | 0–15 | — | — |
| 4 | 5 | 2–18 | 1–19 | 0–20 | — |
| 5 | 5 | 4–21 | 2–23 | 1–24 | 0–25 |
| 2 | 6 | 0–12 | — | — | — |
| 3 | 6 | 2–16 | 1–17 | — | — |
| 4 | 6 | 3–21 | 2–22 | 1–23 | 0–24 |
| 5 | 6 | 5–25 | 3–27 | 2–28 | 1–29 |
| 6 | 6 | 7–29 | 5–31 | 3–33 | 2–34 |
| 2 | 7 | 0–14 | — | — | — |
| 3 | 7 | 2–19 | 1–20 | 0–21 | — |
| 4 | 7 | 4–24 | 3–25 | 1–27 | 0–28 |
| 5 | 7 | 6–29 | 5–30 | 3–32 | 1–34 |
| 6 | 7 | 8–34 | 6–36 | 4–38 | 3–39 |
| 7 | 7 | 11–38 | 8–41 | 6–43 | 4–45 |
| 2 | 8 | 1–15 | 0–16 | — | — |
| 3 | 8 | 3–21 | 2–22 | 0–24 | — |
| 4 | 8 | 5–27 | 4–28 | 2–30 | 1–31 |
| 5 | 8 | 8–32 | 6–34 | 4–36 | 2–38 |
| 6 | 8 | 10–38 | 8–40 | 6–42 | 4–44 |
| 7 | 8 | 13–43 | 10–46 | 7–49 | 6–50 |
| 8 | 8 | 15–49 | 13–51 | 9–55 | 7–57 |
| 2 | 9 | 1–17 | 0–18 | — | — |
| 3 | 9 | 4–23 | 2–25 | 1–26 | 0–27 |
| 4 | 9 | 6–30 | 4–32 | 3–33 | 1–35 |
| 5 | 9 | 9–36 | 7–38 | 5–40 | 3–42 |
| 6 | 9 | 12–42 | 10–44 | 7–47 | 5–49 |
| 7 | 9 | 15–48 | 12–51 | 9–54 | 7–56 |
| 8 | 9 | 18–54 | 15–57 | 11–61 | 9–63 |
| 9 | 9 | 21–60 | 17–64 | 14–67 | 11–70 |
| 2 | 10 | 1–19 | 0–20 | — | — |
| 3 | 10 | 4–26 | 3–27 | 1–29 | 0–30 |
| 4 | 10 | 7–33 | 5–35 | 3–37 | 2–38 |
| 5 | 10 | 11–39 | 8–42 | 6–44 | 4–46 |
| 6 | 10 | 14–46 | 11–49 | 8–52 | 6–54 |
| 7 | 10 | 17–53 | 14–56 | 11–59 | 9–61 |
| 8 | 10 | 20–60 | 17–63 | 13–67 | 11–69 |
| 9 | 10 | 24–66 | 20–70 | 16–74 | 13–77 |
| 10 | 10 | 27–73 | 23–77 | 19–81 | 16–84 |
| 2 | 11 | 1–21 | 0–22 | — | — |
| 3 | 11 | 5–28 | 3–30 | 1–32 | 0–33 |
| 4 | 11 | 8–36 | 6–38 | 4–40 | 2–42 |
| 5 | 11 | 12–43 | 9–46 | 7–48 | 5–50 |
| 6 | 11 | 16–50 | 13–53 | 9–57 | 7–59 |
| 7 | 11 | 19–58 | 16–61 | 12–65 | 10–67 |

| $n_1$ | $n_2$ | .10 | .05 | .02 | .01 |
|-------|-------|------|------|------|------|
| 8 | 11 | 23–65 | 19–69 | 15–73 | 13–75 |
| 9 | 11 | 27–72 | 23–76 | 18–81 | 16–83 |
| 10 | 11 | 31–79 | 26–84 | 22–88 | 18–92 |
| 11 | 11 | 34–87 | 30–91 | 25–96 | 21–100 |
| 2 | 12 | 2–22 | 1–23 | — | — |
| 3 | 12 | 5–31 | 4–32 | 2–34 | 1–35 |
| 4 | 12 | 9–39 | 7–41 | 5–43 | 3–45 |
| 5 | 12 | 13–47 | 11–49 | 8–52 | 6–54 |
| 6 | 12 | 17–55 | 14–58 | 11–61 | 9–63 |
| 7 | 12 | 21–63 | 18–66 | 14–70 | 12–72 |
| 8 | 12 | 26–70 | 22–74 | 17–79 | 15–81 |
| 9 | 12 | 30–78 | 26–82 | 21–87 | 18–90 |
| 10 | 12 | 34–86 | 29–91 | 24–96 | 21–99 |
| 11 | 12 | 38–94 | 33–99 | 28–104 | 24–108 |
| 12 | 12 | 42–102 | 37–107 | 31–113 | 27–117 |
| 2 | 13 | 2–24 | 1–25 | 0–26 | — |
| 3 | 13 | 6–33 | 4–35 | 2–37 | 1–38 |
| 4 | 13 | 10–42 | 8–44 | 5–47 | 3–49 |
| 5 | 13 | 15–50 | 12–53 | 9–56 | 7–58 |
| 6 | 13 | 19–59 | 16–62 | 12–66 | 10–68 |
| 7 | 13 | 24–67 | 20–71 | 16–75 | 13–78 |
| 8 | 13 | 28–76 | 24–80 | 20–84 | 17–87 |
| 9 | 13 | 33–84 | 28–89 | 23–94 | 20–97 |
| 10 | 13 | 37–93 | 33–97 | 27–103 | 24–106 |
| 11 | 13 | 42–101 | 37–106 | 31–112 | 27–116 |
| 12 | 13 | 47–109 | 41–115 | 35–121 | 31–125 |
| 13 | 13 | 51–118 | 45–124 | 39–130 | 34–135 |
| 2 | 14 | 3–25 | 1–27 | 0–28 | — |
| 3 | 14 | 7–35 | 5–37 | 2–40 | 1–41 |
| 4 | 14 | 11–45 | 9–47 | 6–50 | 4–52 |
| 5 | 14 | 16–54 | 13–57 | 10–60 | 7–63 |
| 6 | 14 | 21–63 | 17–67 | 13–71 | 11–73 |
| 7 | 14 | 26–72 | 22–76 | 17–81 | 15–83 |
| 8 | 14 | 31–81 | 26–86 | 22–90 | 18–94 |
| 9 | 14 | 36–90 | 31–95 | 26–100 | 22–104 |
| 10 | 14 | 41–99 | 36–104 | 30–110 | 26–114 |
| 11 | 14 | 46–108 | 40–114 | 34–120 | 30–124 |
| 12 | 14 | 51–117 | 45–123 | 38–130 | 34–134 |
| 13 | 14 | 56–126 | 50–132 | 43–139 | 38–144 |
| 14 | 14 | 61–135 | 55–141 | 47–149 | 42–154 |
| 2 | 15 | 3–27 | 1–29 | 0–30 | — |
| 3 | 15 | 7–38 | 5–40 | 3–42 | 2–43 |
| 4 | 15 | 12–48 | 10–50 | 7–53 | 5–55 |
| 5 | 15 | 18–57 | 14–61 | 11–64 | 8–67 |
| 6 | 15 | 23–67 | 19–71 | 15–75 | 12–78 |
| 7 | 15 | 28–77 | 24–81 | 19–86 | 16–89 |
| 8 | 15 | 33–87 | 29–91 | 24–96 | 20–100 |
| 9 | 15 | 39–96 | 34–101 | 28–107 | 24–111 |
| 10 | 15 | 44–106 | 39–111 | 33–117 | 29–121 |
| 11 | 15 | 50–115 | 44–121 | 37–128 | 33–132 |
| 12 | 15 | 55–125 | 49–131 | 42–138 | 37–143 |
| 13 | 15 | 61–134 | 54–141 | 47–148 | 42–153 |

| $n_1$ | $n_2$ | .10 | .05 | .02 | .01 |
|---|---|---|---|---|---|
| 14 | 15 | 66–144 | 59–151 | 51–159 | 46–164 |
| 15 | 15 | 72–153 | 64–161 | 56–169 | 51–174 |
| | | | | | |
| 2 | 16 | 3–29 | 1–31 | 0–32 | — |
| 3 | 16 | 8–40 | 6–42 | 3–45 | 2–46 |
| 4 | 16 | 14–50 | 11–53 | 7–57 | 5–59 |
| 5 | 16 | 19–61 | 15–65 | 12–68 | 9–71 |
| 6 | 16 | 25–71 | 21–75 | 16–80 | 13–83 |
| 7 | 16 | 30–82 | 26–86 | 21–91 | 18–94 |
| 8 | 16 | 36–92 | 31–97 | 26–102 | 22–106 |
| 9 | 16 | 42–102 | 37–107 | 31–113 | 27–117 |
| 10 | 16 | 48–112 | 42–118 | 36–124 | 31–129 |
| 11 | 16 | 54–122 | 47–129 | 41–135 | 36–140 |
| 12 | 16 | 60–132 | 53–139 | 46–146 | 41–151 |
| 13 | 16 | 65–143 | 59–149 | 51–157 | 45–163 |
| 14 | 16 | 71–153 | 64–160 | 56–168 | 50–174 |
| 15 | 16 | 77–163 | 70–170 | 61–179 | 55–185 |
| 16 | 16 | 83–173 | 75–181 | 66–190 | 60–196 |
| | | | | | |
| 2 | 17 | 3–31 | 2–32 | 0–34 | — |
| 3 | 17 | 9–42 | 6–45 | 4–47 | 2–49 |
| 4 | 17 | 15–53 | 11–57 | 8–60 | 6–62 |
| 5 | 17 | 20–65 | 17–68 | 13–72 | 10–75 |
| 6 | 17 | 26–76 | 22–80 | 18–84 | 15–87 |
| 7 | 17 | 33–86 | 28–91 | 23–96 | 19–100 |
| 8 | 17 | 39–97 | 34–102 | 28–108 | 24–112 |
| 9 | 17 | 45–108 | 39–114 | 33–120 | 29–124 |
| 10 | 17 | 51–119 | 45–125 | 38–132 | 34–136 |
| 11 | 17 | 57–130 | 51–136 | 44–143 | 39–148 |
| 12 | 17 | 64–140 | 57–147 | 49–155 | 44–160 |
| 13 | 17 | 70–151 | 63–158 | 55–166 | 49–172 |
| 14 | 17 | 77–161 | 69–169 | 60–178 | 54–184 |
| 15 | 17 | 83–172 | 75–180 | 66–189 | 60–195 |
| 16 | 17 | 89–183 | 81–191 | 71–201 | 65–207 |
| 17 | 17 | 96–193 | 87–202 | 77–212 | 70–219 |
| | | | | | |
| 2 | 18 | 4–32 | 2–34 | 0–36 | — |
| 3 | 18 | 9–45 | 7–47 | 4–50 | 2–52 |
| 4 | 18 | 16–56 | 12–60 | 9–63 | 6–66 |
| 5 | 18 | 22–68 | 18–72 | 14–76 | 11–79 |
| 6 | 18 | 28–80 | 24–84 | 19–89 | 16–92 |
| 7 | 18 | 35–91 | 30–96 | 24–102 | 21–105 |
| 8 | 18 | 41–103 | 36–108 | 30–114 | 26–118 |
| 9 | 18 | 48–114 | 42–120 | 36–126 | 31–131 |
| 10 | 18 | 55–125 | 48–132 | 41–139 | 37–143 |
| 11 | 18 | 61–137 | 55–143 | 47–151 | 42–156 |
| 12 | 18 | 68–148 | 61–155 | 53–163 | 47–169 |
| 13 | 18 | 75–159 | 67–167 | 59–175 | 53–181 |
| 14 | 18 | 82–170 | 74–178 | 65–187 | 58–194 |
| 15 | 18 | 88–182 | 80–190 | 70–200 | 64–206 |
| 16 | 18 | 95–193 | 86–202 | 76–212 | 70–218 |
| 17 | 18 | 102–204 | 93–213 | 82–224 | 75–231 |
| 18 | 18 | 109–215 | 99–225 | 88–236 | 81–243 |
| | | | | | |
| 1 | 19 | 0–19 | — | — | — |
| 2 | 19 | 4–34 | 2–36 | 1–37 | 0–38 |

| $n_1$ | $n_2$ | .10 | .05 | .02 | .01 |
|---|---|---|---|---|---|
| 3 | 19 | 10–47 | 7–50 | 4–53 | 3–54 |
| 4 | 19 | 17–59 | 13–63 | 9–67 | 7–69 |
| 5 | 19 | 23–72 | 19–76 | 15–80 | 12–83 |
| 6 | 19 | 30–84 | 25–89 | 20–94 | 17–97 |
| 7 | 19 | 37–96 | 32–101 | 26–107 | 22–111 |
| 8 | 19 | 44–108 | 38–114 | 32–120 | 28–124 |
| 9 | 19 | 51–120 | 45–126 | 38–133 | 33–138 |
| 10 | 19 | 58–132 | 52–138 | 44–146 | 39–151 |
| 11 | 19 | 65–144 | 58–151 | 50–159 | 45–164 |
| 12 | 19 | 72–156 | 65–163 | 56–172 | 51–177 |
| 13 | 19 | 80–167 | 72–175 | 63–184 | 57–190 |
| 14 | 19 | 87–179 | 78–188 | 69–197 | 63–203 |
| 15 | 19 | 94–191 | 85–200 | 75–210 | 69–216 |
| 16 | 19 | 101–203 | 92–212 | 82–222 | 74–230 |
| 17 | 19 | 109–214 | 99–224 | 88–235 | 81–242 |
| 18 | 19 | 116–226 | 106–236 | 94–248 | 87–255 |
| 19 | 19 | 123–238 | 113–248 | 101–260 | 93–268 |
| 1 | 20 | 0–20 | — | — | — |
| 2 | 20 | 4–36 | 2–38 | 1–39 | 0–40 |
| 3 | 20 | 11–49 | 8–52 | 5–55 | 3–57 |
| 4 | 20 | 18–62 | 14–66 | 10–70 | 8–72 |
| 5 | 20 | 25–75 | 20–80 | 16–84 | 13–87 |
| 6 | 20 | 32–88 | 27–93 | 22–98 | 18–102 |
| 7 | 20 | 39–101 | 34–106 | 28–112 | 24–116 |
| 8 | 20 | 47–113 | 41–119 | 34–126 | 30–130 |
| 9 | 20 | 54–126 | 48–132 | 40–140 | 36–144 |
| 10 | 20 | 62–138 | 55–145 | 47–153 | 42–158 |
| 11 | 20 | 69–151 | 62–158 | 53–167 | 48–172 |
| 12 | 20 | 77–163 | 69–171 | 60–180 | 54–186 |
| 13 | 20 | 84–176 | 76–184 | 67–193 | 60–200 |
| 14 | 20 | 92–188 | 83–197 | 73–207 | 67–213 |
| 15 | 20 | 100–200 | 90–210 | 80–220 | 73–227 |
| 16 | 20 | 107–213 | 98–222 | 87–233 | 79–241 |
| 17 | 20 | 115–225 | 105–235 | 93–247 | 86–254 |
| 18 | 20 | 123–237 | 112–248 | 100–260 | 92–268 |
| 19 | 20 | 130–250 | 119–261 | 107–273 | 99–281 |
| 20 | 20 | 138–262 | 127–273 | 114–286 | 105–295 |

# Appendix I

# Two-Tailed Critical Values for the Sign Test (n < 26)

Locate *n*, the number of untied cases, in the first column. Let *c* be the number of positive differences. To determine the approximate *p* value, note the largest value that *c* or $n - c$ equals or exceeds. Read up to the column heading for the approximate *p* value.

## Sign Test Critical Values

| n | .10 | .05 | .02 | .01 | .002 | .001 |
|---|-----|-----|-----|-----|------|------|
| 5  | 5  | —  | —  | —  | —  | —  |
| 6  | 6  | 6  | —  | —  | —  | —  |
| 7  | 7  | 7  | 7  | —  | —  | —  |
| 8  | 7  | 8  | 8  | 8  | —  | —  |
| 9  | 8  | 8  | 9  | 9  | —  | —  |
| 10 | 9  | 9  | 10 | 10 | 10 | —  |
| 11 | 9  | 10 | 10 | 11 | 11 | 11 |
| 12 | 10 | 10 | 11 | 11 | 12 | 12 |
| 13 | 10 | 11 | 12 | 12 | 13 | 13 |
| 14 | 11 | 12 | 12 | 13 | 13 | 14 |
| 15 | 12 | 12 | 13 | 13 | 14 | 14 |
| 16 | 12 | 13 | 14 | 14 | 15 | 15 |
| 17 | 13 | 13 | 14 | 15 | 16 | 16 |
| 18 | 13 | 14 | 15 | 15 | 16 | 17 |
| 19 | 14 | 15 | 15 | 16 | 17 | 17 |
| 20 | 15 | 15 | 16 | 17 | 18 | 18 |
| 21 | 15 | 16 | 17 | 17 | 18 | 19 |
| 22 | 16 | 17 | 17 | 18 | 19 | 19 |
| 23 | 16 | 17 | 18 | 19 | 20 | 20 |
| 24 | 17 | 18 | 19 | 19 | 20 | 21 |
| 25 | 18 | 18 | 19 | 20 | 21 | 21 |

# Appendix J

# Two-Tailed Critical Values for Spearman's Rho

To determine the approximate $p$ value, locate the number of pairs of scores in the left-hand column. Find the largest value that the test statistic, ignoring sign, equals or exceeds. Read the column heading for the approximate $p$ value.

## Spearman's Rho Critical Values

| Number of Pairs | .10 | .05 | .02 | .01 | .002 |
|---|---|---|---|---|---|
| 4 | 1.000 | — | — | — | — |
| 5 | .900 | 1.000 | 1.000 | — | — |
| 6 | .829 | .886 | .943 | 1.000 | — |
| 7 | .714 | .786 | .893 | .929 | 1.000 |
| 8 | .643 | .738 | .833 | .881 | .952 |
| 9 | .600 | .700 | .783 | .833 | .917 |
| 10 | .564 | .648 | .745 | .794 | .879 |
| 11 | .536 | .619 | .709 | .764 | .845 |
| 12 | .503 | .587 | .678 | .734 | .825 |
| 13 | .484 | .560 | .648 | .703 | .797 |
| 14 | .464 | .538 | .626 | .679 | .771 |
| 15 | .446 | .521 | .604 | .657 | .750 |
| 16 | .429 | .503 | .585 | .635 | .729 |
| 17 | .414 | .488 | .566 | .618 | .711 |
| 18 | .401 | .474 | .550 | .600 | .692 |
| 19 | .391 | .460 | .535 | .584 | .675 |
| 20 | .380 | .447 | .522 | .570 | .660 |
| 21 | .370 | .436 | .509 | .556 | .647 |
| 22 | .361 | .425 | .497 | .544 | .633 |
| 23 | .353 | .416 | .486 | .532 | .620 |
| 24 | .344 | .407 | .476 | .521 | .608 |
| 25 | .337 | .398 | .466 | .511 | .597 |
| 26 | .331 | .390 | .475 | .501 | .586 |
| 27 | .324 | .383 | .449 | .492 | .576 |
| 28 | .318 | .375 | .441 | .483 | .567 |
| 29 | .312 | .369 | .433 | .475 | .557 |
| 30 | .306 | .362 | .426 | .467 | .548 |

# Answers to Selected Problems

## Chapter 1: Introduction

1. a. number: 76; object: John; variable: midterm grade
   b. number: 6.98; object: Rolling Stones album; variable: record cost
   c. number: 28; object: 1986 Ford Tempo; variable: miles per gallon
   d. number: brown-eyed; object: Mary; variable: eye color

2. a. interval    b. nominal    c. ordinal

3. a. 4.20    b. $-.60$

5. a. .52    b. $-.32$    c. .84    d. .53
   e. $-.48$    f. $-.13$    g. $-.13$    h. .36

6. a. 39    b. 251    c. 1521    d. 31    e. 38    f. 181

7.

| Category | Proportion | Odds |
|----------|-----------|------|
| A | .40 | .67 |
| B | .15 | .18 |
| C | .11 | .12 |
| D | .09 | .10 |

## Chapter 2: The Distribution of Scores

1.

| Class Interval | Frequency | Relative Frequency |
|----------------|-----------|--------------------|
| 46 to 50 | 1 | 4.2 |
| 51 to 55 | 0 | 0.0 |
| 56 to 60 | 1 | 4.2 |
| 61 to 65 | 3 | 12.5 |
| 66 to 70 | 3 | 12.5 |
| 71 to 75 | 3 | 12.5 |
| 76 to 80 | 5 | 20.8 |

1. *(continued)*

| Class Interval | Frequency | Relative Frequency |
|---|---|---|
| 81 to 85 | 5 | 20.8 |
| 86 to 90 | 1 | 4.2 |
| 91 to 95 | 1 | 4.2 |
| 96 to 100 | 1 | 4.2 |

2. a.

b.

c.

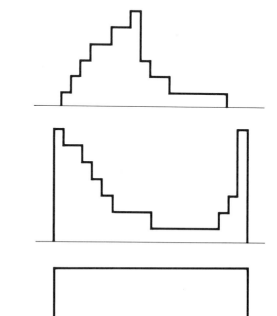

3. Assume that the peak is not in the middle. Because the distribution is symmetric, it would have two peaks, one on each side. So the distribution must be peaked in the middle if there is a single peak.

5.

| Class Interval | Frequency | Smoothed Frequency |
|---|---|---|
| −1.40 to −1.21 | 0 | .25 |
| −1.20 to −1.01 | 1 | .50 |
| −1.00 to −.81 | 0 | .25 |
| −.80 to −.61 | 0 | .25 |
| −.60 to −.41 | 1 | 2.50 |
| −.40 to −.21 | 8 | 4.75 |
| −.20 to −.01 | 2 | 6.25 |
| 0 to .19 | 13 | 8.00 |
| .20 to .39 | 4 | 5.75 |
| .40 to .59 | 2 | 2.25 |
| .60 to .79 | 1 | 1.00 |
| .80 to .99 | 0 | .25 |

6. a. The unit is one, but most scores are multiples of five so the class width should be a multiple of five. The range is 425 minus 230, which equals 195. The maximum class width is $195/8 = 24.375$ and the minimum is $195/15 = 13$. Reasonable class widths are 15 and 20. The lowest lower limit should be some value less than or equal to 230, the lowest score.

b.

| Rent | Frequency | Relative Frequency |
|---|---|---|
| 226–250 | 2 | 5 |
| 251–275 | 7 | 18 |
| 276–300 | 12 | 32 |
| 301–325 | 8 | 21 |
| 326–350 | 2 | 5 |
| 351–375 | 2 | 5 |
| 376–400 | 4 | 11 |
| 401–425 | 1 | 3 |

c. The distribution is positively skewed.

7.

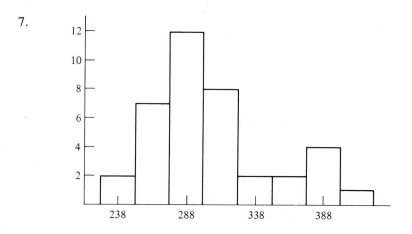

8. a. Males

| Class Interval | Frequency | Relative Frequency |
|---|---|---|
| 35.0 to 39.9 | 2 | 7 |
| 40.0 to 44.9 | 3 | 10 |
| 45.0 to 49.9 | 4 | 13 |
| 50.0 to 54.9 | 4 | 13 |
| 55.0 to 59.9 | 5 | 17 |
| 60.0 to 64.9 | 3 | 10 |
| 65.0 to 69.9 | 8 | 27 |
| 70.0 to 74.9 | 1 | 3 |

Females

| Class Interval | Frequency | Relative Frequency |
|---|---|---|
| 35.0 to 39.9 | 2 | 7 |
| 40.0 to 44.9 | 1 | 3 |
| 45.0 to 49.9 | 5 | 17 |
| 50.0 to 54.9 | 4 | 13 |
| 55.0 to 59.9 | 2 | 7 |
| 60.0 to 64.9 | 4 | 13 |
| 65.0 to 69.9 | 2 | 7 |
| 70.0 to 74.9 | 6 | 20 |
| 75.0 to 79.9 | 4 | 13 |

    b. Both distributions are negatively skewed. Females live longer than males.

9.

| | Males | | Females |
|---|---|---|---|
| | 35\|67 | | 35\|69 |
| | 40\|113 | | 40\|0 |
| | 45\|5789 | | 45\|56678 |
| | 50\|1333 | | 50\|1333 |
| | 55\|67799 | | 55\|78 |
| | 60\|234 | | 60\|0123 |
| | 65\|56788999 | | 65\|67 |
| | 70\|2 | | 70\|134444 |
| | | | 75\|5667 |

14.   a. .25    b. 20

# Chapter 3: Central Tendency

1. a. mode = 6; median = 6; mean = 5.29
   b. mode = 2; median = 3.5; mean = 3.5
   c. modes = 2, 8; median = 4.5; mean = 5
   d. mode = 3; median = 4.5; mean = 16

2. sample c, because of the two modes;
   sample d, because of the outlier of 96

3. a. mean = 3.90; median = 3.5; mode = 3
   b. mean = 12.64; median = 4; mode = 3
   The mean changed the most.
   c. mean = 4.45; median = 4; mode = 3
   The mean is most affected.

5. a. mean = 11101.33; median = 8780.5
   b. the median

7. negatively skewed

9. a.

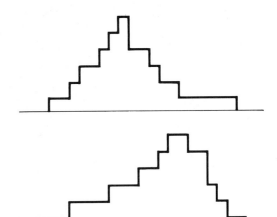

b.

14. a. mean: $28.49; median: $22; mode: $8 and $45
    b. the median because of the outlier of $120

# Chapter 4: Variability

1. a. range = 10; interquartile range = 6; $s$ = 3.54
   b. range = 13; interquartile range = 8.5; $s$ = 4.54
   c. range = 13; interquartile range = 9; $s$ = 4.72

2. a. 42.25    b. 84.50

3. a. $s$ = 46.11; range = 144; interquartile range = 10
   b. $s$ = 5.01; range = 16; interquartile range = 6
   c. the interquartile range

4. a. Marvel Motors: range = 8; interquartile range = 6; $s$ = 3.22
      Amazing Auto: range = 2; interquartile range = 2; $s$ = .89
   b. Amazing Auto

6. a. yes    b. no    c. no    d. no

7. a. Control:        $\bar{X}$ = 87.00; $s$ = 7.24; $s^2$ = 52.44
      Experimental:   $\bar{X}$ = 94.57; $s$ = 4.06; $s^2$ = 16.48
   b. The control group is more variable.

9. The median is 19 and the interquartile range is 8.5. To be an outlier a score
   must be less than 2.00 or greater than 36.00. Using these criteria, the score
   37 qualifies as an outlier.

# *Chapter 5: Transformation*

1. a. 2.303    b. 6.245    c. 2.094    d. .021    e. .412    f. .160

3.

| Original Score | Z Score |
|:---:|:---:|
| 1 | −1.14 |
| 2 | −.91 |
| 3 | −.68 |
| 4 | −.45 |
| 5 | −.23 |
| 6 | .00 |
| 7 | .23 |
| 8 | .45 |
| 9 | .68 |
| 15 | 2.05 |
| 18 | 2.73 |

4. Multiply each score by .09 and then add 1.0.

5. a. mean: 53; $s$: 30
   b. mean: 16.8; $s$: .2

6.

| Number | Percentile Rank |
|:---:|:---:|
| 8 | 27.5 |
| 12 | 52.5 |
| 17 | 82.5 |

7. a. mean = 30; $s$ and $s^2$ remain the same
   b. mean = 75; $s$ = 9.6; $s^2$ = 92.16

# *Chapter 6: Measuring Association:*
# *The Regression Coefficient*

1. slope: 1.0; intercept: 40

2. a. no cigarettes: 2.23 days; 20 cigarettes: 3.85 days; 40 cigarettes: 5.47 days
   b. 114.21 days (15 × 94 × .081)

3. a. Predictor: similarity; criterion: marital satisfaction
   b. Predictor: effort; criterion: performance
   c. Predictor: sleep; criterion: efficiency
   d. Predictor: mood; criterion: health

4. a. positive    b. positive    c. negative    d. positive

5. a. slope: 3.8793; intercept: −89.138; variance of errors: 534.48

   b.

   | Score | Predicted Score | Error |
   |-------|-----------------|-------|
   | 140 | 143.62 | −3.62 |
   | 170 | 159.14 | 10.86 |
   | 210 | 190.17 | 19.83 |
   | 180 | 174.66 | 5.35 |
   | 150 | 182.41 | −32.41 |

7. a. causal    b. predictive

8. a.

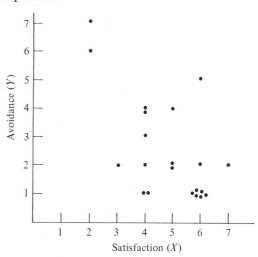

   b. The relationship appears negative. The slope equals −.77. It shows that less satisfaction with privacy was associated with more avoidance behaviors. The intercept equals 6.18. This is the predicted avoidance score for someone with a score of zero on satisfaction. Because the lowest possible score on satisfaction is one, the intercept is an extrapolation.

   c. variance of $X$: 2.03; variance of $Y$: 3.31; variance of errors: 2.23

# Chapter 7: Relationship: The Correlation Coefficient

1. $r_{XY} = 1.000$

3. $r_{XY} = .404$

4. a. $b_{XY} = .20$    b. $b_{XY} = −1.0$    c. $r_{XY} = .10$

5. $r_{XY} = .463$. There is a fairly large positive correlation between typing and preparing stencils, such that those who perform one of these tasks will tend to perform the other task well. The 2sd advantage is about .762. So someone who is one standard deviation above the mean on typing is

about 75% more likely to better at preparing stencils than someone who is one standard deviation below the mean on typing.

7.
$$\frac{8511 - (749)(195)/16}{\sqrt{(37011 - 749^2/16)(2853 - 195^2/16)}} = -.641$$

The relationship between memory and age is a large negative one. So as persons age, their memory declines. The 2sd advantage is about .887. So someone who is one standard deviation above the mean in age is about 89% more likely to have a worse memory than someone who is one standard deviation below the mean in age.

12. a. Because school and not person is the unit in the correlation, aggregation is likely to increase the size of the correlation.
   b. Because the child's initial height is used to measure how much the child grew (growth equals current height minus initial height), there is a part-whole problem. The most likely outcome is that the correlation will be negative.
   c. Because almost all air-traffic controllers experience high levels of stress, there is likely to be a restriction in range of the stress variable. The size of the correlation will likely be underestimated.
   d. The likely form of this relationship is nonlinear. Persons who eat nothing will not likely be very happy and persons who overconsume will also be unhappy. The resulting pattern will mean that only those who eat moderate amounts will have relatively high scores, resulting in convex curvilinearity. A linear measure of association such as a correlation coefficient will understate the true size of the relationship.
   e. A single item on an intelligence test is not a very reliable measure of intelligence and so the strength of the correlation will be lessened by unreliability.

# Chapter 8: Measures of Association: Ordinal and Nominal Variables

1. a. Members of religions: Protestant, Catholic, Jewish
      Nonmembers: agnostic, atheist
   b. Precipitation: rainy, snowy
      No precipitation: clear, cloudy

2. a. $\overline{X} = .347$, $s = .479$

3.

| Political Party | Capital Punishment Approve | Disapprove | |
|---|---|---|---|
| Democrat | 76 | 73 | 149 |
| Republican | 108 | 111 | 219 |
| | 184 | 184 | 368 |

Percentage difference 2%; phi .02; logit difference .07

4. Percentage difference: 43%; the difference between the percentage of women who smoke minus the percentage of men who smoke is 43%. Phi: .43; the correlation between gender and smoking is .43. This is a moderate to large correlation.
Logit difference: 1.86. The odds of a woman smoking are about six (the antilog of the logit difference) times greater than for a man in this sample.

5. Percentage difference = –39%; people over 30 are 39% less likely to agree than persons under 30.
Phi: –.32; the correlation between age and opinion is –.32. The correlation is moderate in size.
Logit difference: –1.66. The odds of a younger person agreeing are about five times more than an older person agreeing.

6. a.

| Solitude | Intimacy Primary | Secondary | Public |
|---|---|---|---|
| Primary | 28 | 20 | 34 |
| Secondary | 3 | 0 | 4 |
| Public | 69 | 80 | 62 |
| | 100 | 100 | 100 |

b.

| Solitude | Intimacy Primary | Secondary | Public | |
|---|---|---|---|---|
| Primary | 39 | 3 | 57 | 99 |
| Secondary | 43 | 0 | 57 | 100 |
| Public | 45 | 6 | 49 | 100 |

10. Rho equals −.571.

# Chapter 9: *Statistical Principles*

1. a. nonrandom    b. nonrandom    c. nonrandom    d. random

2. Although statistic $p$ is unbiased, statistic $q$ is to be preferred. Its standard error is so much smaller than $q$'s that $q$ is a better statistic.

5. a. yes    b. $k$    c. For $k$ the standard error is 4.47 and for $q$ the standard error is 5.00.

6. a. yes    b. $\sqrt{1/(n-2)}$

9. It is more efficient.

11.

| $\overline{X}$ | Frequency |
|---|---|
| 6.0 | 1 |
| 6.5 | 0 |
| 7.0 | 2 |
| 7.5 | 2 |
| 8.0 | 1 |
| 8.5 | 2 |
| 9.0 | 3 |
| 9.5 | 0 |
| 10.0 | 2 |
| 10.5 | 2 |
| 11.0 | 0 |
| 11.5 | 0 |
| 12.0 | 1 |

Mean of the random sampling distribution: 8.75; standard deviation: 1.58.

15.
$$\frac{6!}{3!3!}\left(\frac{1}{6}\right)^3\left(\frac{5}{6}\right)^3 = .054$$

16.
$$\frac{5!}{4!1!}\left(\frac{5}{36}\right)^4\left(\frac{31}{36}\right)^1 = .0016$$

# Chapter 10: The Normal Distribution

1. a. .2794    b. .0438    c. .4115 − .1293 = .2822
   d. .5000 − .2642 = .2358

2. a. .3413    b. .4082 − .2486 = .1596    c. .5000 − .1293 = .3707

4. a. .412    b. −1.282

5. Percentile Ranks:  6, 17, 28, 39, 50, 61, 72, 83, 94
   Normalized ranks:  −1.555, −.954, −.583, −.279, .000, .279, .583, .954, 1.555

7. The ranks are first transformed into percentile ranks. The percentile ranks for one through ten are 5, 15, 25, 35, 45, 55, 65, 75, 85, and 95, respectively. The normalized ranks for the ten cities are

| | Trans. | Econ. | Average | Ave. Rank |
|---|---|---|---|---|
| Atlanta | −1.036 | −.126 | −.581 | 3.5 |
| Boston | .126 | −.385 | −.130 | 5 |
| Chicago | −.126 | 1.645 | .760 | 7.5 |
| Cincinnati | .674 | .674 | .674 | 8 |
| Dallas | .385 | −1.645 | −.630 | 4 |
| Denver | −.385 | −1.036 | −.710 | 3 |
| New York | −1.645 | .385 | −.630 | 4 |
| Phoenix | 1.645 | −.674 | .486 | 6.5 |
| Pittsburgh | 1.036 | 1.036 | 1.036 | 9 |
| San Francisco | −.674 | .126 | −.274 | 4.5 |

Note that a lower score means that the city is ranked ahead of the other city. Now comparing the average rank (fourth column of numbers) to the average of the normalized ranks (third column of numbers), it is found that five cities' ranks do not change. The following changes do occur. First, using normalized ranks Cincinnati is ranked ahead of Chicago. Second, using normalized ranks both Dallas and New York are ranked ahead of Atlanta.

8. The sampling distribution of $\overline{X}$ with $n = 36$ has a mean of 50 and variance of 81/36. The standard error of $\overline{X}$ is 1.5. So the probability that $\overline{X}$ is between 50 and 51 equals the probability that Z is between zero and

(51 – 50)/1.5 or .67. This probability is .2486. Because the probability that $\bar{X}$ is between 49 and 50 is also .2486, the probability that $\bar{X}$ is between 49 and 51 is .4972.

12. a. .253      b. –.496      c. –.553

# Chapter 11: Special Sampling Distributions

1. a. normal      b. chi square      c. $t$      d. chi square

2. a. .70      b. 2.21

3. 64

5. a. .5000 – .3413 = .1587      b. .5000 – .4207 = .0793

8. First, because $t(df)^2 = F(1,df)$, then $t(\infty)^2 = F(1,\infty)$. Second, because $F(1,\infty) = \chi^2(1)$, then $t(\infty)^2 = \chi^2(1)$.

# Chapter 12: Testing a Model

1. a. 3.055      b. 2.069      c. 3.435 (not 3.416, because one rounds down the $df$)

2. a. .20      b. .10      c. .01

3.
$$t(8) = \frac{29.556 - 25}{12.72/\sqrt{9}} = 1.075$$

This $t$ with eight degrees of freedom is not statistically significant. There is then insufficient evidence to claim that the inmates score any differently from the national norm of 25.

4.
$$t(5) = \frac{12.667 - 10}{3.011/\sqrt{6}} = 2.170$$

This $t$ with five degrees of freedom is marginally significant at the .10 level. Using the conventional .05 level of significance, there is not sufficient evidence to claim that the subjects are performing above chance.

5. a. 2.131      b. 2.004

6. A Type I error is concluding that the restricted model is false when in fact it is not. A Type II error is concluding that there is insufficient evidence to reject the restricted model when in fact the complete model is true.

8.
$$t(7) = \frac{112.125 - 100}{13.250/\sqrt{8}} = 2.588$$

This $t$ with seven degrees of freedom is statistically significant at the .05 level. So the couples believe that they do significantly more than 100% of the housework.

9. The first model is the complete model and the second is the restricted model. The restriction is that the independent variable has no effect.

# Chapter 13: The Two-Group Design

1. a. 2.056    b. 3.707    c. 1.684

2. The mean for method A is 19.40 and the mean for method B is 14.40. Method A is superior to method B. The test of whether this difference is statistically significant is $t(8) = 3.356$, which is statistically significant at the .01 level. (The pooled variance is 5.55.)

3. The six subjects lost an average of 8.5 pounds.

$$t(5) = \frac{-8.5}{7.007/\sqrt{6}} = -2.971$$

This $t$ with five degrees of freedom is statistically significant at the .05 level. The weight loss then is statistically significant and cannot be explained by chance.

4. a. .59    b. .42

5. a. .33    b. .29    c. .99

6. The mean for the treatment group is 122.50 and the mean for the control group is 109.14. To test this difference a $t(13) = 2.846$ is obtained which is statistically significant at the .02 level. The program does significantly increase IQ. (The pooled variance is 82.220.)

8. The use of individual therapy as the control group would control for therapeutic experience because all subjects would receive some form of therapy. It would then contrast two different types of talk therapy. Its weakness is that it is a low power test because presumably both types of therapy are effective.

   Having a hypnosis control group would control for receiving some intervention that was attempting to reduce smoking. Because the two types of treatments are so very different, it would be difficult to determine why it is that one treatment is more effective than the other.

   The film condition would control for information and for the motivation to quit smoking. It would also provide the most powerful test. It would provide no evidence about why group therapy is effective.

13. The standard deviation of the treatment group is 8.06. The standard deviation of the control group is 4.75. Although the difference is large because the sample sizes are equal, the unequal variances should not have much effect on the $p$ values.

15. a. 32     b. 26

# Chapter 14: One-Way Analysis of Variance

1. a. 4.49     b. 2.61     c. 3.23     d. 11.97

2.

| Source of Variation | Sum of Squares | Degrees of Freedom | Mean Square | F |
|---|---|---|---|---|
| Factor A | 140.0 | 4 | 35 | 3.5 |
| Subjects within A (S/A) | 380.0 | 38 | 10 | |
| Total | 520.0 | 42 | | |

The $F(4,38) = 3.5$ is statistically significant at the .05 level.

3. The means for the three groups are 13.2, 9.8, and 10.4. The test that they are significantly different is $F(2,12) = 16.467/7.233 = 2.28$, which is not a significant difference. The value of $\omega^2$ is .15, which is small to moderate in size.

4. a. One cannot do the Tukey lsd test because the overall $F$ test is not statistically significant.
   b. The mean square for the constant equals $15(11.1333 - 10)^2 = 19.266$. The $F(1,12)$ is 2.66, which is not significant at the .05 level.

5. The result of the $t$ test is $t(15) = 2.794$, which is statistically significant at the .02 level. The result of the one-way ANOVA is $F(1, 15) = 114.421/14.654 = 7.81$, which is significant at the .05 level. The mean for group I is 18.7 and for group II is 13.43. Note that $2.794^2$ equals, within the level of rounding error, 7.81.

6. The contrast weights are 1, 1, −1, −1, and 0. (Also acceptable are −1, −1, 1, 1, and 0). Each mean is multiplied by twelve to obtain a group total: 24, 38.4, 49.2, 62.4, and 61.2. The mean square for the contrast is

$$\frac{(24 + 38.4 - 49.2 - 62.4)^2}{(12)(4)} = 50.43$$

7. 
$$\frac{55.33 - (3)(2.49)}{194.77 + 2.49} = .24$$

This indicates a moderate effect size.

# Chapter 15: Two-Way Analysis of Variance

1.

| Source of Variation | Sum of Squares | Degrees of Freedom | Mean Square | F |
|---|---|---|---|---|
| A | 6.3 | 3 | 2.1000 | 3.13 |
| B | 4.3 | 2 | 2.1500 | 3.21 |
| A × B | 6.0 | 6 | 1.0000 | 1.49 |
| Subjects within AB (S/AB) | 72.4 | 108 | .6704 | |
| Total | 89.0 | 119 | | |

The *F* tests for A and B are statistically significant at the .05 level while the *F* for A × B is not significant.

2.

| Source of Variation | Sum of Squares | Degrees of Freedom | Mean Square | F |
|---|---|---|---|---|
| Meat (M) | 105.8 | 1 | 105.8 | 3.04 |
| Region (R) | 57.8 | 1 | 57.8 | 1.66 |
| M × R | 1.8 | 1 | 1.8 | .05 |
| Pigs within MR (P/MR) | 556.8 | 16 | 34.8 | |
| Total | 722.2 | 19 | | |

None of the *F* tests is statistically significant at the .05 level. The main effect for meat is marginally significant at the .10 level. The fat content of bacon, 33.6, is higher than ham, 29.0.

3.

| Source of Variation | Sum of Squares | Degrees of Freedom | Mean Square | F |
|---|---|---|---|---|
| A | 7.225 | 1 | 7.225 | 1.57 |
| B | 25.750 | 4 | 6.438 | 1.40 |
| A × B | 78.150 | 4 | 19.538 | 4.24 |
| Subjects within AB (S/AB) | 138.250 | 30 | 4.608 | |
| Total | 249.375 | 39 | | |

The interaction is significant at the .05 level. The table of cells means is:

| | B 1 | 2 | B 3 | 4 | 5 |
|---|---|---|---|---|---|
| A 1 | 6.75 | 3.25 | 6.75 | 4.75 | 2.50 |
| A 2 | 3.50 | 3.25 | 3.75 | 2.50 | 6.75 |

The A1 cell mean is higher than the A2 cell mean for B1, B3, and B4. However, the A2 cell mean is higher than the A1 cell mean for B5. Finally, the two cell means do not differ for B2.

5. There is evidence for a main effect of A. The A1 means are lower than the A2 means. There is also an indication of a main effect of B. The B3 means are the highest, then the B1 means, and finally the B2 means. There is little evidence for interaction because the A effect across levels of B is always about 3.0.

6.

| Source of Variation | Sum of Squares | Degrees of Freedom | Mean Square | F |
|---|---|---|---|---|
| Day (D) | 555.4 | 3 | 185.13 | 12.51 |
| Subject (S) | 664.8 | 4 | 166.20 | |
| D × S | 177.6 | 12 | 14.80 | |
| Total | 1397.8 | 19 | | |

The effect for day is statistically significant at the .01 level of significance. Performance is highest at day 4 and lowest at day 1.

7.

| Source of Variation | Sum of Squares | Degrees of Freedom | Mean Square | F |
|---|---|---|---|---|
| A | 124 | 2 | 62.000 | 23.25 |
| B | 25 | 3 | 8.333 | 3.12 |
| C | 48 | 1 | 48.000 | 18.00 |
| A × B | 12 | 6 | 2.000 | .75 |
| A × C | 19 | 2 | 9.500 | 3.56 |
| B × C | 14 | 3 | 4.667 | 1.75 |
| A × B × C | 12 | 6 | 2.000 | .75 |
| S/ABC | 64 | 24 | 2.667 | |
| Total | 318 | 47 | | |

Significant results: A main effect, $p < .001$; B main effect, $p < .05$; C main effect, $p < .001$; A × C interaction, $p < .05$.

# Chapter 16: Testing Measures of Association

1.
$$t(146) = \frac{.742\sqrt{146}}{\sqrt{1 - .742^2}} = 13.374$$

The positive correlation between age and susceptibility to glare is statistically significant at the .001 level.

2. a. −.1307     b. .0701     c. 1.5275     d. .9287

3. a. −.6963     b. −.4053     c. .7211     d. .0599

4.
$$t(42) = \frac{.31\sqrt{31.93}}{\sqrt{\dfrac{22.41 - (.31^2)(31.93)}{42}}} = 2.581$$

The positive regression coefficient is statistically significant at the .02 level.

5.
$$Z = \frac{.2342 - .5230}{\sqrt{\dfrac{1}{209} + \dfrac{1}{133}}} = -2.60$$

The difference between the two correlations is statistically significant at the .01 level.

6.
$$t(82) = \frac{.39\sqrt{82}}{\sqrt{1 - .39^2}} = 3.835$$

The positive correlation is statistically significant at the .001 level.

8. a. $t(142) = 5.186$, $p < .001$ ($K = .563$)
   b. $Z = 3.00$, $p = .0026$ ($Q = .464$)

9. Test of the difference between slopes: $t(106) = -1.434$, which is not statistically significant. The pooled slope is .331 and its $t(104) = 4.529$, which is statistically significant at the .001 level. So the slopes do not differ, and the pooled slope does differ from zero.

10. The average of the four correlations yields a Fisher's $z$ of 336.2733/480 = .7006, which when converted back to $r$ is .6044. The test that the pooled correlation is different from zero yields $Z = .7006/\sqrt{1/480} = 15.35$, which is statistically significant at the .001 level. The test that the correlations differ is $\chi^2(3) = 8.79$, which is statistically significant at the .05 level. The four correlations differ significantly.

12. a. .06     b. .92

13. a. 384     b. 8

# Chapter 17: Models for Nominal Dependent Variables

1. a. 3.84     b. 15.09     c. 16.27

2. Men are less likely to agree (29%) than women (52%). The test of independence yields a $\chi^2(2) = 15.48$, which is statistically significant at the .001 level.

3.

|           | Female  | Male    |
|-----------|---------|---------|
| Observed  | 212     | 246     |
| Expected  | 238.16  | 219.84  |

$\chi^2(1) = 5.99$, which is statistically significant at the .02 level. So women are underrepresented in the juries of the county.

4. Persons are more likely to buy a product when they have seen the ad (54%) than when they have not seen the ad (21%). The test of independence yields a $\chi^2(1) = 94.28$, which is statistically significant at the .001 level.

5. Blacks prefer less to live in the North (47%) than whites (54%). The test of independence yields $\chi^2(1) = 38.18$, which is statistically significant at the .001 level. When the data are split by area of birth, the result reverses. Of persons who were born in the North, blacks are more likely to prefer the North (81%) than whites (73%). This difference is statistically significant $\chi^2(1) = 37.43$ at the .001 level. Of persons who were born in the South, blacks are more likely to prefer the North (28%) than whites (16%). This difference is statistically significant $\chi^2(1) = 68.41$ at the .001 level.

11. Fathers are less likely to recognize their infants' cries (22%) than are mothers (52%). To test this difference, McNemar's test is used. The result is $\chi^2(1) = (|9 - 1| - 1)^2/10 = 4.90$, which is statistically significant at the .05 level.

# Chapter 18: Models for Ordinal Dependent Variables

1.
$$1 - \frac{(6)(146)}{(8)(63)} = -.738$$

This negative rank-order coefficient is statistically significant at the .05 level.

2. The sum of the ranks in group A is 70.5 and the value of $U$ is $49 + 28 - 70.5 = 6.5$. This value of $U$ with group sizes both equal to seven is

statistically significant at the .05 level. Using group B, the sum of the ranks is 34.5 and $U$ is 42.5. This value of $U$ is also significant at the .05 level, as it should be.

4. $$H = \left[\frac{12}{(15)(16)}\right]\left[\frac{46.5^2 + 33^2 + 40.5^2}{5}\right] - (3)(16) = .915$$

A $\chi^2$ test statistic of .915 with two degrees of freedom is not statistically significant. There is no evidence to conclude that the groups differ.

6. Eight values are higher in the after measure, three are lower, and one is tied. Using the sign test with $n = 11$, an eight is not significant.

7. Let heaviest have a rank of one, middle a rank of two, and lightest a rank of three. The sum of the ranks for the spherical condition is $3 + 2(9) + 3(8) = 45$; for the conical condition the sum is $5 + 2(4) + 3(11) = 46$; and for the cubical condition the sum is $12 + 2(7) + 3(1) = 29$. The Friedman test is

$$\left[\frac{12}{(20)(3)(4)}\right](45^2 + 46^2 + 29^2) - (3)(20)(4) = 9.1$$

A $\chi^2$ with two degrees of freedom is statistically significant at the .02 level. The groups' distributions significantly differ.

8. The sum of the ranks in group B is 15.5 and the value of $U$ is $25 + 15 - 15.5 = 24.5$. This value of $U$ with group sizes both equal to five is statistically significant at the .02 level of significance. Using group A, the sum of the ranks is 39.5 and $U$ is .5. This value of $U$ is also significant at the .02 level, as it should be.

9. The sum of the ranks in control group is 34.5 and the value of $U$ is $56 + 28 - 34.5 = 49.5$. This value of $U$ with a control group $n$ equal to seven and experimental group $n$ equal to eight is statistically significant at the .02 level of significance. One cannot use the sum of the treated group ranks because the sample sizes of the two groups differ.

17. a. $$U = (20)(25) + \frac{(21)(20)}{2} - 248 = 462$$

$$\frac{462 - (20)(25)/2}{\sqrt{(20)(25)(46)/12}} = -4.84$$

This value of $Z$ is statistically significant at the .001 level.

18. a. $t(76) = -1.872, p < .10$
    b. $t(40) = -3.187, p < .01$
    c. $p < .02$
    d. not statistically significant

# Glossary of Symbols

## Mathematical Symbols

| | |
|---|---|
| $b > c$ | $b$ greater than $c$ |
| $b < c$ | $b$ less than $c$ |
| $b = c$ | $b$ equal to $c$ |
| $b \approx c$ | $b$ approximately equal to $c$ |
| $+$ | Plus |
| $-$ | Minus |
| $\pm$ | Plus or minus |
| $\infty$ | Infinity |
| $\sqrt{\phantom{x}}$ | Square root |
| $\times$ | Multiplication or interaction in ANOVA |
| $\lvert c \rvert$ | Absolute value of $c$; negative signs are ignored |
| $n!$ | Factorial; $n(n - 1)(n - 2) \ldots (3)(2)(1)$ |
| $e$ | The number 2.718. . . |
| $\ln(c)$ | Natural logarithm; logarithm to base $e$ |
| $\log(c)$ | Common logarithm; logarithm to base 10 |

## Statistical Symbols

| | |
|---|---|
| $a$ | Intercept |
| $a$, $b$, $c$, and $d$ | Frequencies in a $2 \times 2$ table |
| A, B, and C | Factors in ANOVA |
| $b$ | Regression coefficient |
| $C$ | Correction term for the mean |
| $d$ | Cohen's measure of effect size |
| $D$ | Difference between ranks or scores |
| $df$ | Degrees of freedom |
| $f$ | Frequency |
| $H$ | Test statistic for Kruskal-Wallis test |
| $H_0$ | Null hypothesis |

| | |
|---|---|
| $k$ | Number of levels in one-way ANOVA |
| lsd | Least significant difference |
| MS | Mean square |
| $n$ | Sample size |
| $N$ | Sample size in analysis of variance |
| $p$ | $p$ value; also proportion or probability |
| $r$ | Correlation coefficient |
| $R$ | Sum of the ranks of the group with the smaller $n$ |
| $R_i$ | Score $i$'s rank |
| $r_S$ | Rank-order correlation or Spearman's rho |
| $s$ | Sample standard deviation |
| $s^2$ | Sample variance |
| $s_p^2$ | Pooled variance |
| $s_{y.x}^2$ | Error variance |
| $2 \times 2$ table | Table with two rows and columns |
| 2sd advantage | Two standard deviation advantage |
| S | Subject or person |
| S/A | Subjects within levels of A |
| SS | Sum of squares |
| $T$ | Sum of scores or total |
| TOT | Total variability |
| $U$ | Test statistic for Mann-Whitney test |
| $x$ | Number of successes in $n$ trials |
| $X$ and $Y$ | Variables |
| $\bar{X}$ | Sample mean |
| $\hat{Y}$ | Predicted score of $Y$ |
| $z$ | Fisher's $z$ transformation |
| $Z$ | Standard normal distribution |

# Greek Letters

| | |
|---|---|
| $\alpha$ | Alpha: probability of making a Type I error |
| $\beta$ | Beta: probability of making a Type II error |
| $\mu$ | Mu: population mean |
| $\phi$ | Phi: correlation between two dummy coded dichotomies |
| $\rho$ | Rho: population correlation coefficient |
| $\sigma$ | Sigma: population standard deviation |
| $\sigma^2$ | Sigma squared: population variance |
| $\Sigma$ | Summation sign |
| $\chi^2$ | Chi-square distribution |
| $\omega^2$ | Omega squared |

# Glossary of Terms

*Aggregation:*  Creating a score that is an average or sum of other scores.

*Alpha:*  Probability of making a Type I error.

*Alternative hypothesis:*  Hypothesis that is true if the null hypothesis is false.

*Analysis of variance:*  Procedure for testing the differences between means.

*Analysis of variance table:*  Table with sums of squares, mean squares, degrees of freedom, and $F$ ratios.

*ANOVA:*  Analysis of variance.

*Antilog:*  For $x = \log(y)$, $y$ is the antilog of $x$; inverse logarithm function.

*Arcsin transformation:*  Two-stretch transformation of proportions that stretches less than probit and logit.

*Asymmetric distribution:*  Distribution whose shape changes when its mirror image is examined.

*Bar graph:*  Graph of the frequencies of a nominal variable.

*Bimodal distribution:*  Distribution with two peaks.

*Binomial distribution:*  Distribution that describes the probability of $x$ successes in $n$ independent trials.

*Cell:*  Particular row and column combination.

*Central limit theorem:*  With increasing sample size, the distribution of the mean approaches a normal distribution, regardless of the shape of the original distribution of the scores.

*Central tendency:*  Typical value of an observation from the sample.

*Chi square distribution:*  Sampling distribution with a lower limit of zero and no upper limit; sum of independent $Z^2$ values; $\chi^2$.

*Chi square test of independence:*  Test to evaluate whether two nominal variables are associated.

*Circle diagram:*  Representation of the partitioning of sums of squares and degrees of freedom in analysis of variance.

*Class interval:*  Range of possible scores that can be a member of a given class.

*Class midpoint:*   One-half the sum of a class's lower and upper limits.

*Class width:*   Difference between adjacent lower limits.

*Coefficient of variation:*   Standard deviation divided by the mean.

*Cohen's* **d***:*   Measure of effect size in a two-group study; difference between the means divided by the pooled within-groups standard deviation.

*Complete model:*   Model that contains the term that is to be tested.

*Concave curvilinearity:*   Relationship that begins negative and becomes positive; U shape.

*Constant in model:*   Term added to every score; often the population mean of the dependent variable.

*Contrast:*   Set of weights assigned to levels of the independent variable in ANOVA; weights that are chosen for theoretical reasons and must sum to zero.

*Convex curvilinearity:*   Relationship that begins positive and becomes negative; inverted U shape.

*Correction term of the mean:*   Squared sum of all the observations which is divided by the total number of observations; symbolized by $C$.

*Correlated correlations:*   Two or more correlations computed using the same sample of objects.

*Correlation coefficient:*   Regression coefficient between $Z$ scored variables that varies from $-1$ to $+1$; $r$.

*Criterion variable:*   Outcome or dependent variable in a regression equation.

*Critical value:*   Value that the test statistic must meet or exceed to be deemed statistically significant.

*Cummulative frequency:*   Sum of the frequencies of all classes that are less than or equal to the class's upper limit.

*Curvilinearity:*   Nonlinear relationship in which the relationship changes direction.

*Data:*   Numerical values given to objects.

*Datum:*   Single score.

*Degrees of freedom for a contrast:*   One.

*Degrees of freedom for $\chi^2$ goodness of fit test:*   Number of levels of the nominal variable less one.

*Degrees of freedom for $\chi^2$ test of independence:*   $(r - 1)(c - 1)$.

*Degrees of freedom for error variance in a regression equation:*   $n - 2$.

*Degrees of freedom for* **F** *in one-way ANOVA:*   $k - 1$ in the numerator and $N - k$ in the denominator.

*Degrees of freedom for interaction in two-way ANOVA:*   $(a - 1)(b - 1)$.

*Degrees of freedom for pooled variance:*   $n_1 + n_2 - 2$.

*Degrees of freedom for* **t***:*   For one-sample test, $n - 1$; for two-sample test, $n_1 + n_2 - 2$; for a test of a single correlation or regression coefficient, $n - 2$.

*Degrees of freedom of the standard deviation:*   Sample size minus one, or $n - 1$.

*Dependent variable:*   Outcome or variable caused by the independent variable.

*Descriptive statistics:*   Numerical values that summarize sample data.

*Dichotomy:*   Nominal variable with two levels.

*Distribution:*   Shape of a sample or population; usually represented by a histogram.

*Distribution-free test:*   Procedure for testing a model that makes no distributional assumptions.

*Distribution-tied test:*   Test that assumes a normal distribution that is analogous to a distribution-free test.

*Dummy coding:*   Numbers used to create a dummy variable.

*Dummy variable:*   Numerical variable that is created by assigning arbitrary numbers to the levels of a nominal variable.

*Ecological fallacy:*   Inferring individual relations from aggregate relations.

*Effect size:*   Measure of the strength of effect as opposed to its $p$ value.

*Efficient statistic:*   Statistic with a relatively small standard error.

*Error in a regression equation:*   Observed score minus the predicted score; the vertical distance in the scatterplot from the regression line to the point.

*Factor:*   Nominal independent variable in ANOVA.

*Factorial design:*   In two-way ANOVA, the creation of all possible combinations of two independent variables.

*F distribution:*   Sampling distribution that is the ratio of two independent variances.

*Fisher's z transformation:*   Transformation of a correlation that makes its distribution approximately normal.

*Flat distribution:*   Distribution in which all scores are equally likely.

*Flat transformation:*   Transformation that changes the shape of a distribution into flat one; rank order and percentile rank.

*Frequency:*   Number of observations that fall in a cell of a table or the number of observations in a class interval.

*Frequency table:*   Table with the classes and their frequencies.

*Friedman two-way ANOVA:*   Test used to evaluate the medians and other aspects of two or more nonindependent groups.

*Goodness of fit $\chi^2$ test:*   Test to compare the observed distribution of a nominal variable to a predicted distribution.

*Histogram:*   Graph of the frequency table of a distribution with the $X$ axis being the classes and the $Y$ axis being the frequency.

*Hotelling test:*   Test of the equality of two nonindependent correlations in which two of the variables are in common.

*Independent groups:*   Two or more samples that contain different persons who do not influence one another.

*Independent sampling:*   If one object is sampled, every other object in the population has the same probability of being sampled.

*Independent variable:* Causal variable in a model.

*Inferential statistics:* Using sample data to draw conclusions about the population; tests of models.

*Interaction:* Effect of an independent variable changes as a function of a second variable.

*Intercept:* Predicted value of $Y$ when $X$ is zero in a regression equation in which $X$ is the predictor and $Y$ the criterion.

*Interquartile range:* Difference between the upper median and the lower median.

*Interval level of measurement:* Measurement level at which numbers can be used to quantify differences between objects.

*Kruskal-Wallis ANOVA:* Test used to compare the medians and other aspects of two or more independent groups.

*Leaf:* In a stem and leaf display, the next digit after the stem.

*Least significant difference test:* Post hoc test of means in one-way AN-OVA; Tukey lsd.

*Leptokurtic distribution:* Distribution that has a high peak in the center and skinny tails.

*Linearity:* One-unit change in $X$ produces the same change in $Y$ regardless of where the change in $X$ comes.

*Logarithm:* If $x^y = b$, $y$ is the logarithm of $b$ to base $x$.

*Logit difference:* In a $2 \times 2$ table the difference between logits; also the natural logarithm of the odds ratio.

*Logit transformation:* Natural logarithm of the odds.

*Log linear model:* Model for multiple nominal independent variables and a nominal dependent variable.

*Lower median:* Median of scores below the median of the sample.

*Lowest lower limit:* Lower limit of the lowest class interval.

*McNemar's test:* Test of the effect of a dichotomous independent variable on a dichotomous dependent variable when groups are nonindependent.

*Main effect:* In two-way ANOVA the effect of an independent variable averaged across levels of the other independent variable.

*Mann-Whitney test:* Distribution-free test that compares the medians and other aspects of two independent groups; $U$.

*Margin:* In a table, sum of frequencies across a row or a column.

*Mean:* Sum of the observations divided by the sample size.

*Mean square:* In ANOVA the sum of squares divided by degrees of freedom.

*Measurement:* Assignment of numbers to objects by a rule.

*Median:* Middle observation in a sample.

*Mode:* Most frequent observation in a sample.

*Model:* Mathematical equation specified by a theory.

*Negative association:* As one variable increases, the other decreases.

*Negative skew:*    Distribution with a long, skinny tail on the left side.

*Nominal level of measurement:*    Measurement level at which only differentiation of objects is possible.

*Nonindependent groups:*    Two or more samples that contain the same persons or sampling units.

*Nonlinearity:*    Relationship between two variables that varies in strength as a function of one variable.

*Normal distribution:*    Unimodal, symmetric, bell-shaped distribution with limits of positive and negative infinity.

*Normalized ranks transformation:*    Transformation that alters a variable's distribution to make the distribution more normal.

*No-stretch transformation:*    Constant multiplied or added to each score; basic shape of the distribution not altered.

*Null hypothesis:*    Constraint on the complete model that is present in the restricted model; $H_0$.

*Odds:*    Proportion divided by the quantity one minus the proportion.

*Odds ratio:*    In a $2 \times 2$ table (ad)/(bc).

*Omega squared:*    Measure of variance explained in one-way ANOVA.

*One-stretch transformation:*    Transformation to remove positive skew, which stretches the left side of the distribution; square root, logarithm, and reciprocal.

*One-tailed test:*    Test in which only one alternative hypothesis is considered.

*One-way analysis of variance:*    Method used to test for differences between independent means.

*Operational definition:*    Set of procedures used to measure a construct.

*Ordinal level of measurement:*    Measurement level at which objects can be rank ordered.

*Outlier:*    Extremely large or small score.

*Paired* t *test:*    Test of the difference between two nonindependent means.

*Parameter:*    Quantity computed using all objects in the population, often symbolized by a Greek letter.

*Part-whole problem:*    Two variables, one of which is derived from the other.

*Pearson Filon test:*    Test of the equality of two nonindependent correlations in which none of the variables are in common.

*Percentage difference:*    In a $2 \times 2$ table, the difference between percentages computed across either rows or columns.

*Percentile rank:*    Percentage of scores that the object is greater than.

*Phi:*    Correlation between two dummy-coded dichotomies.

*Platykurtic distribution:*    Distribution with a low peak in the center and fat tails.

*Pooled variance:*    Weighted average of variances used in two-group *t* test, where the weights are sample size minus one for each group.

*Population:*  All possible observations.

*Positive association:*  As one variable increases, the other increases.

*Positive skew:*  Distribution with a long, skinny tail on the right side.

*Post hoc test of means:*  Test in which all possible pairs of means are compared.

*Power:*  Probability of rejecting the restricted model when the restricted model is false; one minus the probability of making a Type II error.

*Power efficiency:*  Ratio of sample size needed for a distribution-tied test to the sample size needed for a distribution-free test in which the same power is achieved and the assumptions of the distribution-tied test hold.

*Predicted score:*  In a regression equation, the intercept plus the predictor score times the regression coefficient.

*Probit transformation:*  Two-stretch transformation of proportions based on the standard normal distribution.

**p** *value:*  The probability of obtaining a value of the test statistic at least as large as the one obtained.

*Random assignment:*  Each object having the same probability of being assigned to a level of the independent variable.

*Random sample:*  Each object equally likely to be chosen from the population.

*Range:*  Crude measure of variability; largest score minus the smallest score.

*Rank-order correlation:*  Spearman's rho; correlation between ranks; $r_S$.

*Rank-order transformation:*  Scores rank ordered from smallest to largest and the smallest score assigned a 1, the next a 2, and so on.

*Reciprocal transformation:*  One-stretch transformation in which one is divided by the score; $1/X$.

*Rectangular distribution:*  Flat distribution.

*Regression coefficient:*  Measure of association of how much a one-unit change in the predictor variable creates in the criterion variable.

*Regression equation:*  Criterion equals the intercept plus the regression coefficient times the predictor.

*Regression toward the mean:*  Predicted scores in a regression equation are less variable than the scores of the criterion.

*Relative frequency:*  One hundred times the frequency divided by sample size.

*Reliability:*  Proportion of true variance in a variable.

*Repeated measures design:*  All subjects measured at each level of the independent variable.

*Residual variable:*  All other sources of variation in the dependent variable besides that due to the independent variables.

*Restricted model:*  Model that is a constrained version of the complete model, the constraint being the null hypothesis.

*Restriction in range:*  Variable with limited variability.

*Robust statistic:*   Statistic not influenced much by outliers.

*Sample:*   Set of scores that refer to different objects.

*Sample size:*   Number of observations in the sample; $n$.

*Sampling distribution:*   Distribution of a statistic that is created by drawing repeated samples and recomputing the statistic.

*Sampling error:*   The fact that a statistic changes when it is recomputed using a different sample.

*Scatterplot:*   Graph to represent the association between two variables; variables form the axes and points are the data.

*Significance level:*   Alpha or the probability of making a Type I error.

*Sign test:*   Distribution-free test for evaluating the difference between the medians and other aspects of two nonindependent groups.

*Skew:*   Long, skinny tail on just one side of a distribution.

*Slope:*   Regression coefficient; linear measure of association.

*Smoothed frequency:*   One-half the class's frequency plus one-quarter the sum of the adjacent class frequencies.

*Smoothing:*   Procedure to make a frequency table less influenced by choice of lowest lower limit and class width.

*Spearman's rho:*   Correlation coefficient of ranks; $r_S$.

*Standard deviation:*   Measure of variability that uses all observations; square root of the variance; $s$.

*Standard error of the mean:*   Standard deviation divided by the square root of the sample size.

*Standard normal distribution:*   $Z$ distribution; normal distribution with a mean of zero and a variance of one.

*Standard score:*   $Z$ score; score minus the mean and the difference divided by the standard deviation.

*Statistic:*   Quantity computed from sample data.

*Statistically significant:*   $p$ value is equal to or less than the significance level.

*Stem:*   Lower limit of a class used to represent a class interval in a stem and leaf display.

*Stem and leaf display:*   Vertical histogram that essentially preserves the raw data.

*Sum of squares:*   In ANOVA the numerator of a mean square.

*Summation sign:*   Symbol that represents the sum of all the scores; $\Sigma$.

*Symmetric distribution:*   Distribution that when folded vertically perfectly coincides.

*Tail of a distribution:*   Frequency of very large or very small values.

*t distribution:*   Sampling distribution used to test hypotheses about means and to test correlation coefficients.

*Test statistic:*   Quantity computed from sample data used to evaluate the plausibility of a restricted model.

*Tukey least significant difference test:*   Post hoc test in one-way ANOVA.

*Two-standard deviation advantage:*   Measure of effect size of $r$ that equals how much more likely someone who is one standard deviation above the mean on $X$ will outscore on $Y$ someone who is one standard deviation below the mean on $X$.

*Two-stretch transformation:*   Transformation that is used to remove lower and upper limits, commonly used on proportions; arcsin, probit, and logit.

*Two-tailed test:*   Test in which the two alternative hypotheses are considered.

*Two-way analysis of variance:*   Procedure to evaluate models with two nominal independent variables and an interval dependent variable.

*Type I error:*   Rejecting the restricted model when it is true; alpha, or $\alpha$.

*Type II error:*   Retaining the restricted model when it is false.

*Unbiased statistic:*   Statistic whose mean of the sampling distribution equals the population parameter that the statistic is estimating.

*Unimodal distribution:*   Distribution with one peak.

*Unit in a distribution:*   Smallest possible difference between a pair of scores.

*Unit of measurement:*   Term that defines the meaning of a one-point difference between two scores.

*Upper median:*   Median of scores above the median of the sample.

*Variability:*   How much the observations differ from one another.

*Variance:*   Measure of variability that is based on deviations from the mean; $s^2$.

**X** *axis:*   Horizontal (left to right) axis in a graph.

**Y** *axis:*   Vertical (up and down) axis in a graph.

**Z** *distribution:*   Standard normal distribution; a normal distribution with a mean of zero and a variance of one.

**Z** *score:*   Score in which the mean has been subtracted and this difference is divided by the standard deviation.

*z transformation:*   Transformation to make the distribution of $r$ more normal, commonly called Fisher's $r$ to $z$ transformation.

**Z** *transformation:*   Scores in which the mean has been subtracted and this difference is divided by the standard deviation.

# *Index*

# Formulas for Computation

(Continued from front endleaf.)

**Pearson-Filon Test**

$$Z = \frac{\sqrt{(n-3)}(z_{12} - z_{34})}{\sqrt{2 - Q(1 - r^2)^2}}$$

280

where

$$Q = (r_{13} - r_{23}r)(r_{24} - r_{23}r) + (r_{14} - r_{13}r)(r_{23} - r_{13}r) + (r_{13} - r_{14}r)(r_{24} - r_{14}r) + (r_{14} - r_{24}r)(r_{23} - r_{24}r)$$

and

$$r = \frac{r_{12} + r_{34}}{2}$$

**Percentage Difference**

$$100\left[\frac{a}{a+b} - \frac{c}{c+d}\right]$$

135

**Percentile Rank**

$$100\left[\frac{R - .5}{n}\right]$$

84

**Phi**

$$\phi = \frac{ad - bc}{\sqrt{(a+b)(c+d)(a+c)(b+d)}}$$

133

**Pooled Variance**

$$s_p^2 = \frac{\sum X_1^2 - (\sum X_1)^2/n_1 + \sum X_2^2 - (\sum X_2)^2/n_2}{n_1 + n_2 - 2}$$

205

**Predicted Score**

$$\hat{Y} = a + bX$$

98

**Regression Coefficient**

$$b = \frac{\sum XY - (\sum X)(\sum Y)/n}{\sum X^2 - (\sum X)^2/n}$$

99

**Sign Test**

$$Z = \frac{|2c - n| - 1.0}{\sqrt{n}}$$

315

**Spearman's Rho**

$$r_S = 1 - \frac{6\sum D_i^2}{n(n^2 - 1)}$$

138

**Standard Deviation**

$$s = \sqrt{\frac{\sum X^2 - \frac{(\sum X)^2}{n}}{n - 1}}$$

66